T0173044

Microbial Nanotechnology

Editors

Mahendra Rai

Department of Biotechnology
SGB Amravati University
Amravati, Maharashtra, India

Patrycja Golińska

Department of Microbiology
Nicolaus Copernicus University
Toruń, Poland

CRC Press
Taylor & Francis Group
Boca Raton   London   New York

CRC Press is an imprint of the
Taylor & Francis Group, an **informa** business

A SCIENCE PUBLISHERS BOOK

Cover photos: Cover illustrations are by kind courtesy of Dr Patrycja Golińska, Department of Microbiology, Nicolaus Copernicus University, 87100 Torun, Poland and Dr Gabriela Kratošová, Nanotechnology Centre, VŠB – Technical University of Ostrava, 17. listopadu 15/2172, Ostrava, Czech Republic

CRC Press
Taylor & Francis Group
6000 Broken Sound Parkway NW, Suite 300
Boca Raton, FL 33487-2742

First issued in paperback 2022

© 2020 by Taylor & Francis Group, LLC
CRC Press is an imprint of Taylor & Francis Group, an Informa business

No claim to original U.S. Government works

Version Date: 20200226

ISBN-13: 978-0-367-22676-3 (hbk)
ISBN-13: 978-0-367-51710-6 (pbk)

DOI: 10.1201/9780429276330

**Visit the Taylor & Francis Web site at
http://www.taylorandfrancis.com**

**and the CRC Press Web site at
http://www.crcpress.com**

Library of Congress Cataloging-in-Publication Data

Names: Rai, Mahendra, editor. | Golińska, Patrycja, 1978- editor.
Title: Microbial nanotechnology / editors, Mahendra Rai, Patrycja Golińska.
Description: Boca Raton : CRC Press, [2020] | Includes bibliographical references and index.
Identifiers: LCCN 2020009295 | ISBN 9780367226763 (hardback)
Subjects: MESH: Nanotechnology | Nanostructures--microbiology | Biocompatible Materials--chemical synthesis
Classification: LCC R857.N34 | NLM QT 36.5 | DDC 610.28--dc23
LC record available at https://lccn.loc.gov/2020009295

Preface

The development of green processes for the synthesis of nanomaterials is evolving into an important branch of nanotechnology. Nanotechnology involves creating and manipulating organic and inorganic matter at the nanoscale. Microbial nanotechnology is based on the use of microorganisms, including fungi, bacteria, actinobacteria, and algae, for synthesis of metallic and non-metallic nanomaterials.

Presently, researchers are looking into the development of cost-effective procedures for the production of reproducible, stable and biocompatible nanomaterials. Microbe-mediated synthesis of metal nanomaterials is gaining more importance because it allows for cost-effective, eco-friendly and rapid rate of synthesis of nanomaterials of attractive and diverse morphologies and bioactivity.

The ecologically friendly nature of microbiological synthesis of nanomaterials without the use of toxic reagents needed for chemical synthesis or high temperature and pressure required in the case of physical synthesis has made it interesting to nanotechnologists and microbiologists. This aspect is strongly highlighted in this book. Moreover, bionanoparticles are considered more biocompatible because they are covered by microbe-derived capping agents. The association of biomolecules with silver nanoparticles has been shown to modify the stability of the nanoparticles as well as their behavior in the physiological environment. These complex biomolecule–nanoparticle conjugates may be used in various medical applications.

The size and shape of fabricated bionanoparticles can also be completely regulated by controlling the environment in which the nanocrystal growth occurs. Moreover, biological methods have the greater advantage of easy bulk synthesis that can be exploited for industrial-scale production.

This book is a comprehensive review of the state of the art in microbial nanotechnology with an emphasis on the mechanisms of biosynthesis, diverse applications in food and nutrition sciences, biomedicine, agriculture and other fields, and their toxicity. It contains numerous references to the primary literature, making it the perfect guide for scientists who want to explore the fascinating world of microbial nanotechnology.

We believe this book will be useful for microbiologists, biotechnologists, chemists, and researchers who are interested in eco-friendly microbial synthesis of nanoparticles. Postgraduate students and researchers should also find it useful.

MR thankfully acknowledges the financial support by the CNPq (National Council for Scientific and Technological Development, Brazil; process number 403888/2018-2) and the Polish National Agency for Academic Exchange (NAWA), Poland, for financial support to visit the Department of Microbiology, Nicolaus Copernicus University, Toruń, Poland.

Mahendra Rai, India
Patrycja Golińska, Poland

Contents

Chapter 1

Synthesis of Nanoparticles by Actinomycetes

Mechanism and Applications

Patrycja Golińska

Introduction

Metal nanoparticles have attracted considerable attention from researchers for their unique physical, chemical and biological properties (Bhattacharyya and Singh 2009, Rizvi and Saleh 2018) that result from shape, size, and composition of metallic nanomaterials (Lee and Jun 2019). It is claimed that the most important property of nanoparticles is the ratio of their surface area to volume, which easily allows them to interact with other molecules (Harish et al. 2018). Metal nanoparticles have exhibited their potential in biotechnology, bioremediation of environmental pollutants, medicine for gene delivery in treatment or prevention of genetic disorders, as delivery of antigen for vaccination or better drug-delivery methods. However, there is a need to develop environment-friendly procedures for the synthesis of nanoparticles (Golinska et al. 2014). Biological methods of nanoparticle synthesis are more useful than physical and chemical methods that involve the use of hazardous chemicals, different kinds of radiations and high cost (Ingle et al. 2008, Wolska et al. 2017, Yusof et al. 2019).

Biosynthesis of metal nanoparticles, which involves the use of biological systems such as microorganisms and plants, is a type of "bottom-up" approach where nanoparticles are formed because of the reduction/oxidation of metal by enzymes and secondary metabolites secreted by microbes and plants, respectively (Prabhu and Poulose 2012).

There is a long list of microorganisms, fungi and bacteria and also algae, which have been used for the synthesis of metal nanoparticles over the past two decades (Kowshik et al. 2003, Prasad et al. 2007, Luangpipat et al. 2011, Rai et al. 2015,

Department of Microbiology, Nicolaus Copernicus University, 87100 Torun, Poland.
Email: golinska@umk.pl

Quester et al. 2016, Ghiută et al. 2018, Li et al. 2018). However, to date, researchers have paid less attention to biosynthesis of nanoparticles using actinomycetes.

Different genera of actinobacteria have been used for the synthesis of different types of metal nanoparticles, such as copper, gold, manganese, silver and zinc (Ahmad et al. 2003a, Ahmad et al. 2003b, Sastry et al. 2003, Sadhasivam et al. 2010, Rajamanickam et al. 2012, Selvakumar et al. 2012, Usha et al. 2010, Deepa et al. 2013, Abdeen et al. 2014, Waghmare et al. 2011, Waghmare et al. 2014, Golińska et al. 2015, Rathod et al. 2016, Składanowski et al. 2017, Wypij et al. 2018a, Avilala and Golla 2019), but mainly for silver nanoparticles (Table 1.1). The biogenic nanoparticles were synthesized extra- and/or intracellularly (Sukanya et al. 2013, Golinska et al. 2014, Wypij et al. 2018a, Wypij et al. 2018b, Avilala and Golla 2019).

Actinobacteria-mediated nanoparticles have been found to have antibacterial, antifungal, antiviral, anticancer and other activities (Usha et al. 2010, Balagurunathan et al. 2011, Rajamanickam et al. 2012, Chauhan et al. 2013, Abd-Elnaby et al. 2016, Anasane et al. 2016, Składanowski et al. 2016, Składanowski et al. 2017, Wypij et al. 2018a, Wypij et al. 2018b, Avilala and Golla 2019). Detailed information on synthesis of metal nanoparticles using different actinomycetes and their potential applications in the medical field are given in the later part of this chapter.

Biogenic synthesis of metal nanoparticles using actinomycetes

It is well known that the reduction in an ionic concentration using biological systems leads to the formation of size-controlled, stable and dispersed nanoparticles, which possess attractive physico-chemical properties (Salata 2004, Morones et al. 2005, He et al. 2007, Kim et al. 2007, Shah et al. 2015); hence, green synthesis has received particular attention.

The biosynthesis of metal nanoparticles by actinomycetes may take place intra- or extracellularly (Golinska et al. 2014). There are a few proposed mechanisms of intracellular synthesis. Intracellular reduction of metal ions occurs on the surface of mycelia along with cytoplasmic membrane leading to the formation of nanoparticles (Ahmad et al. 2003b). A similar mechanism was proposed by Abdeen and co-workers (2014), who claimed that intracellular synthesis of silver nanoparticles (AgNPs) was performed by trapping of the Ag^+ ions on the surface of the actinomycete cells via electrostatic interactions between the Ag^+ and negatively charged carboxylate groups in enzymes present in the cell wall of mycelia. The silver ions are reduced by enzymes present in the cell wall leading to the formation of the silver nuclei, which subsequently grow by further reduction of Ag^+ ions and accumulation on these nuclei (Abdeen et al. 2014). Intracellular synthesis of metal nanoparticles using actinomycetes has been reported less often than extracellular synthesis (Usha et al. 2010, Balagurunathan et al. 2011, Prakasham Shetty et al. 2012, Sukanya et al. 2013).

The hypothetical mechanism of extracellular synthesis of silver nanoparticles in *Streptomyces* sp. LK3 has been reported by Karthik et al. (2014). The possible mechanism, which may involve the reduction of silver ions, is probably the electron shuttle enzymatic metal reduction process (Fig. 1.1). It is claimed that NADH-dependent nitrate reductase enzyme, involved in the nitrogen cycle, is an important factor in biogenic synthesis of silver nanoparticles (Duran et al. 2005). This enzyme

Table 1.1. Synthesis of metal nanoparticles from actinomycetes and their antimicrobial activity.

Type of NPs	Actinomycetes used for synthesis	Type of synthesis	Size (nm)	Biological activity against	References
silver					
	Streptomyces rochei MHM13	Extracellular	22–85	*Bacillus subtilis,* *Staphylococcus aureus,* *Pseudomonas aeruginosa,* *Bacillus cereus,* *Salmonella typhimurium,* *Escherichia coli,* *Vibrio fluvialis,* *Vibrio damsela*	Abd-Elnaby et al. 2016
	Pilimelia columellifera subsp. *pallida* SF23 *Pilimelia columellifera* subsp. *pallida* C9	Extracellular Extracellular	4–36 8–60	*Candida albicans,* *Candida tropicalis,* *Malassezia furfur,* *Trichophyton robrum*	Anasane et al. 2016
	Nocardiopsis alba		20–60	*Staphylococcus aureus,* *Pseudomonas aeruginosa,* *Escherichia coli,* *Klebsiella pneumoniae*	Avilala and Golla 2019
	Streptomyces sp. JAR1	Extracellular	60–70	*Enterococcus faecalis,* *Staphylococcus aureus,* *Escherichia coli,* *Salmonella typhimurium,* *Proteus mirabilis,* *Pseudomonas aeruginosa,* *Fusarium sp.,* *Aspergillus terreus* JAS1	Chauhan et al. 2013

Table 1.1 contd. ...

... Table 1.1 contd.

Type of NPs	Actinomycetes used for synthesis	Type of synthesis	Size (nm)	Biological activity against	References
silver					
	Pilimelia columellifera subsp. *pallida* SL19 *Pilimelia columellifera* subsp. *pallida* SL24	Extracellular	12.7 15.9	*Bacillus subtilis* PCM2021, *Escherichia coli* ATCC8739, *Klebsiella pneumoniae* ATCC700603, *Pseudomonas aeruginosa* ATCC 10145, *Staphylococcus aureus* ATCC6538, Uropathogens of: *Enterobacter* sp., *Escherichia coli*, *Klebsiella pneumoniae*, *Pseudomonas aeruginosa*, *Staphylococcus aureus*	Golinska et al. 2015
	Streptomyces sp. LK3	Extracellular	5	*Rhipicephalus microplus*, *Haemaphysalis bispinosa*	Karthik et al. 2014
	Nocardiopsis sp. MBRC-1	Extracellular	45	*Escherichia coli*, *Bacillus subtilis*, *Enterococcus hirae*, *Pseudomonas aeruginosa*, *Shigella lexneri*, *Staphylococcus aureus*, *Aspergillus niger*, *Aspergillus brasiliensis*, *Aspergillus fumigatus*, *Candida albicans*	Manivasagan et al. 2013
	Actinomycetes	Extracellular	5–50	*Escherichia coli*, *Staphylococcus* sp., *Pseudomonas* sp., *Bacillus* sp.	Narasimha et al. 2013

Streptomyces albidoflavus CNP10	Extracellular/Intracellular	10–40	*Bacillus subtilis, Micrococcus luteus, Escherichia coli, Klebsiella pneumoniae*	Prakasham Shetty et al. 2012
Nocardiopsis valliformis OT1	Extracellular	5–50	*Bacillus subtilis* PCM2021, *Escherichia coli* ATCC8739, *Klebsiella pneumoniae* ATCC700603, *Pseudomonas aeruginosa* ATCC10145, *Staphylococcus aureus* ATCC6538	Rathod et al. 2016
Streptomyces sp. VITPK1	Extracellular	20–45	*Candida albicans* MTCC227, *Candida tropicalis* MTCC184, *Candida krusei* MTCC9215	Sanjenbam et al. 2014
Streptomyces kasugaensis NH28	Extracellular	4.2–65	*Bacillus subtilis* PCM2021, *Escherichia coli* ATCC8739, *Klebsiella pneumoniae* ATCC700603, *Proteus mirabilis, Pseudomonas aeruginosa* ATCC 0145, *Salmonella infantis, Staphylococcus aureus* ATCC6538	Składanowski et al. 2016
Streptomyces sp. BDUKAS10	Extracellular	21–48	*Pseudomonas aeruginosa* MTCC1688, *Bacillus cereus* MTCC1272, *Staphylococcus aureus* MTCC96	Sivalingam et al. 2012
Streptomyces sp. VITBT7	Extracellular	20–70	*Staphylococcus aureus* MTCC739, *Pseudomonas aeruginosa* MTCC424, *Aspergillus niger* MTCC1344, *Aspergillus fumigatus* MTCC3002	Subashini and Kannabiran 2013
Streptomyces sp. I, *Streptomyces* sp. II, *Rhodococcus* sp.	Extracellular/Intracellular	65–80	*Staphylococcus aureus, Escherichia coli, Pseudomonas aeruginosa, Klebsiella pneumoniae*	Sukanya et al. 2013

Table 1.1 contd. ...

... Table 1.1 contd.

Type of NPs	Actinomycetes used for synthesis	Type of synthesis	Size (nm)	Biological activity against	References
silver					
	Streptomyces sp. JF741876	Extracellular	80–100	*Staphylococcus aureus* MTCC3160, *Escherichia coli* MTCC1302, *Trichophyton rubrum*, *Trichophyton tonsurans*	Vidyasagar et al. 2012
	Streptomyces sp. SH11	Extracellular	13.2	*Bacillus subtilis* PCM2021, *Escherichia coli* ATCC8739, *Staphylococcus aureus* ATCC6538	Wypij et al. 2017a
	Pilimelia columellifera subsp. *pallida* SL19	Extracellular		*Candida albicans* ATCC10231, *Malassezia furfur* DSM 6170, *Trichophyton erinacei* DSM 25374	Wypij et al. 2017b
	Streptomyces xinghaiensis OF1	Extracellular	5–20	*Bacillus subtilis* PCM2021, *Escherichia coli* ATCC8739, *Pseudomonas aeruginosa* ATCC10145, *Staphylococcus aureus* ATCC 6538, *Candida albicans* ATCC10231, *Malassezia furfur* DSM 6170	Wypij et al. 2018a
	Streptomyces calidiresistens IF11 *Streptomyces calidiresistens* IF17	Extracellular	5–50 5–20	*Bacillus subtilis* PCM2021, *Escherichia coli* ATCC8739, *Staphylococcus aureus* ATCC6538, *Candida albicans* ATCC10231, *Malassezia furfur* DSM 6170	Wypij et al. 2018b
gold					
	Streptomycetes viridogens HM10	Intracellular	18–20	*Staphylococcus aureus*, *Escherichia coli*	Balagurunathan et al. 2011

zinc				
Streptomyces sp.	Extracellular	—	*Staphylococcus aureus*, *Escherichia coli*, *Salmonella* sp.	Rajamanickam et al. 2012
Streptomyces sp.	Extracellular	—	*Escherichia coli* ATCC8739, *Staphylococcus aureus* ATCC6538	Usha et al. 2010
copper				
Streptomyces sp.			*Escherichia coli* ATCC8739, *Staphylococcus aureus* ATCC6538	Usha et al. 2010

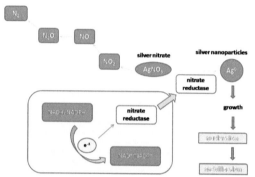

Figure 1.1. The hypothetical mechanism of extracellular synthesis of silver nanoparticles in microorganisms.

might be responsible for the bioreduction of silver ions to metallic silver and the subsequent formation of silver nanoparticles (Karthik et al. 2014).

The mechanism of gold nanoparticle synthesis is similar to the mechanism of silver nanoparticle formation and based on reduction of gold ions from aqueous $AgCl_4^-$ (Kalishwaralal et al. 2009). Bioreduction of gold ions seems to be initiated by electron transfer from NADH by NADH-dependent reductase as electron carrier. The gold ions are reduced to gold (Au^0) and then to gold nanoparticles (He et al. 2007, Sadowski 2010).

Biosynthesis of metal nanoparticles using microorganisms is rather slow as compared with the chemical or physical synthesis process. A comparatively longer time period of 24–96 hours or even 120 hours is required to synthesize the nanoparticles in actinomycetes (Ahmad et al. 2003a, Ahmad et al. 2003b, Sastry et al. 2003, Alani et al. 2012, Selvakumar et al. 2012, Manivasagan et al. 2013). However, biogenic synthesis was found to be effective, cheap and eco-friendly, as it does not require addition of stabilizing agents (Golinska et al. 2014, Golinska et al. 2015, Składanowski et al. 2016, Wypij et al. 2018a, Wypij et al. 2019). In chemical methods stabilizer is added to prevent agglomeration of metal nanoparticles and the metal nanoparticles obtained possess a high toxicity. The biogenic metal nanoparticles from microorganisms, including actinomycetes, are very stable and well dispersed as they are covered by biomolecules named capping agents, which are considered to be responsible for the formation and stabilization of metal nanoparticles (Duran et al. 2005, Mukherjee et al. 2008, Parikh et al. 2008, Bar et al. 2009, Sanghi and Verma 2009, Prakasham Shetty et al. 2012, Narasimha et al. 2013, Sanjenbam et al. 2014, Wypij et al. 2018a). Such biomolecules from biological systems can bind to metal nanoparticles either through free amino groups or cysteine residues in proteins (Mandal et al. 2005, Avilala and Golla 2019). Controlled synthesis of nanoparticles with desirable properties by using actinomycetes is easy to achieve (Golinska et al. 2014, Wypij et al. 2019). Optimization of the growth conditions, such as media components, pH, temperature, substrate concentration and inoculum size, will not only support the growth but also enhance the productivity and monitor the rate of enzyme activity that affects the synthesis of silver nanoparticles (El-Naggar and Abdelwahed 2014, Iravani 2014, Singh et al. 2014, Mohamedin et al. 2015, Wypij et al. 2019).

Most available studies on actinomycete-mediated synthesis of metal nanoparticles concern synthesis of silver nanoparticles with antibacterial properties. These studies have shown that actinomycetes produced silver nanoparticles ranging from 5 to 100 nm, mainly 10–50 nm, and with a spherical shape (Table 1.1).

Biomedical applications of nanoparticles

Antimicrobial activity of actinomycete metal nanoparticles, mainly silver but also gold, zinc and copper, against bacterial pathogens, including multidrug-resistant strains and fungi, was proved by many researchers (Usha et al. 2010, Balagurunathan et al. 2011, Rajamanickam et al. 2012, Chauhan et al. 2013, Składanowski et al. 2016, Składanowski et al. 2017, Wypij et al. 2018a, Wypij et al. 2018b, Avilala and Golla 2019), as presented in Table 1.1. Silver nanoparticles from actinomycetes also displayed activity against acaricides (Karthik et al. 2014) and viruses (Avilala and Golla 2019) as well as cytotoxicity towards cancer cells (Chauhan et al. 2013, Abd-Elnaby et al. 2016, Wypij et al. 2018a, Wypij et al. 2018b). Because of the effective potential of metal nanoparticles against a broad range of microorganisms, they are considered the new generation of antimicrobials (Rai et al. 2009).

Biogenic synthesis of metal nanoparticles and its bioactivities depend on a few factors, mainly size, shape and stability (Pal et al. 2007, Sharma et al. 2009, Rai et al. 2012, Golinska et al. 2014, Wypij et al. 2018a). Based on available reports it can be concluded that actinomycetes produce mainly spherical metal nanoparticles (Golinska et al. 2014). However, rod-shaped gold nanoparticles were also formed by *Streptomycetes viridogens* HM10 (Balagurunathan et al. 2011).

Researchers have proved that triangular silver nanoparticles showed strongest biocidal properties against *E. coli*, when compared with spherical and rod-shaped nanoparticles. The triangular nanoparticles inhibited the growth of bacteria in lower concentration than nanoparticles of other shapes (Pal et al. 2007).

The antimicrobial properties of metal nanoparticles are related to the total surface area–to–volume ratios of the nanoparticles. Thus, smaller nanoparticles display higher antimicrobial activity (Baker et al. 2005, Choi and Hu 2008). Many studies have demonstrated that smaller nanoparticles have higher toxicity on bacterial pathogens, as these nanoparticles probably diffuse more easily by cell envelopes than the larger ones (Panacek et al. 2006, Mohan et al. 2007). It is also claimed that nanoparticles smaller than 10 nm interact with bacterial cells and produce electronic effects, which enhance the reactivity of metal nanoparticles (Morones et al. 2005, Raimondi et al. 2005). Detailed mechanism of nanoparticle action on microbial cells is presented later in this chapter.

Antibacterial activity of NPs

The bactericidal activity of silver nanoparticles synthesized from actinomycetes, both intra- and extracellularly, against different species of bacteria, including clinically important pathogens of *Bacillus subtilis, Enterobacter* sp., *Enterococcus faecalis, Escherichia coli, Klebsiella pneumoniae, Proteus mirabilis, Pseudomonas aeruginosa, Staphylococcus aureus, Salmonella infantis, Streptococcus aureus, Vibrio* sp. and many more has been reported, as presented in Table 1.1 (Chauhan et al. 2013, Manivasagan et al. 2013, Sukanya et al. 2013, Golinska et al. 2015, Abd-Elnaby et al. 2016, Składanowski et al. 2016, Wypij et al. 2018a, Avilala and Golla 2019). Antibacterial activity of AgNPs has been estimated by using well/disc diffusion method (Chauhan et al. 2013, Manivasagan et al. 2013, Narasimha et al. 2013, Sukanya et al. 2013, Abd-Elnaby et al. 2016, Avilala and Golla 2019) and/ or evaluation of minimum inhibitory concentration (MIC) and minimum biocidal concentration (MBC) (Golinska et al. 2015, Rathod et al. 2016, Składanowski et al. 2016, Wypij et al. 2018a, Wypij et al. 2018b). Many authors reported that AgNPs synthesized from actinobacteria inhibited growth of both Gram-positive and Gram-negative bacteria, including multidrug-resistant ones (Dayanand et al. 2010, Golinska et al. 2015, Rathod et al. 2016, Wypij et al. 2018a, Avilala and Golla 2019). High activity of AgNPs from *Streptomyces* sp. against Gram-positive bacteria (*S. aureus, Staphylococcus epidermidis, E. faecalis, B. subtilis*) and Gram-negative bacteria (*E. coli, K. pneumoniae, P. mirabilis, P. aeruginosa*) was found in studies by Abdeen et al. (2010), Selvakumar et al. (2012), Subashini and Kannabiran (2013), Sukanya et al. (2013) and Avilala and Golla (2019). The inhibition zones observed were found to be in the range of 7–40 mm. Składanowski et al. (2016, 2017) using microdilution method observed that growth of Gram-positive and Gram-negative bacteria was inhibited in concentration range of AgNPs between 1.25 and 10 µg ml^{-1}. Other authors reported that MIC of silver nanoparticles synthesized from *Nocardiopsis valliformis* OT1, *Pilimelia columellifera* subsp. *pallida* SL19 and SL24, *Streptomyces xinghaensis* OF1 and *Streptomyces calidiresistens* IF11 and IF17 against *B. subtilis, E. coli, K. pneumoniae, P. aeruginosa* and *S. aureus* were found to be in concentration range between 8 and 256 µg ml^{-1} (Golinska et al. 2015, Rathod et al. 2016, Wypij et al. 2018a, Wypij et al. 2018b).

Furthermore, Golinska et al. (2015) and Rathod et al. (2016) reported that bacterial strains that were resistant to antibiotics were highly susceptible to silver nanoparticles. The authors in their studies showed that *K. pneumoniae* and *P. aeruginosa* were resistant to ampicillin and kanamycin but sensitive to bionanoparticles synthesized from two strains of *Pilimelia columellifera* subsp. *pallida*. From the obtained results it was concluded that silver nanoparticles synthesized from actinomycetes seem to be promising and effective antibacterial agents against multidrug-resistant bacteria (Golinska et al. 2015, Rathod et al. 2016).

Selvakumar et al. (2012) showed that silver nanoparticles biosynthesized from *Streptomyces rochei* in the presence of 0.1 and 1 mM concentration of silver nitrate displayed different antimicrobial activity. Silver nanoparticles obtained in the presence of 1 mM AgNO$_3$ were active against *P. aeruginosa, E. coli, K. pneumoniae, E. faecalis* and *S. aureus* with formation of 23, 24, 20, 18, and 27 mm inhibition zones, respectively. Those received in the presence of 0.1 mM silver nitrate showed higher antibacterial activity giving 28, 22, 21, 25 and 31 mm inhibition zones, respectively. However, most authors used 1 mM AgNO$_3$ for efficient biosynthesis of silver nanoparticles (Golinska et al. 2015, Rathod et al. 2016, Wypij et al. 2018a, Wypij et al. 2018b, Avilala and Golla 2019, Wypij et al. 2019).

Although synthesis of metal nanoparticles other than silver (e.g., gold, copper and zinc) using various genera of actinomycetes, namely *Rhodococcus* sp., *Streptomyces* sp., *Thermoactinomycete* spp., *Nocardia* sp., and *Thermomonospora* sp., was reported extensively (Ahmad et al. 2003a, Ahmad et al. 2003b, Sastry et al. 2003, Balagurunathan et al. 2011, Oza et al. 2012, Kalabegishvili et al. 2013, Waghmare et al. 2014, Składanowski et al. 2017), their antibacterial activity was found less often. Balagurunathan et al. (2011), Rajamanickam et al. (2012) and Usha et al. (2010) showed antibacterial activity of gold, zinc and copper nanoparticles from actinomycetes of the genus *Streptomyces* against *S. aureus, E. coli* (gold, zinc, copper NPs) or *Salmonella* sp. (zinc NPs). Spherical zinc nanoparticles showed higher activity against *S. aureus*, followed by *E. coli* and *Salmonella* sp., and displayed inhibition zone 23, 21 and 19 mm, respectively (Rajamanickam et al. 2012).

Antifungal activity of NPs

Nanoparticles produced by actinomycetes were also found to be active against various fungi such as *Candida albicans, C. tropicalis, C. krusei*, different species of the genus *Aspergillus* (*A. niger, A. fumigatus, A. flavus, A. terreus, A. brasiliensis*), and dermatophytes such as *Malassezia furfur, Trichophyton rubrum, T. tonsurans*, and *T. erinacei* (Sadhasivam et al. 2010, Vidyasagar et al. 2012, Chauhan et al. 2013, Manivasagan et al. 2013, Subashini and Kannabiran 2013, Sanjenbam et al. 2014, Anasane et al. 2016, Wypij et al. 2017a, Wypij et al. 2018a, Wypij et al. 2018b), as presented in Table 1.1. Silver nanoparticles produced by actinomycetes, mainly *Streptomyces* sp., tested by disc diffusion method were found to be active against different species of *Candida, Aspergillus* and *Trichophyton*, giving inhibition zones of their growth in the range of 16–28.4 mm (Vidyasagar et al. 2012, Chauhan et al. 2013, Subashini and Kannabiran 2013, Sanjenbam et al. 2014). However, silver nanoparticles synthesized extracellularly by *Streptomyces* sp. showed selective inhibition over fungal pathogens tested. Silver nanoparticles exhibited 20 and 22 mm inhibition zone for *A. fumigatus* and *A. niger*, respectively, but growth inhibition of *A. flavus* and *C. albicans* was not observed (Subashini and Kannabiran 2013). Similarly, Anasane and co-authors (2016) reported high antifungal activity (inhibition zone in the range of 14–26 mm) of silver nanoparticles synthesized by *Pilimelia columellifera* subsp. *pallida* strains SF23 and C9 against *C. albicans, C. tropicalis, M. furfur* and *T. rubrum*, with the highest antifungal activity against threatening human pathogen of *T. rubrum*. They also estimated minimal inhibitory concentrations of AgNPs against tested fungi, which were found to be in the range of 20–40 µg ml^{-1} (Anasane et al. 2016). However, Wypij et al. (2018b) found that AgNPs from two alkalophilic strains of *Streptomyces calidiresistens* effectively inhibited growth of *M. furfur* at concentration equal to 8 and 16 µg ml^{-1} and *Candida albicans* at concentration equal to 64 and 96 µg ml^{-1}, respectively. Similarly, high antifungal activity of AgNPs from alkalophilic *Streptomyces xinghaiensis* strain OF1 against *M. furfur* and *C. albicans* was reported in studies of Wypij et al. (2018a). Authors found that growth of both yeasts was inhibited at concentration of AgNPs equal to 32 µg ml^{-1}. Wypij et al. (2017a) also studied antifungal activity of AgNPs from acidophilic actinobacteria of *Pilimelia columellifera* subsp. *pallida* strain SL19. These AgNPs inhibited growth of *M. furfur, C. albicans* and *T. erinacei* at concentration of 16, 32 and 192 µg ml^{-1}, respectively.

Other bioactivities

Antibacterial and antifungal activity of AgNPs reduces the density of microbial cells and acts as an anti-biofouling agent (Abd-Elnaby et al. 2016). Significant inhibition of biofilm formation with AgNPs from actinomycetes was reported against *E. coli* (Abd-Elnaby et al. 2016) and against *E. coli, S. aureus* and *C. albicans* (Wypij et al. 2018b). The AgNPs from alkalophilic actinomycete strains, namely IF11 and IF17, showed different antimicrobial activity. The AgNPs from IF17 strain showed higher antimicrobial activity and prevented biofilm formation at concentrations of 2, 2 and 16 μg ml^{-1}, and those from strain IF11 at concentrations of 4, 64 and 96 μg ml^{-1}, respectively.

Apart from the antibacterial and antifungal potential of silver nanoparticles synthesized from actinomycetes, their antiviral activity against Newcastle disease virus (Avilala and Golla 2019) and antiparasitic activity against *Rhipicephalus microplus* and *Haemaphysalis bispinosa* (Karthik et al. 2014) were also reported. Antiviral activity of AgNPs was noted after 48–96 hours but not after 24 hours of incubation and was stronger after application of higher dosage (Avilala and Golla 2019). Karthik and co-authors (2014) observed that concentrations of AgNPs equal to 16.1 and 16.45 mg l^{-1}, respectively, were efficiently active against test parasites (both values were found to be LC$_{50}$). On the basis of these results, researchers suggested that biogenic silver nanoparticles could provide a safer alternative to conventional acaricidal agents in the form of a topical antiparasitic formulation (Karthik et al. 2014).

Silver nanoparticles synthesized from actinomycetes were reported to be active against various types of cancer cell lines (Manivasagan et al. 2013, Abd-Elnaby et al. 2016, Rathod et al. 2016, Subbaiya et al. 2017, Wypij et al. 2017a, Wypij et al. 2018a, Wypij et al. 2018b). Manivasagan et al. (2013) found that the IC$_{50}$ value of AgNPs biosynthesized from *Nocardiopsis* sp. against HeLa cells was equal to 200 μg ml^{-1}. In contrast, other authors reported that AgNPs synthesized from actinomycetes displayed high cytotoxicity toward cancer cells at much lower concentrations used (Rathod et al. 2016, Wypij et al. 2017a, Wypij et al. 2018a, Wypij et al. 2018b). Rathod and coworkers (2016) found that AgNPs from *Nocardiopsis valliformis* OT1 were toxic towards human cervical cancer cell line (HeLa) and demonstrated a dose–response activity. The IC$_{50}$ value of AgNPs was found to be 100 μg/mL (Rathod et al. 2016). Moreover, Wypij and co-authors (2017a, 2018a, 2018b) observed much higher cytotoxicity of AgNPs from *Pilimelia columellifera* subsp. *pallida* SL19, *Streptomyces calidiresistens* strains IF11 and IF17 and *S. xinghaiensis* OF1 toward HeLa cells. The IC$_{50}$ values were found to be 55, 28.5, 53.8 and 4 μg ml^{-1}, respectively. Subbaiya et al. (2017) reported that AgNPs from *Streptomyces atrovirens* show profound inhibition of breast cancer cells (MCF-7) at a concentration of 44.51 μg ml^{-1}. Further, the authors concluded that bioAgNPs might be emerging alternative biomaterials for human breast cancer therapy (Subbaiya et al. 2017). Other authors showed anticancer potential of AgNPs from *Streptomyces rochei* HMM13 towards five different cancer cell lines, namely hepatocellular carcinoma cells (HepG-2), breast carcinoma cells (MCF-7), colon carcinoma cells (HCT-116), prostate carcinoma cells (PC-3), and lung carcinoma cells (A-549). The IC$_{50}$ values reported were 32.9, 40.0, 9.05, 48.5, 42.1 μg/Well. However, anticancer activity of test AgNPs against intestinal carcinoma cells (CACO), larynx carcinoma cells (HEP-2) and cervical carcinoma cells (HELA) was found to be lower (Abd-Elnaby et al. 2016).

Synergism of NPs with antibiotics

Beyond antimicrobial activity of actinomycete silver nanoparticles itself, the enhancement or synergistic effect of AgNPs in combination with commercial antibiotics against Gram-positive and Gram-negative bacteria (ampicillin, amoxycillin, vancomycin, erythromycin, kanamycin, ofloxacin, tetracycline or tigecycline) and fungi (nystatin, amphotericin B, fluconazole, ketoconazole) were widely reported (Chauhan et al. 2013, Manivasagan et al. 2013, Wypij et al. 2017a, Wypij et al. 2018a, Wypij et al. 2018b). Chauhan et al. (2013) studied antimicrobial activity of four antibiotics (erythromycin, vancomycin, tigecycline and ofloxacin) in combination with silver nanoparticles. They found that antibacterial activities of tested antibiotics increased in the presence of silver nanoparticles against the clinically important pathogenic microorganisms (*E. faecalis, S. aureus, E. coli, Salmonella typhimurium, Shigella* sp., *P. mirabilis, K. pneumoniae* and *P. aeruginosa*). The enhancement effect of silver nanoparticles representing the highest percentage of increase in inhibition was found with vancomycin, followed by erythromycin, ofloxacin and tigecycline against the tested strains.

The combined formulation of silver nanoparticles from *Streptomyces rochei* with standard antibiotic discs, from different groups and having a different mode of action (ciprofloxacin, ampicillin, streptomycin, gentamicin, tetracycline and lincomycin) was studied against *Vibrio fluvialis, P. aeruginosa, S. typhimurium, Vibrio damsela, E. coli, B. subtilis, Bacillus cereus* and *S. aureus* (Abd-Elnaby et al. 2016). They showed that combined use of AgNPs and antibiotics significantly enhanced antibacterial effect against all tested pathogens. Similarly, an increase of antifungal activity of ketoconazole after combined use with AgNPs from two strains of *Pilimelia columellifera* subsp. *pallida* (C9 and SF23) against *C. albicans, C. tropicalis, M. furfur* and *T. rubrum* was reported by Anasane et al. (2016). Rathod et al. (2016) used dilution plate method for estimation of synergistic effect of ampicillin, kanamycin and tetracycline in combination with AgNPs from *Nocardiopsis valliformis* OT1 against *E. coli, S. aureus, K. pneumoniae, P. aeruginosa* and *B. subtilis*. They observed maximum synergistic effect for ampicillin and kanamycin when combined with AgNPs against *P. aeruginosa* (41.92% and 35.8% higher bacterial growth inhibition, respectively, than AgNPs used individually). A similar method was used by Wypij et al. (2017b), who studied antibacterial effect of AgNPs from acidophilic *Streptomyces* sp. strain SH11 in combination with ampicillin, kanamycin and tetracycline against *E. coli, S. aureus* and *B. subtilis*. They also showed that the antimicrobial efficacy of antibiotics was enhanced in the presence of AgNPs (Wypij et al. 2017b). However, the most accurate method for estimation of activity of both antimicrobial agents (AgNPs and antibiotics) seems to be those performed by Wypij et al. (2017a, 2018a, 2018b). Authors determined the effect of AgNP and antibiotic combination against bacteria or fungi by estimation of fractional inhibitory concentration (FIC) index proposed by Doern (2014). Synergistic interactions of AgNPs and amphotericin B or ketoconazole were observed against *C. albicans* and *M. furfur* (FIC index $\leq$ 0.5). Synergistic interaction of AgNPs and fluconazole was found only against *M. furfur*, while non-synergistic effect was found when AgNPs were combined with fluconazole against *C. albicans* (FIC index of 1) (Wypij et al. 2017a). A similar effect was observed when combined antimicrobials were tested against bacteria, namely *E. coli, P. aeruginosa,*

S. aureus and *B. subtilis*, and yeasts, namely *C. albicans* and *M. furfur* (Wypij et al. 2018a, Wypij et al. 2018b). Combination of AgNPs and antibiotics significantly decreased concentrations of both antimicrobials used and retained their high antibacterial and antifungal activity (Wypij et al. 2018a, Wypij et al. 2018b). Wypij and co-workers (2018a) reported that 16 out of 21 antimicrobial agent combinations (AgNPs from *Streptomyces xinghaiensis* OF1 strain and various antibiotics) showed high synergistic effect against tested bacteria and fungi. The FIC index for those combinations was found to be 0.12. Such low FIC index value indicated that microbial growth was inhibited in the presence of 1/16 of MIC of both antimicrobials (AgNPs and antimicrobial agent), when compared to antimicrobials used alone. However, combination of AgNPs and antibiotics against *P. aeruginosa* revealed indifferent effect on bacterial growth (FIC equal to 2.0), while combination of amphotericin B and fluconazole with AgNPs against *M. furfur* demonstrated non-synergistic or additive effect (FIC equal to 1.0). Similar effects were noticed when combined use of AgNPs from two alkalophilic *Streptomyces calidiresistens* strains (IF11 and IF17) and antibiotics was tested against Gram-positive and Gram-negative bacteria and yeasts. However, the authors observed that synergism of AgNPs synthesized from IF17 strain with test antibiotics against bacteria and yeasts was found to be higher than those described for AgNPs synthesized from IF11 strain (Wypij et al. 2018b).

All the above researchers suggested that antibiotic molecules, which contain many active groups such as hydroxyl and amide groups, may easily react with nanosilver by chelation, which increases growth inhibition of bacterial cells.

Mechanisms of action of NPs against microbial cells

The proposed mechanisms of action of metal nanoparticles on microbial cells have been described for silver nanoparticles (Ghosh et al. 2012, Chauhan et al. 2013, Manivasagan et al. 2013, Abbaszadegan et al. 2015). Microorganisms exposure to AgNPs causes adhesion of nanoparticles onto the cell wall and the membrane (Abbaszadegan et al. 2015). Antimicrobial activity of biogenic AgNPs results from the interaction of AgNPs with the sulfur-containing proteins present in the cell wall. It causes irreversible changes in cell wall structure resulting in its disruption (Ghosh et al. 2012). This in turn affects the integrity of lipid bilayer and permeability of the cell membrane, which affects the ability of cells to properly regulate transport activity through the plasma (Panacek et al. 2006, Ghosh et al. 2012, Chauhan et al. 2013). Silver ions, as a secondary oxidation process of silver nanoparticles, contribute to the biocidal properties of silver nanoparticles (Morones et al. 2005, Manivasagan et al. 2013). Silver ions can affect membrane transport and the release of potassium (K^+) ions from the microbial cells. The increase of membrane permeability may lead to leakage of cellular contents, including ions, proteins, reducing sugars and sometimes cellular energy reservoir (ATP) (Lok et al. 2006, Kim et al. 2011, Chauhan et al. 2013, Li et al. 2013). AgNPs can further penetrate inside the microbial cell and interact with cellular structures (e.g., ribosomes) and biomolecules such as proteins, lipids, and DNA, which have damaging effect on microbial cells. Silver ions affect the losses in ability to replicate DNA, translation process in ribosomes and protein activities (Morones et al. 2005, Chauhan et al. 2013, Jung et al. 2008, Rai et al. 2012).

Conclusion and future perspectives

A crucial need in the field of nanotechnology is the development of an eco-friendly and reliable process for nanoparticle synthesis. Based on available reports it can be concluded that actinomycetes have been used efficiently for nanoparticle synthesis, mainly for silver nanoparticle synthesis, by means of a low-cost, natural and renewable bio-reducing agent. Optimization of the culture conditions for actinomycetes made it possible to obtain small and stable nanoparticles. Actinomycete-mediated silver nanoparticles showed good antibacterial, antifungal, antibiofilm, antiparasitic and anticancer activities. Thus, activities of AgNPs from actinomycetes indicate their potential application in the field of nanomedicine.

Biological approaches to nanoparticle synthesis are still in the development stages. The synthesis of nanoparticles using microorganisms is a slow process compared to physical and chemical approaches. Thus, the reduction of synthesis time will make this biosynthesis route much more attractive. Stability and aggregation of the biosynthesized NPs, control of crystal growth, shape, size, and size distribution are the most significant factors. Therefore, optimization of biosynthesis parameters such as microorganism type, growth stage of microbial cells, growth medium, synthesis conditions, pH, substrate concentrations, source compound of target nanoparticle, temperature, reaction time, and addition of non-target ions for effective control of the particle size and monodispersity should be more extensively investigated, as strictly controlled particle morphology would offer considerable advantage to chemical and physical approaches. Moreover, a better understanding of the synthesis mechanism on a cellular and molecular level, including isolation and identification of the compounds responsible for the reduction of metal ions to nanoparticles and molecules acting as capping agents, is still in its infancy. These studies are required to involve biosynthesized nanoparticles in clinical trials in the future.

References

Abbaszadegan, A., Ghahramani, Y., Gholami, A., Hemmateenejad, B., Dorostkar, S., Nabavizadeh, M. 2015. The effect of charge at the surface of silver nanoparticles on antimicrobial activity against gram-positive and gram-negative bacteria: a preliminary study. J. Nanomater. 2015: 720654; http://dx.doi.org/10.1155/2015/720654.

Abdeen, S., Geo, S., Sukanya, S., Praseetha, P.K., Dhanya, R.P. 2014. Biosynthesis of silver nanoparticles from *Actinomycetes* for therapeutic applications. Int. J. Nano Dimens. 5: 155–162.

Abd-Elnaby, H.M., Abo-Elala, G.M., Abdel-Raouf, U.M., Hamed, M.M. 2016. Antibacterial and anticancer activity of extracellular synthesized silver nanoparticles from marine *Streptomyces rochei* MHM13. Egyptian J. Aquatic Res. 42: 301–312.

Ahmad, A., Senapati, S., Khan, M.I., Kumar, R., Ramani, R., Srinivas, V., Sastry, M. 2003a. Extracellular biosynthesis of monodisperse gold nanoparticles by a novel extremophilic actinomycete *Thermomonospora* sp. Langmuir 19: 3550–3553.

Ahmad, A., Senapati, S., Khan, M.I., Kumar, R., Ramani, R., Srinivas, V., Sastry, M. 2003b. Intracellular synthesis of gold nanoparticles by a novel alkalotolerant actinomycete, *Rhodococcus* species. Nanotechnology 14: 824.

Alani, F., Moo-Young, M., Anderson, W. 2012. Biosynthesis of silver nanoparticles by a new strain of *Streptomyces* sp. compared with *Aspergillus fumigatus*. World J. Microbiol. Biotechnol. 28: 1081–1086.

Anasane, N., Golińska, P., Wypij, M., Rathod, D., Dahm, H., Rai, M. 2016. Acidophilic actinobacteria synthesised silver nanoparticles showed remarkable activity against fungi-causing superficial mycoses in humans. Mycoses 59: 157–166.

Avilala, J., Golla, N. 2019. Antibacterial and antiviral properties of silver nanoparticles synthesized by marine actinomycetes. Int. J. Pharm. Res. 10: 1223–1228.

Baker, C., Pradhan, A., Pakstis, L., Pochan, D.J., Shah, S.I. 2005. Synthesis and antibacterial properties of silver nanoparticles. J. Nanosci. Nanotechnol. 2: 244–249.

Balagurunathan, R., Radhakrishnan, M., Rajendran, R.B., Velmurugan, D. 2011. Biosynthesis of gold nanoparticles by actinomycete *Streptomyces viridogens* strain HM10. J. Biochem. Biophys. 48: 331–335.

Bar, H., Bhudi, D.K., Sahoo, G.P., Sarkar, P., De, S.P., Misra, A. 2009. Green synthesis of silver nanoparticles using latex of *Jatropha curcas*. Coll. Surf. A Physicochem. Eng. Asp. 339: 134–139.

Bhattacharyya, D., Singh, S. 2009. Nanotechnology, big things from a tiny world: a review. Int. J. u- and e-Serv. Sci. Technol. 2: 29–38.

Chauhan, R., Kumar, A., Abraham, J. 2013. A biological approach to the synthesis of silver nanoparticles with *Streptomyces* sp. JAR1 and its antimicrobial activity. Sci. Pharm. 81: 607–621.

Choi, O.K., Hu, Z.Q. 2008. Size dependent and reactive oxygen species related nanosilver toxicity to nitrifying bacteria. Environ. Sci. Technol. 42: 4583–4588.

Dayanand, S.A., Sreedhar, B., Dastager, S.G. 2010. Antimicrobial activity of silver nanoparticles synthesized from novel *Streptomyces* species. Digest J. Nanomater. Biostruc. 5: 447–451.

Deepa, S., Kanimozhi, K., Panneerselvam, A. 2013. Antimicrobial activity of extracellularly synthesized silver nanoparticles from marine derived *Actinomycetes*. Int. J. Cur. Microbiol. App. Sci. 2: 223–230.

Doern, C.D. 2014. When does 2 plus 2 equal 5? A review of antimicrobial synergy testing. J. Clinical Microbiol. 52: 4124–4128.

Durán, N., Marcato, P.D., Alves, O.L., De Souza, G.I.H., Esposito, E. 2005. Mechanistic aspects of biosynthesis of silver nanoparticles by several *Fusarium oxysporum* strains. J. Nanobiotechnol. 3: 8.

El-Naggar, N.E., Abdelwahed, N.A.M. 2014. Application of statistical experimental design for optimization of silver nanoparticles biosynthesis by a nanofactory *Streptomyces viridochromogenes*. J. Microbiol. 52: 53–63.

Ghiuță, I., Cristea, D., Croitoru, C., Kost, J., Wenkert, R., Vyrides, I., Anayiotos, A., Munteanu, D. 2018. Characterization and antimicrobial activity of silver nanoparticles, biosynthesized using *Bacillus* species. Appl. Surf. Sci. 438: 66–73.

Ghosh, S., Patil, S., Ahire, M., Kitture, R., Kale, S., Pardesi, K. 2012. Synthesis of silver nanoparticles using *Dioscoreabulbifera* tuberextract and evaluation of its synergistic potential in combination with antimicrobial agents. Int. J. Nanomed. 7: 483–496.

Golinska, P., Wypij, M., Ingle, A.P., Gupta, I., Dahm, H., Rai, M. 2014. Biogenic synthesis of metal nanoparticles from actinomycetes: biomedical applications and cytotoxicity. Appl. Microbiol. Biotechnol. 98: 8083–8097.

Golinska, P., Wypij, M., Rathod, D., Tikar, S., Dahm, H., Rai, M. 2015. Synthesis of silver nanoparticles from two acidophilic strains of *Pilimelia columellifera* subsp. *pallida* and their antibacterial activities. J. Basic Microbiol. 55: 1–16.

Harish, K.K., Nagasamy, V., Himangshu, B., Anuttam, K. 2018. Metallic nanoparticle: a review. Biomed. J. Sci. Tech. Res. 4: 3765–3775.

He, S., Guo, Z., Zhang, Y., Zhang, S., Wang, J., Gu, N. 2007. Biosynthesis of gold nanoparticles using the bacteria *Rhodopseudomonas capsulata*. Mater. Lett. 61: 3984–3987.

Ingle, A.P., Gade, A.K., Pierrat, S., Sönnichsen, C., Rai, M.K. 2008. Mycosynthesis of silver nanoparticles using the fungus *Fusarium acuminatum* and its activity against some human pathogenic bacteria. Curr. Nanosci. 4: 141–144.

Iravani, S., Korbekandi, H., Mirmohammadi, S.V., Zolfaghari, B. 2014. Synthesis of silver nanoparticles: chemical, physical and biological methods. Resin. Pharma. Sci. 9: 385–406.

Jung, W.K., Koo, H.C., Kim, K.W., Shin, S., Kim, S.H., Park, Y.H. 2008. Antibacterial activity and mechanism of action of the silver ion in *Staphylococcus aureus* and *Escherichia coli*. Appl. Environ. Microbiol. 74: 2171–2178.

Kalabegishvili, T., Kirkesali, E., Ginturi, E., Rcheulishvili, A., Murusidze, I., Pataraya, D., Gurielidze, M., Bagdavadze, N., Kuchava, N., Gvarjaladze, D., Lomidze, L. 2013. Synthesis of gold nanoparticles by new strains of thermophilic *Actinomycetes*. Nano Studies 7: 255–260.

Kalishwaralal, K., Deepak, V., Ram Kumar Pandian, S., Gurunathan, S. 2009. Biological synthesis of gold nanocubes from *Bacillus licheniformis*. Bioresour. Technol. 100: 5356–5358.

Karthik, L., Kumar, G., Vishnu Kirthi, A., Rahuman, A.A., Bhaskara Rao, K.V. 2014. *Streptomyces* sp. LK3 mediated synthesis of silver nanoparticles and its biomedical application. Bioprocess Biosyst. Eng. 37: 261–267.

Kim, J.S., Kuk, E., Yu, K.N., Kim, J.H., Park, S.J., Lee, H.J., Jeong, D.H., Cho, M.H. 2007. Antimicrobial effects of silver nanoparticles. Nanomed. Nanotechnol. Biol. Med. 3: 95–101.

Kim, S.H., Lee, H.S., Ryu, D.S., Choi, S.J., Lee, D.S. 2011. Antibacterial activity of silver-nanoparticles against *Staphylococcus aureus* and *Escherichia coli*. Korean J. Microbiol. Biotechnol. 39: 77–85.

Kowshik, M., Ashtaputre, S., Kharrazi, S., Vogel, W., Urban, J., Kulkarni, S.K. Paknikar, K.M. 2003. Extracellular synthesis of silver nanoparticles by a silver-tolerant yeast strain MKY3. Nanotechnology 14: 95–100.

Lee, S.H., Jun, B.H. 2019. Silver nanoparticles: synthesis and application for nanomedicine Int. J. Mol. Sci. 20: 865.

Li, J., Tian, B., Li, T., Dai, S., Weng, Y., Lu, J., Xu, X., Jin, Y., Pang, R., Hua, Y. 2018. Biosynthesis of Au, Ag and Au–Ag bimetallic nanoparticles using protein extracts of *Deinococcus radiodurans* and evaluation of their cytotoxicity. Int. J. Nanomed. 13: 1411–1424.

Li, J., Rong, K., Zhao, H., Li, F., Lu, Z., Chen, R. 2013. Highly selective antibacterial activities of silver nanoparticles against *Bacillus subtilis*. J. Nanosci. Nanotechnol. 13: 6806–6813.

Lok, C.N., Ho, C.M., Chen, R., He, Q.Y., Yu, W.Y., Sun, H. 2006. Proteomic analysis of the mode of antibacterial action of silver nanoparticles. J. Proteome. Res. 5: 916–924.

Luangpipat, T., Beattie, I.R., Chisti, Y., Haverkamp, R.G. 2011. Gold nanoparticles produced in a microalga. J. Nanoparticle Res. 13: 6439–6445.

Mandal, S., Phadtare, S., Sastry, M. 2005. Interfacing biology with nanoparticles. Curr. Appl. Physics 5: 118–127.

Manivasagan, P., Venkatesan, J., Senthilkumar, K., Sivakumar, K., Kim, S. 2013. Biosynthesis, antimicrobial and cytotoxic effect of silver nanoparticles using a novel *Nocardiopsis* sp. MBRC-1. Bio. Med. Res. Int. 2013; http://dx.doi.org/10.1155/2013/287638.

Mohamedin, A., El-Naggar, N.E., Hamza, S.S., Sherief, A.A. 2015. Green synthesis, characterization and antimicrobial activities of silver nanoparticles by *Streptomyces viridodiastaticus* SSHH-1 as a living nanofactory: statistical optimization of process variables. Curr. Nanosci. 11: 640–654.

Mohan, Y.M., Lee, K., Premkumar, T., Geckeler, K.E. 2007. Hydrogel networks as nanoreactors: a novel approach to silver nanoparticles for antibacterial applications. Polymer 48: 158–164.

Morones, J.R., Elechiguerra, J.L., Camacho, A., Ramirez, J.T. 2005. The bactericidal effect of silver nanoparticles. Nanotechnology 16: 2346–2353.

Mukherjee, P., Roy, M., Mandal, P., Dey, G.K., Mukherjee, P.K., Ghatak, J., Tyagi, A.K., Kale, S.P. 2008. Green synthesis of highly stabilized nanocrystalline silver particles by a non-pathogenic and agriculturally important fungus *T. asperellum*. Nanotechnology 19: 1–7.

Narasimha, G., Janardhan Alzohairy, M., Khadri, H., Mallikarjuna, K. 2013. Extracellular synthesis, characterization and antibacterial activity of silver nanoparticles by *Actinomycetes* isolative. Int. J. Nano Dimens. 4: 77–83.

Oza, G., Pandey, S., Gupta, A., Kesarkar, R., Sharon, M. 2012. Biosynthetic reduction of gold ions to gold nanoparticles by *Nocardia farcinica*. J. Microbiol. Biotechnol. Res. 4: 511–515.

Pal, S., Tak, Y.K., Song, J.M. 2007. Does the antibacterial activity of silver nanoparticles depend on the shape of the nanoparticle? A study of the gram-negative bacterium *Escherichia coli*. Appl. Environ. Microbiol. 27: 1712–1720.

Panacek, A., Kvitek, L., Prucek, R., Kolar, M., Vecerova, R., Pizurova, N., Sharma, V.K., Nevecna, T. 2006. Silver colloid nanoparticles: synthesis, characterization, and their antibacterial activity. J. Phys. Chem. 110: 16248–16253.

Parikh, R.Y., Singh, S., Prasad, B.L.V., Patole, M.S., Sastry, M., Shouche, Y.S. 2008. Extracellular synthesis of crystalline silver nanoparticles and molecular evidence of silver resistance from *Morganella* sp.: towards understanding biochemical synthesis mechanism. Chem-biochem. 9: 34–41.

Prabhu, S., Poulose, E.K. 2012. Silver nanoparticles: mechanism of antimicrobial action, synthesis, medical applications, and toxicity effects. Int. Nano Lett. 2: 32.

Prakasham Shetty, R., Kumar, B.S., Kumar, Y.S., Shankar, G.G. 2012. Characterization of silver nanoparticles synthesized by using marine isolate *Streptomyces albidoflavus*. J. Microbiol. Biotechnol. 22: 614–621.

Prasad, K., Jha, A.K., Kulkarni, A.R. 2007. *Lactobacillus* assisted synthesis of titanium nanoparticles. Nanoscale Res. Lett. 2: 248–250.

Quester, K., Avalos-Borja, M., Castro-Longoria, E. 2016. Controllable biosynthesis of small silver nanoparticles using fungal extract. J. Biomaterials Nanobiotechnol. 7: 118–125.

Rai, M., Deshmukh, S.D., Ingle, A.P., Gade, A.K. 2012. Silver nanoparticles: the powerful nanoweapon against multidrug-resistant bacteria. J. Appl. Microbiol. 112: 841–852.

Rai, M., Ingle, A.P., Gade, A.K., Duarte, M.C., Duran, N. 2015. Three *Phoma* spp. synthesised novel silver nanoparticles that possess excellent antimicrobial efficacy. IET Nanobiotechnol. 9: 280–287.

Rai, M., Yadav, A., Gade, A. 2009. Silver nanoparticles as a new generation of antimicrobials. Biotechnol. Adv. 27: 76–83.

Raimondi, F., Scherer, G.G., Kotz, R., Wokaun, A. 2005. Nanoparticles in energy technology: examples from electrochemistry and catalysis. Angew. Chem. Int. Ed Engl. 44: 2190–2209.

Rajamanickam, U., Mylsamy, P., Viswanathan, S., Muthusamy, P. 2012. Biosynthesis of zinc nanoparticles using *Actinomycetes* for antibacterial food packaging. International Conference on Nutrition and Food Sciences IPCBEE vol. 39 IACSIT.

Rathod, D., Golińska, P., Wypij, M., Dahm, H., Rai, M. 2016. A new report of *Nocardiopsis valliformis* strain OT1 from alkaline Lonar crater of India and its use in synthesis of silver nanoparticles with special reference to evaluation of antibacterial activity and cytotoxicity. Med. Microbiol. Immunol. 205: 435–447.

Rizvi, S.A.A., Saleh, A.M. 2018. Applications of nanoparticle systems in drug delivery technology. Saudi Pharm. J. 26: 64–70.

Sadhasivam, S., Shanmugam, P., Yun, K. 2010. Biosynthesis of silver nanoparticles by *Streptomyces hygroscopicus* and antimicrobial activity against medically important pathogenic microorganisms. Coll. Surf. B Biointerfaces 81: 358–362.

Sadowski, Z. 2010. Biosynthesis and application of silver and gold nanoparticles, silver nanoparticles, David Pozo Perez (Ed.), ISBN: 978-953-307-028-5, InTech, http://www.intechopen.com/books/silver-nanoparticles/biosynthesis-and-application-of-silver-and-goldnanoparticles.

Salata, O. 2004. Applications of nanoparticles in biology and medicine. J. Nanobiotechnol. 2: 1–6.

Sanghi, R., Verma, P. 2009. Biomimetric synthesis and characterization of protein capped silver nanoparticles. Bioresour. Technol. 100: 501–504.

Sanjenbam, P., Gopal, J.V., Kannabiran, K. 2014. Anticandidal activity of silver nanoparticles synthesized using *Streptomyces* sp. VITPK1. J. Mycol. Méd. http://dx.doi.org/10.1016/j.mycmed.2014.03.004.

Sastry, M., Ahmed, A., Khan, M.I., Kumar, R. 2003. Biosynthesis of metal nanoparticles using fungi and *Actinomycetes*. Curr. Nanosci. 85: 162–170.

Selvakumar, P., Viveka, S., Prakash, S., Jasminebeaula, S., Uloganathan, R. 2012. Antimicrobial activity of extracellularly synthesized silver nanoparticles from marine derived *Streptomyces rochei*. Int. J. Pharma Bio. Sci. 3: 188–197.

Shah, M., Fawcett, D., Sharma, S., Tripathy, S.K., Poinern, G.E.J. 2015. Green synthesis of metallic nanoparticles via biological entities. Materials (Basel). 8: 7278–7308.

Sharma, V.K., Yngard, R.A., Lin, Y. 2009. Silver nanoparticles: green synthesis and their antimicrobial activities. Adv. Coll. Interface Sci. 145: 83–96.

Singh, D., Rathod, V., Ninganagouda, S., Hiremath, J., Kumar, A.S., Mathew, J. 2014. Optimization and characterization of silver nanoparticle by endophytic fungi *Penicillium* sp. isolated from

Curcuma longa (Turmeric) and application studies against MDR *E. coli* and *S. aureus*. Bioinorg. Chem. Appl. 2014: 1–8. Article ID408021.

Sivalingam, P., Antony, J.J., Siva, D., Achiraman, S., Anbarasu, K. 2012. Mangrove Streptomyces sp. BDUKAS10 as nanofactory for fabrication of bactericidal silver nanoparticles. Colloids Surf. B: Biointerfaces 98: 12–17.

Składanowski, M., Golinska, P., Rudnicka, K., Dahm, H., Rai, M. 2016. Evaluation of cytotoxicity, immune-compatibility and antibacterial activity of biogenic silver nanoparticles. Medical Microbiol. Immunol. 205: 603–613.

Składanowski, M., Wypij, M., Laskowski, D., Golińska, P., Dahm, H., Rai, M. 2017. Silver and gold nanoparticles synthesized from *Streptomyces* sp. isolated from acid forest soil with special reference to its antibacterial activity against pathogens. J. Cluster Sci. 28: 59–79.

Subashini, J., Kannabiran, K. 2013. Antimicrobial activity of *Streptomyces* sp. VITBT7 and its synthesized silver nanoparticles against medically important fungal and bacterial pathogens. Der. Pharmacia Lett. 5: 192–200.

Subbaiya, R., Saravanan, M., Priya, A.R., Shankar, K.R., Selyam, M., Ovais, M., Ballajee, R., Barabadi, M. 2017. Biomimetic synthesis of silver nanoparticles from *Streptomyces atrovirens* and their potential anticancer activity against human breast cancer cells. IET Nanobiotechnol. 11: 965–972.

Sukanya, M.K., Saju, K.A., Praseetha, P.K., Sakthivel, G. 2013. Potential of biologically reduced silver nanoparticles from actinomycete cultures. J. Nanosci. 1–8.

Usha, R., Prabu, E., Palaniswamy, M., Venil, C.K., Rajendran, R. 2010. Synthesis of metal oxide nanoparticles by *Streptomyces* sp. for development of antimicrobial textiles. Global J. Biotechnol. Biochem. 5: 153–160.

Vidyasagar, G.M., Shankaravva, B., Begum, R., Imrose Raibagkar, R.L. 2012. Antimicrobial activity of silver nanoparticles synthesized by *Streptomyces* species JF714876. Int. J. Pharm. Sci. Nanotechnol. 5: 1638–1642.

Waghmare, S.S., Deshmukh, A.M., Kulkarni, W., Oswaldo, L.A. 2011. Biosynthesis and characterization of manganese and zinc nanoparticles. Uni J. Environ. Res. Technol. 1: 64–69.

Waghmare, S.S., Deshmukh, A.M., Sadowski, Z. 2014. Biosynthesis, optimization, purification and characterization of gold nanoparticles. African J. Microbiol. Res. 8: 138–146.

Wolska, K.I., Markowska, K., Wypij, M., Golińska, P., Dahm, H. 2017. Nanocząstki srebra, synteza i biologiczna aktywność. Kosmos 66: 125–138.

Wypij, M., Czarnecka, J., Dahm, H., Rai, M., Golinska, P. 2017a. Silver nanoparticles from *Pilimelia columellifera* subsp. *pallida* SL19 strain demonstrated antifungal activity against fungi causing superficial mycoses. J. Basic Microbiol. 57: 793–800.

Wypij, M., Golińska, P., Dahm, H., Rai, M. 2017b. Actinobacterial-mediated synthesis of silver nanoparticles and their activity against pathogenic bacteria. IET Nanobiotechnol. 11: 336–342.

Wypij, M., Czarnecka, J., Świecimska, M., Dahm, H., Rai, M., Golińska, P. 2018a. Synthesis, characterization and evaluation of antimicrobial and cytotoxic activities of biogenic silver nanoparticles synthesized from *Streptomyces xinghaiensis* OF1 strain. World J. Microbiol. Biotechnol. 34: 23; doi: 10.1007/s11274-017-2406-3.

Wypij, M., Świecimska, M., Czarnecka, J., Dahm, H., Rai, M., Golińska, P. 2018b. Antimicrobial and cytotoxic activity of silver nanoparticles synthesized from two haloalkaliphilic actinobacterial strains alone and in combination with antibiotics. J. Appl. Microbiol. 124(6): 1411–1424.

Wypij, M., Świecimska, M., Dahm, H., Rai, M., Golińska, P. 2019. Controllable biosynthesis of silver nanoparticles using actinobacterial strains. Green Process. Synth. 8: 207–214.

Yusof, H.M., Mohamad, R., Zaidan, U.H., Rahman, N.A.A. 2019. Microbial synthesis of zinc oxide nanoparticles and their potential application as an antimicrobial agent and a feed supplement in animal industry: a review. J. Animal Sci. Biotechnol. 10: 57.

Chapter 2

Toxic Metal Removal Using Microbial Nanotechnology

Sougata Ghosh

Introduction

Metals play a very significant role in cellular metabolism. Many essential heavy metals are cofactors: copper (Cu), zinc (Zn), Iron (Fe), and cobalt (Co) play vital roles in respiration, electron transport chain, cell division, and diverse types of enzymatic reactions, bimolecular synthesis and immunological responses of the body. Fe is specifically found in hemoglobin, myoglobin, cytochromes (a, b, c), catalase, aconitase, succinate dehydrogenase, aldehyde oxidase, peroxidases, tryptophan 2,3-dioxygenase and many more enzymes (Cassat and Skaar 2013). Cu is required for tyrosinase, superoxide dismutase, cytochrome c oxidase, ceruloplasmin, and dopamine-β-hydoxylase (Desai and Kaler 2008). Similarly, proper protein folding, conformational and configurational alterations and activity, DNA replication, male fertility and growth hormone synthesis and activity are mediated by Zn (Foresta et al. 2014). Co is required for vitamin B_{12} synthesis (Lindsay and Kerr 2011). Specific protein transporters are responsible for metals homeostasis in the body as they actively take part in the uptake, distribution, storage, and excretion of metal ions inside the body. Similarly, the vacuolar systems of eukaryotic cells store metal ions and transport to various sites of the cell, particularly to the cell membranes via the secretory pathway. Metal ions are also abundantly found in cell organelles such as peroxisome, chloroplasts, and mitochondria. Transporter proteins such as Ctr1, Ctr2, Atox1, CCS, Cox1, Ceruloplasmin, ATP7A, ATP7B, and metallothioneins are responsible for regulation of cellular copper content. Disturbance in the metal homeostasis may result in severe clinical contraindications and critical metabolic disorder leading to morbidity and mortality. Mutation in *ATP7A* and *ATP7B* gene are reported to cause Menkes and Wilson diseases (WD) respectively. Deficiency in Cu is observed in Menkes disease, whereas WD is characterized by the accumulation of Cu in the liver, brain and other organs (Kozlowski et al. 2009, Gaetke et al. 2014)

Department of Microbiology, School of Science, RK University, Rajkot 360020, Gujarat, India.
Email: ghoshsibb@gmail.com

Transferrin, ceruloplasmin, hephaestin, ferroportin, and divalent metal transporter 1 (DMT1) are responsible for homeostasis of Fe, impairment of which may lead to Fe-deficiency anemia, hemochromatosis, Huntington's chorea, Parkinson's disease, Alzheimer's disease and amyotrophic lateral sclerosis (Britton et al. 2002, Carocci et al. 2018). Zinc transporter families (ZnT and ZIP) together with metallothioneins help in Zn homeostasis as the ZnT family decreases intracellular zinc level, while the ZIP family plays a vital role in increasing intracellular Zn levels by importing zinc ions from the extracellular space into the cytoplasm (Fukada et al. 2011). Heavy metal concentrations are rapidly increasing in water and soils at an alarming rate during the last few decades owing to the increased levels of electronic waste, fossil fuel burning, disposal of municipal wastes, mining and smelting, and application of fertilizer, pesticides, and sewage. Heavy metals are lethal because of their non-biodegradable nature and high cytotoxicity. Even low concentrations of nonessential heavy metals (As, Hg, Pb, and Cd) can lead to severe health risks, while essential metals such as Zn, Cu, and Fe may become toxic if they exceed their threshold levels (Hu and Cheng 2013, Li et al. 2014, Wu et al. 2015a, Kapusta and Sobczyk 2015). Heavy metal pollution poses a global threat due to its deleterious effects on metabolic processes in biological systems. Industrial effluents are considered the key source of hazardous metals that can cause severe toxicity depending on solubility, concentration, pH and other associated suspended solids. The closing and flooding of mines may also generate large volumes of water contaminated with metals (U, Sn, Cu, As, Fe) (Diels et al. 2003). Conformational alteration of nucleic acids and proteins due to these metals can cause long term patho-physiological defects. Heavy metals that are a prime cause of environmental concern include chromium (Cr), manganese (Mn), iron (Fe), cobalt (Co), copper (Cu), zinc (Zn), molybdenum (Mo), silver (Ag), mercury (Hg), cadmium (Cd) and nickel (Ni). Other metallic elements such as aluminum (Al), tin (Sn), lead (Pb) and bismuth (Bi) and metalloids such as arsenic (As), antimony (Sb) and selenium (Se) are also considerably toxic. Cu is abundant in acid mine drainage (Kim et al. 2019). For these reasons, when these metals are discharged in the effluents of metal plating, tannery, paint, dyeing, and smelting industries and enter a wastewater treatment plant, they should be effectively removed during the treatment processes.

Presently, the most widely used techniques for heavy metal remediation involve supplementation of different inorganic and organic agents such as phosphate salts, biochar, zeolites, manure, and lime, which are efficient, practical, simple and economical. These methods aim to prevent the translocation of heavy metals to the food chain via precipitation, complex formation and adsorbing properties. However, many metal immobilizers are toxic to water and soil. Thus, low-cost, easily available, highly efficient and environmentally benign metal immobilizers are continuously being explored and developed to remediate sites contaminated with many metals and restore soil properties and plant health (Khanna et al. 2019). Bacteria and other biological resources have been actively explored in the past few decades because they are simple, economic, efficient, and environment-friendly. Biosorption eliminates hazardous sludge generation and biosorbents can be easily regenerated and recycled for multiple uses. The biosorption process represents an effective sequestration of metal pollutants by diverse mechanisms. Various microbes have evolved highly efficient cellular metabolic processes to cope with metal toxicity enabling them to survive even in the presence of high metal concentrations. *Bacillus subtilis* senses

Zn using zinc uptake regulator (Zur) protein, which binds DNA and represses the Zur regulon under higher Zn concentration, leading to Zn uptake (Chandrangsu et al. 2017). Ferric uptake regulator (Fur) protein represses Fe acquisition genes in a similar way, leading to regulation of Fe uptake. Further, P_{1b}-type ATPases (also known as E1-E2 ATPases) pumps, conserved in many bacteria, are reported as typical metal exporters, apart from the resistance-nodulation-division (RND) transporters superfamily, which includes the heavy metal efflux (HME)-RND family. Cu efflux from the cytoplasm or the periplasm to the extracellular space is mediated by CusABC pump in *E. coli*. Another major and noteworthy type of bacterial efflux pumps is the cation diffusion facilitators (CDF) family. It is important to note that in *B. subtilis* and other bacteria MneP and MneS CDF proteins are under the control of Mn sensor and regulator MntR (Gillet et al. 2019). Among various transporters playing a critical role in cellular metal trafficking, ABC transporters family is the most conserved transport system in microbes. Fe integral inner membrane MbfA exporter of *Bradyrhizobium japonicum* and RcnA inner membrane Ni/Co pump in *E. coli* are reported as well. Chaperones are small cytoplasmic or periplasmic proteins acting as metal carriers that isolate toxic free metal ions and mediate targeted delivery to their cognate protein targets. CusF binds Cu^+ ions and delivers them for export to CusB within the CusABC RND efflux pump in the periplasm of *E. coli* (Bagai et al. 2008). Similarly, MerP binds Hg^{2+} cations and eventually transports them to MerT transporter, associated to cytoplasmic MerA mercuric reductase for conversion of Hg^{2+} to Hg^0 leading to detoxification (Das 2017, Moore et al. 1990). Chaperones like ZinT are also reported to bind multiple metals Co, Cd, and Hg, apart from Zn (Colaco et al. 2016). Small cytoplasmic or periplasmic proteins called metallothioneins bind Zn, Cu, Hg, Cd, As, or Ag via thiolate bounds within cysteine residues (Chandrangsu et al. 2017). Ferritin, bacterioferritin, and ferritin-like proteins are multimeric proteins selectively responsible for storage and detoxification of free Fe ions (Carrondo 2003, Das 2017). Likewise, multicopper oxidases use Cu ions as cofactors for oxidation of various substrates (Komori and Higuchi 2015). In view of the background, it can be assumed that microbes can be among the most promising agents for removal of toxic metals.

This chapter encompasses the different sources of metal pollutants and their associated toxicity. Further, the mechanism of microbe-assisted bioconversion of various toxic metals to their corresponding nanostructures is discussed in detail. Finally, an account of applications of biogenic metal nanoparticles is presented.

Metal toxicity

Cadmium

Cadmium is a naturally occurring metal that is commercially used in electronic appliances, batteries, lasers, welding, paint, cosmetics, electroplating, nuclear reactors, fertilizers and pesticides. It is highly toxic and causes tissue injury, oxidative stress, and epigenetic alterations in DNA expression, inhibition or up-regulation of transport/metabolic pathways, kidney damage, inhibition of heme synthesis, impairment of mitochondrial function, apoptosis induction and glutathione depletion. Cd affects S1 segment of the proximal tubule of kidney leading to defects in protein, amino acid, glucose, bicarbonate, and phosphate re-absorption (Fanconi syndrome) resulting from Cd-induced oxidative damage to transport proteins and mitochondria that induces apoptosis of tubular cells. Cd impairs vitamin D metabolism and calcium (Ca) absorption in gut, and collagen metabolism, leading to osteomalacia and/or osteoporosis. Cd also induces hypertension, diabetes, peripheral arterial disease, enhanced vascular intima media thickness and myocardial infarction. The underlying mechanism is thought to be disruption of Ca channels, vasoconstriction and inhibition of vasodilators (Bernhoft 2013).

Chromium

Chromium is used in antifreeze, cement, chrome alloy production, electroplating, copier servicing, glassmaking, leather tanning, paints/pigments, photoengraving, porcelain/ceramics manufacturing, production of high-fidelity magnetic audio tapes, tattooing, textile manufacturing, welding of alloys or steel, and wood preservatives. It is also found in cement-producing plants, the wearing down of asbestos linings, emissions of Cr-based automotive catalytic converters, and tobacco smoke. The most notable toxic effects after contact inhalation or ingestion of hexavalent Cr compounds include dermatitis, allergic and eczematous skin reactions, skin and mucous membrane ulcerations, perforation of the nasal septum, allergic asthmatic reactions, bronchial carcinomas, gastroenteritis, hepatocellular deficiency, and renal oligoanuric deficiency (Baruthio 1992).

Cobalt

Cobalt and its associated compounds are found in the industries related to production of hard metals, grinding, mining and paint. It is used in medical implants and also as a colouring agent in glass, pottery and jewellery. It is found in several electronic appliances. The surface of diamond polishers is composed of microdiamonds, cemented in ultrafine Co powder. Co blue dyes are widely used for painting porcelain pottery (Glade and Meguid 2018). Therefore, there is a possibility of occupational exposure through inhalation and ingestion. Widespread immunological response is observed along with adverse tissue reaction on exposure to Co. Inhalation of Co dust results in adverse respiratory effects, while Co nanoparticles (CoNPs) generated by wearing of metal on metal hip implants lead to inflammatory fluid collection or osteolysis. Other clinical implications of Co toxicity include metal allergy, contact dermatitis, generation of reactive oxygen species (ROS) and lipid peroxidation, mitochondrial dysfunction, alteration of Ca and Fe homeostasis, triggering of erythropoiesis, interruption of thyroid iodine uptake, and induction of genotoxic effects and possible perturbation of DNA repair processes (Leyssens et al. 2017).

Copper

Although Cu is an essential element, its level in the body largely varies with food intake. Higher levels of Cu are often observed due to consumption of Cu-rich foods like oysters, liver, nuts, legumes, whole grains, and dried fruit (Shokunbi et al. 2019). Windblown dust, volcanoes, and forest fires, Cu smelters, iron and steel production industries, and municipal incinerators release Cu into the air (Cai et al. 2019). Cu is bound to amino acids like histidine, methionine, and cysteine that facilitate its absorption through an amino acid transport system. Cu forms ligands with reduced glutathione, citric, gluconic, lactic, and acetic acids that are readily absorbed. Other sources of Cu include machinery, construction, transportation, military weapons, and jewellery. Dental products, intrauterine devices and cosmetics also use Cu (Gutiérrez et al. 2019). Elevated levels of Cu are associated with hepatotoxicity, liver cirrhosis, and hemolysis and lead to damage of renal tubules, brain, and other organs. Severe patho-physiological conditions may result in coma, hepatic necrosis, vascular collapse, and death. Cu-contaminated water or foods may result in Cu poisoning indicated by weakness, lethargy, anorexia, erosion of epithelial lining of the gastrointestinal tract, hepatocellular necrosis in the liver, acute tubular necrosis in the kidney, obstructive bile excretion, primary biliary cirrhosis, obstructive hepatobiliary disease, extrahepatic biliary atresia, neonatal hepatitis, choledochal cysts and α-1-antitrypsin deficiency (Gaetke and Chow 2003).

Iron

In spite of Fe being an essential component for hemoglobin, a higher level of Fe poses a toxic threat as excess Fe is deposited in the cells of various organs and tissues, including liver, pancreas, heart, endocrine glands, skin, and joints. This may lead to severe clinical damage such as micronodular cirrhosis of the liver, atrophy of pancreas, hepatocellular liver failure, diabetes mellitus, arthritis, cardiac dysfunction, and even hypogonadism. On the other hand, secondary Fe overload is a result of disorders of erythropoiesis and chronic liver disease where excessive dietary Fe absorption and tissue deposition is observed. Fe-loading associated with refractory anemia with hypercellular bone marrow and ineffective erythropoiesis may also include severe conditions like thalassemia and sideroblastic anemia. Eventually, Fe overload leads to oxidative stress and ROS-mediated damage to lipids, proteins, carbohydrates, and DNA. Further, Fe overload leads to acquired lysosomal storage disease and may lead to damage of hepatic mitochondria, endoplasmic reticulum, and plasma membrane (Britton et al. 2002).

Lead

Toxic levels of Pb affect the cognitive and behavioral development of the central nervous system (CNS), causing encephalopathy and even ataxia, coma, and convulsions in severe cases. Those who survive face elevated risks of neurological complications, mental retardation, deafness, blindness, impairment of fine motor function, concept formation, and altered behavioral profiles. Severe edema due to proteinaceous transudate from permeable capillaries and intercapillary spaces disrupts brain function as a result of Pb toxicity. Pb activates phospholipid-dependent protein kinase and later produces deficits in neurotransmission through inhibition of cholinergic function. Impairment of dopamine uptake by synaptosomes and inhibitory neurotransmitter α-aminobutyric acid function are among other associated problems. Exposure to Pb may change circulating levels of 1,25-dihydroxyvitamin D3, impair new bone formation, inhibit collagen or bone sialoprotein synthesis, and alter Ca and cAMP messenger systems in cells. Pb nephropathy may develop into renal adenocarcinoma. Moreover, Pb is considered to have genotoxic properties that may affect the functions of enzymes involved in DNA synthesis and repair (Soliman et al. 2019, Giri et al. 2019, Goyer 1993).

Manganese

Mn regulates many enzymes by binding as co-factors for arginase, superoxide dismutase, and pyruvate carboxylase. However, occupational exposure (for example, in welders) and dietary overexposure may lead to Mn toxicity associated with CNS, lung, heart, liver, reproductive system and fetal development. Accumulation of Mn in brain tissue results in a progressive disorder of the extrapyramidal system identical to Parkinson's disease. Mn is distributed from blood into brain tissue either through the blood–brain barrier or blood–cerebrospinal fluid barrier, impairment of which leads to Mn accumulation in brain and neurotoxicity (Crossgrove and Zheng 2004).

Mercury

Severe toxicity associated with Hg is reported (Aschner and Carvalho 2019) that may affect various organs and create long-term contraindications (Kishore et al. 2019). Hg is easily and rapidly absorbed in vapour form through mucus membranes, lungs and gut and gets deposited in many tissues. Distribution and deposition in target tissues mostly depends on the form in which it is absorbed. If inhaled, Hg vapour primarily targets the brain, while mercurous and mercuric salts mainly damage the gut lining and kidney. Hg alters tertiary and quaternary structure of proteins by binding with sulfhydryl and selenohydryl groups. It is known to impair peripheral nerve function, renal function, immune function, and endocrine and muscle function, and induce several types of dermatitis. It is associated to erosive bronchitis and bronchiolitis resulting in respiratory failure apart from CNS symptoms such as tremor or erethism. Chronic exposure to Hg causes neurological dysfunction, weakness, fatigue, anorexia, weight loss, and gastrointestinal disturbances. Some of the severe consequences related to Hg exposure include severe behavior and personality changes, emotional excitability, loss of memory, insomnia, depression, fatigue, delirium, hallucination, gingivitis and copious salivation. Other immunological impacts associated with Hg toxicity include hypersensitivity, asthma, dermatitis, autoimmunity, suppression of natural killer cells, and disruption of lymphocyte subpopulations. It may also lead to thyroid dysfunction, inhibition of spermatogenesis, muscular atrophy and capillary damage (Bernhoft 2012).

Nickel

Nickel-induced toxicity is a major concern due to the omnipresence of Ni in our daily activities. Primarily, the respiratory tract and immune system are affected. Ni toxicity results in skin allergies, lung fibrosis, cancer of the respiratory tract and iatrogenic Ni poisoning. Nephrotoxicity, aminoaciduria and proteinuria are common symptoms. Similarly, Ni-associated dermatitis produces erythema, eczema and lichenification. Hypersensitivity related to Ni overexposure causes asthma, conjunctivitis, and inflammatory reactions in response to Ni-containing prostheses and implants. It enhances lipid peroxidation in the liver, kidney, lung, bone marrow and serum, which results from depletion in the level of hepatic glutathione peroxidase and higher level of Fe that eventually lead to peroxidative damage by hydroxyl radicals (Cempel and Nikel 2006).

Palladium

Exposure to palladium (Pd) may occur from mining and smelting or in those personnel who work in industries related to manufacturing and production of electronic appliances, jewellery, optical instruments, chemicals and catalysts. Many Pd salts cause skin and eye irritations. Studies indicate that overexposure to Pd may lead to histopathological changes in lung, liver, kidney, small intestine, and colon. Symptoms related to acute Pd toxicity include emaciation, ataxia, tiptoe gait, clonic and tonic convulsions, cardiovascular effects, peritonitis, and biochemical changes in activity of hepatic enzymes, haematological parameters, proteinuria and ketonuria. Hemorrhages of lungs and the small intestine are also associated with higher levels of Pd exposure (Kielhorn et al. 2002).

Platinum

Platinum (Pt) drugs are effective chemotherapeutic agents used widely against cancer. The most severe side effect related to Pt is ototoxicity. Activated monoaqua-platin is reported to bind with DNA, forming intra- and inter-strand complexes that inhibit DNA synthesis, suppress RNA transcription, arrest cell cycle, and induce apoptosis. Cisplatin alkylation in mitochondria results in release of pro-apoptotic factors and generation of ROS, which together eventually trigger caspase activation–mediated cell death. ROS-mediated proteins and lipid damage in addition to depletion of cellular intrinsic antioxidant molecules cause further cytotoxicity. They may also lead to genetic variations in two specific genes, thiopurine *S*-methyltransferase (TPMT) and catechol-*O*-methyltransferase (COMT), which are identified as key targets of cisplatin-induced ototoxicity (Brock et al. 2012).

Zinc

Toxic properties of Zn are comparatively less studied, as more attention has been given to Zn as a vital element in diverse metabolic processes. Severe pollution with zinc oxide (ZnO) fumes can cause metal-fumes fever. Further, Zn overdose leads to severe Cu deficiency, characterized by hypocupremia, anemia, leukopenia, and neutropenia. Levels of cholesterol and lipoprotein in serum are altered; low-density-lipoprotein (LDL) cholesterol increases, while high-density-lipoprotein (HDL) cholesterol decreases. Decrease in erythrocyte Cu, Zn-superoxide dismutase, serum ferritin and haematocrit occurs as a result of Zn toxicity (Dekani et al. 2019, Fosmire 1990).

Bioconversion of metals by microbes

Bacteria isolated from activated sludge are found to be more resistant to heavy metals and can effectively be exploited for metal removal as well. *Enterobacter* sp.,

Stenotrophomonas sp., *Providencia* sp., *Chryseobacterium* sp., *Comamonas* sp., *Ochrobactrum* sp. and *Delftia* sp. isolated from activated sludge were reported to be efficient heavy metal removers (Bestawy et al. 2013). A mixture of heavy metal-resistant and bioprecipitating *Ralstonia eutropha* CH34 (former name *Alcaligenes eutrophus*), metal biosorbing *Pseudomonas mendocina* AS302 and *Arthrobacter* sp. BP7/26 was used for efficient removal of Zn, Cu, Co, Fe, Al, Ag, Cr, As and Se (Diels et al. 2003). Initially the metals are converted to less toxic forms by microbes, most commonly nanoparticles (Gahlawat and Choudhury 2019). Thus, microbial nanotechnology has become integral to an understanding of the mechanisms of bioconversion, reduction of toxicity, removal of metals and recovery.

Cadmium nanostructures

Cadmium-resistant and selenite-reducing *Pseudomonas* sp. strain RB was reported as a promising microbial catalyst to synthesize CdSe nanoparticles (Fig. 2.1A). It accumulated small CdSe particles (10–20 nm) in/on the cells and large particles (over 100 nm) on the cell surface. This indicates that the CdSe particles grow inside or on the cells to a certain size and eventually detach from the cells. This bacterial process results in faster removal of selenite and Cd^{2+} ions from the aqueous phase (Ayano et al. 2014). Highly luminescent spherical CdSe quantum dots (9–15 nm) were synthesized by *Fusarium oxysporum* at room temperature when incubated with a mixture of $CdCl_2$ and $SeCl_4$ (Kumar et al. 2007). Similarly, *Saccharomyces cerevisiae* (ATCC9763)

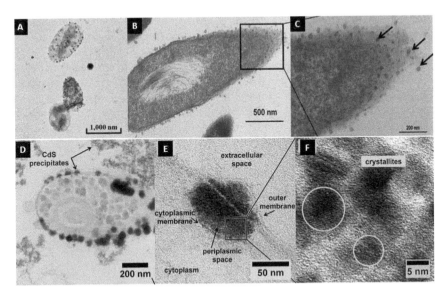

Figure 2.1. Cd nanostructures synthesized by microbes. (A) CdSeNPs synthesized by *Pseudomonas* sp. strain RB. Reproduced with permission from Ayano et al. (2014). (B) CDS QDs synthesized by *Pseudomonas* spp. (C) Magnified image of CDS QDs synthesized by *Pseudomonas* spp. Reproduced with permission from Gallardo et al. (2014). (D) TEM micrograph showing larger CdS NPs in stained *E. coli* (Mag = 53 kx). (E) TEM micrograph of a single CdS NPs in the periplasmic space of the *E. coli* membrane (Mag = 200 kx). (F) High-resolution micrograph of embedded CdS NPs crystallites in *E. coli* (Mag = 700 kx). Reproduced with permission from Marusak et al. (2016).

shows intracellular synthesis of CdSe quantum dots (QDs), which property is largely dependent on the time of adding Na_2SeO_3, the optimal concentrations of selenite and Cd, and the incubating time with yeast cells, respectively (Wu et al. 2015b). Photosynthetic bacteria *Rhodopseudomonas palustris* used cysteine desulfhydrase enzyme for synthesis of cadmium sulfide (CdS) nanocrystals, which were further transported out of the cells and stabilized by protein secreted by the cell (Bai et al. 2009). Various Cd nanostructures synthesized by microbes are presented in Fig. 2.1. *Escherichia coli* was reported to synthesize intracellular cadmium telluride (CdTe) as well (Monrás et al. 2012).

Chromium nanostructures

Some chromate-resistant bacteria such as *Pseudomonas fluorescens*, *Enterobacter cloacae* and *Acinetobacter* sp. (Fig. 2.2) have been shown to remove Cr from industrial effluent (Srivastava and Thakur 2007, Thatheyus and Ramya 2016). Bacteria detoxify chromium mainly by reducing Cr(VI) to Cr(III), through Cr(V) and Cr(IV) intermediates, and it is a potentially useful process in the remediation of Cr(VI)-affected environments. Reduction of Cr(VI) to Cr(III) can be performed by various species of *Pseudomonas*, including *Pseudomonas aeruginosa*, *P. synxantha*, *P. putida*, *P. ambigua*, *P. fluorescens*, *P. dechromaticans* and *P. chromatophila* (Cheung et al. 2006). Bacteria from other genera that can reduce Cr(VI) include *Acinetobacter lwoffii*, *Bacillus megaterium*, *Aeromonas dechromatica* and *Escherichia coli* ATCC 33456 (Srivastava and Thakur 2007, Shen and Wang 1993). Sulfate-reducing bacteria *Desulfovibrio desulfuricans* and *D. vulgaris* were reported to reduce Cr(VI). Some extremophiles such as the radiation-resistant *Deinococcus radiodurans* and *Thermoanaerobacter ethanolicus* isolated from subsurface sediments reduced Cr(VI) (Fredrickson et al. 2000). *Pyrobaculum islandicum* was capable of reducing Cr(VI) at high temperatures (Kashefi and Lovley 2000). Resistance to Cr(VI) was investigated in *Pseudomonas aeruginosa*, which was attributed to the decreased uptake and/ or enhanced efflux of Cr(VI) by the cell membrane (Dogan et al. 2011). A similar resistance mechanism was reported in *Cupriavidus metallidurans* CH34, which was resistant to eight metals including Zn(II), Cd(II), Co(II), Ni(II), Cu(II), CrO_4^{2-},

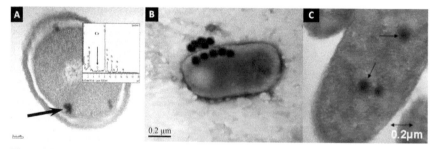

Figure 2.2. TEM images of Cr accumulated by microbes. (A) Cross-section of *B. megaterium* TKW3 with inset illustrating the energy dispersive X-ray spectrum of the electron-dense Cr deposit (arrow) along the intracellular membrane region. Reproduced with permission from Cheung et al. (2006). (B) Magnetotactic bacteria after removal of Cr. Reproduced with permission from Qu et al. (2014). (C) Intracellular accumulation of Cr in *Acinetobacter* sp. Reproduced with permission from Srivastava and Thakur (2007).

Hg(II), and Pb(II) (Monsieurs et al. 2011, Mergeay et al. 2003). *Bacillus circulans, B. megaterium* and *B. coagulans* exhibited efficient Cr removal but *Agrobacterium radiobacter* EPS-916, *Microbacterium liquefaciens* and *Zoogloea ramigera* did not (Thatheyus and Ramya 2016).

Cobalt nanostructures

It was reported that *Providencia rettgeri* MAM-4 can remove Co^{2+} up to 64%, which is efficient compared to *Bacillus cereus* ATCC 11778 showing 59% removal (Abo-Amer et al. 2012). Gram-negative bacteria are capable of preparing the cobalt oxide nanoparticles (CoONPs) in a highly selective manner (Shim et al. 2013, Shim et al. 2015). Various pure strains of *Micrococcus lylae, Bacillus subtilis, Escherichia coli, Paracoccus* sp., and *Haloarcula vallismortis* have been also reported for rapid and efficient synthesis of monodispersed CoONPs with well-defined size and shape. *Micrococcus lylae* and *Haloarcula vallismortis* formed globular nanoparticles, while *B. subtilis* and *E. coli* formed rod-shaped CoONPs (Jang et al. 2015a). Biconcave CoONPs were obtained using *Paracoccus* sp. as bacterial templates (Jang et al. 2015a). The rod-shaped nanostructures prepared using *E. coli* were found to be of average particle size of 473 ± 54 nm, while those synthesized using *M. lylae* were of dimensions equivalent to 356 ± 55 nm (Jang et al. 2015a). Bacteria showing taxonomic affinity to *Flavobacterium* sp. isolated from Co-enriched ferromanganese crusts on the Afanasiy Nikitin Seamounts in the Equatorial Indian Ocean were reported to tolerate and immobilize Co in seawater. These bacteria exhibited an extracellular slime layer, which may play a key role in immobilizing the Co from the liquid phase (Krishnan et al. 2006). Sulfate-reducing *Desulfovibrio desulfuricans* strain Essex6, *Desulfotomaculum gibsoniae, Desulfomicrobium hypogaea*, and *Desulfoarcula baarsi* were reported to show more than 98% Co removal, which was further facilitated by alkaline pH (Krumholz et al. 2003). Similarly, red algae *Porphyra tenera, Palmaria palmata*, and *Chondrus crispus*, brown algae *Sargassum natans, Undaria* sp., *Ascophyllum nodosum, Macrocystis pyrifera*, and *Luminaria* sp., green algae *Halimeda opuntia* and *Codium taylorii*, and fungal biomass *Aspergillus niger* and *Rhizopus arrhizus* were studied for biosorptive Co uptake (González et al. 2019, Kuyucak and Volesky 1989). Genetic engineering has played a very effective

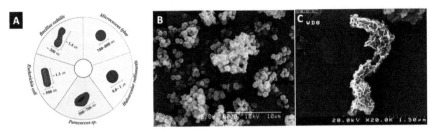

Figure 2.3. CoONPs synthesized using microbes. (A) Schematic representation of synthesis of CoONPs using polymorphic bacterial templates, namely, *M. lylae, B. subtilis, E. coli, H. vallismortis*, and *Paracoccus* sp.; (B) Scanning electron microscopy image of CoONPs prepared from *M. lylae*. Reproduced with permission from Jang et al. (2015a). (C) FESEM images of MTB-KTN90 biomass after exposure to cobalt solutions. Reproduced with permission from Tajer-Mohammad-Ghazvini et al. (2016).

role where special modifications and expression of specific genes may enhance the Co removal/accumulation efficiency. Genetically modified *Escherichia coli* expressing NiCoT genes from *Rhodopseudomonas palustris* CGA009 (RP) and *Novosphingobium aromaticivorans* F-199 (NA) were reported to have increased capacity for Co removal (12 µg/g) in 1 hour (Raghu et al. 2008). *Alphaproteobacterium* MTB-KTN90, which is a novel magnetotactic bacterium, is considered an emerging biosorbent that can remove Co up to 88.55% at pH 6–7 (Tajer-Mohammad-Ghazvini et al. 2016).

Copper nanostructures

Bacillus sp. ATS-1, *Micrococcus luteus* IAM 1056, *Enterobacter* sp. J1, *Pseudomonas aeruginosa* PU21, *Pseudomonas cepacia, Pseudomonas putida, Sphaerotilus natans, Thiobacillus ferrooxidans, Thiobacillus thiooxidans, Streptomyces coelicolor, Zoogloea ramigera, Pseudomonas syringae, Streptomyces noursei, Arthrobacter* sp., *Bacillus firmus, Geobacillus toebii, Geobacillus thermoleovorans, Bacillus* sp. FM1, *Arthrobacter* sp., *Pantoea* sp. TEM18, *Zoogloea ramigera,* and *Bacillus* sp. F19 were reported to efficiently remove Cu from environment (Hansda et al. 2015). Figure 2.4A shows conversion of inexpensive oxidized Cu salts to bimodal-sized elemental copper nanoparticles (CuNPs) by an extracellular metal-reduction employing anaerobic *Thermoanaerobacter* sp. X513 (Jang et al. 2015b). *Micrococcus* sp. isolated from activated sludge was used for removal and recovery of Cu(II) ions from aqueous solutions. Cu uptake by the bacteria was enhanced with increase of pH from 2.0 to 6.0. This biotic process was recyclable and reusable as Cu removal capacity of *Micrococcus* sp. remained unchanged even after five successive sorption and desorption cycles. *Micrococcus* sp. could be immobilized in calcium alginate and polyacrylamide gel beads, which could further increase Cu uptake by 61%. Hence, *Micrococcus* sp. biomass may be applicable to the development of a potentially cost-effective biosorbent for removing and recovering copper from effluents (Wong et al. 2001). Mixed cultures of *Enterobacter* sp., *Stenotrophomonas* sp., *Comamonas* sp. and *Ochrobactrum* sp. could effectively remove Cu up to 31.37% (Bestawy et al. 2013). Similar levels of Cu removal are achieved by using mixed culture of *R. eutrophia, P. mendocina* AS302 and *Arthrobacter* sp. BP7/26 (Diels et al. 2003).

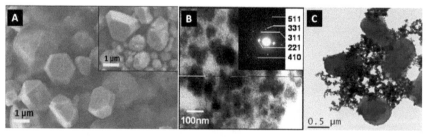

Figure 2.4. Extracellular and cell-bound metal nanoparticles synthesized by bacteria. (A) SEM images of Cu crystallites synthesized by *Thermoanaerobacter* sp. Reproduced with permission from Jang et al. (2015b). (B) TEM image of maghemite nanoparticles synthesized by *Actinobacter* sp. Reproduced with permission from Bharde et al. (2008). (C) TEM image of biogenic manganese oxide nanoparticle produced extracellularly with *Acinetobacter* sp. Reproduced with permission from Hosseinkhani and Emtiazi (2011).

Iron nanostructures

Microbial bioconversion and removal of Fe is achieved because of the unique property of oxidation and precipitation of dissolved Fe that depends on pH and redox potential (*Eh*) (Sharma et al. 2005). Figure 2.4B shows that some bacteria from genera *Actinobacter, Gallionella, Leptothrix, Crenothrix, Clonothrix, Siderocapsa, Sphaerotilus, Ferrobacillus* and *Sideromonas* catalyze exothermic oxidation of Fe(II) by oxido-reductase enzymes secreted as flavins (Bharde et al. 2008). The trivalent Fe is thus converted to insoluble hydroxide form, which is then stored in the mucilaginous secretions (e.g., sheaths, stalks, capsules) of these bacteria. Autotrophic bacteria *Gallionella* and *Leptothrix ochracea* exhibited intracellular oxidation by enzymatic action, while extracellular oxidation by the catalytic action of excreted polymers was exhibited by *Gallionella, Leptothrix, Crenothrix, Clonothrix, Sphaerotilus* and *Siderocapsa*. A pH of 6–8 was suitable for the process, while the range of optimum temperature varied from 10 to 15°C for *Gallionella ferruginea* and 20 to 25°C for *Sphaerotilus–Leptothrix* group (Sharma et al. 2005).

Lead nanostructures

Lactic acid bacteria (LAB) were reported to remove Pb from water in a rapid and metabolism-independent surface process. This pH-dependent process indicated that complex formation, ion exchange, adsorption, chelation and microprecipitation might be the probable underlying mechanisms. *Bifidobacterium longum* 46, *Lactobacillus fermentum* ME3 and *Bifidobacterium lactis* Bb12 were considered as most effective Pb removers. Cell surface of LAB is composed of a thick layer of peptidoglycan, (lipo)teichoic acids, protein and polysaccharides that play a significant role in metal–microbe interactions and thereby Pb removal (Halttunen et al. 2007). Similarly, *Leuconostoc mesenteroides* showed remarkable Pb resistance and removal capacity that can eliminate 60% Pb after 30 min of exposure. It even protected from Pb exposure–associated hepatotoxicity, which was rationalized by lower glutamate oxaloacetate transaminase and glutamate pyruvate transaminase levels in Pb-exposed male mice that received strain L-96 as a probiotic (Yi et al. 2017). Marine yeasts *Rhodosporidium diobovatum* showed intracellular synthesis of stable lead sulfide nanoparticles (PbSNPs) of a size range 2–5 nm on exposure to lead. These PbSNPs were capped by a sulfur-rich peptide, while marked enhancement of nonprotein thiols was observed when yeast was exposed to Pb in the stationary phase (Seshadri et al. 2011). *Aspergillus* species accumulated Pb outside and inside the cell in the form of lead nanoparticles (PbNPs) in the 5–20 nm size range. The particles were distributed on the cell surface, in the periplasmic space and in the cytoplasm (Pavani et al. 2012).

Manganese nanostructures

Manganese-oxidizing bacteria (MOB) play a major role in biological Mn removal, which is considered a cost-effective treatment strategy for Mn-contaminated groundwater (Piazza et al. 2019). Bacteria of the genera *Crenothrix, Hyphomicrobium, Leptothrix, Metallogenium, Siderocapsa, Siderocystis, Pseudomonas, Bacillus, Pedomicrobium* and *Acinetobacter* are potential members of MOB group (Piazza et al. 2019). Biofilter-associated indigenous biofilm of members of *Leptothrix* genus *L. discophora* and *L. ochracea* exhibited Mn removal (Burger et al. 2008). Figure 2.4C shows aerobic Gram-negative bacteria *Acinetobacter* sp. isolated from Persian Gulf water produces extracellular bixbyite-like Mn_2O_3 nanoparticles by enzymatic method (Hosseinkhani and Emtiazi 2011). It was observed that the bacteria-mediated Mn precipitation by oxidation, particularly in case of marine *Bacillus* (strain SG-1), was enhanced by interaction with solid surfaces. On attachment to surfaces like glass beads, frosted glass, sand grains and calcite crystals Mn is rapidly precipitated and removed (Nealson and Ford 1980). Marine bacterium *Saccharophagus degradans* ATCC 43961 (strain 2–40) and yeast *Saccharomyces cerevisiae* were also reported to synthesize MnO_2 NPs that could lead to effective Mn removal and recovery (Salunke et al. 2015).

Mercury nanostructures

Microbes *Bacillus cereus, Lysinibacillus* sp., *Bacillus* sp., *Kocuria rosea, Microbacterium oxydans, Serratia marcescens,* and *Ochrobactrum* sp. precipitate and remove Hg in what is considered an exopolysaccharide-associated process. *S. marcescens* HG19 and *M. oxydans* HG3 showed black precipitates within the bacterial cultures, while amorphous extracellular aggregates adjacent to bacterial surface were observed for *Ochrobactrum* sp. HG16 and *Lysinibacillus* sp. HG17 (François et al. 2012). Mercurial resistance in *Pseudomonas* K-62 is conferred by plasmids pMR26 and pMR68. Bacterial Hg-resistant (*mer*) gene *merA* plays a vital role in biotransformation of Hg by bacterial reduction and volatilization of mercurials for environmental remediation in case of Hg contamination. *Mer* operons in pMR26 consist of six open reading frames (ORFs), five of which are *merR, merT, merP, merA* and *merB1* in pMRA17, while three ORFs in the second *mer* operon (pMRB01) are referred to as *merR, merB2* and *merD. MerR* is a regulatory gene that controls transcription of the structural genes both negatively and positively. *MerD* is associated with transcriptional co-regulatory function. *MerT, merP, merA,* and *merB* code for membrane Hg^{2+} transport protein, periplasmic H^{2+} binding protein, mercuric reductase and organomercurial lyase, respectively. Thus genetic modification of bacteria for Hg remediation has become a very powerful strategy. Deletion of *merG* from pMRA17 makes bacterium more sensitive to phenylmercury than its isogenic strain. Bacterial cells with the *merG*-deleted plasmid (pMU29) took up appreciably more phenyl mercury than the cell with intact pMRA17. Both recovery and accumulation of Hg from mercury-contaminated sites using the *Pseudomonas* K62 *mer* operon could be enhanced by further selective deletion of *merA* and *merG* genes from the operon (Kiyono and Hou 2006).

Nickel nanostructures

Rhodococcus opacus-mediated Ni(II) removal is a pH-dependent process. Ni uptake increased as pH increased from 2.0 to 5.0 but gradually decreased beyond pH 6. It could remove Ni up to 92% owing to its diverse surface-associated compounds, including polysaccharides, carboxylic acids, lipid groups and mycolic acids that impart an amphoteric behavior (Cayllahua et al. 2009). Alginate-immobilized *Bacillus cereus*

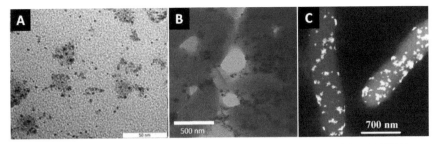

Figure 2.5. TEM images of microbe-assisted synthesis of Ni, Pd and Pt nanoparticles. (A) TEM image of Ni nanoparticles synthesized by *P. aeruginosa* SM1. Reproduced with permission from Srivastava and Constanti (2012). (B) TEM images of palladium reduction with hydrogen as the electron donor on the surface *G. sulfurreducens*. Reproduced with permission from Yates et al. (2013). (C) Dark-field TEM image of *S. algae* cells after exposure to a 1 mM aqueous H_2PtCl_6 solution. Reproduced with permission from Konishi et al. (2007).

M^1_{16} (MTCC 5521) can remove Ni(II) ions most effectively owing to the interactions via carboxyl, sulfur, and amine groups, indicating ionic interaction and complexation/chelation as the underlying mechanism. This adsorption process is strongly pH dependent and maximum adsorption yield takes place at pH 6.0. It is interesting to note that such removal was found to be facilitated by surfactant such as sodium dodecyl sulfate (Naskar and Bera 2017). As seen in Fig. 2.5A, *Pseudomonas aeruginosa* SM1 is reported to synthesize extracellular Ni nanoparticles at room temperature, which might be a powerful strategy for Ni removal (Srivastava and Constanti 2012).

Palladium nanostructures

Anaerobic bacterial community of sludge sample from a municipal wastewater treatment plant located in Montenegro, Faro, in Southern Portugal was reported to remove 60% of Pd(II) from an aqueous solution. This microbial consortium predominantly consisted of several *Clostridium* species apart from members of genera *Bacteroides* and *Citrobacter* where live cells were found to be more efficient removers compared to dead biomass. Sulfate-reducing bacteria (SRB) produce H_2, which may subsequently reduce Pd(II) to Pd(0), which were found to be attached to the cell surfaces (Martins et al. 2013). Similarly, dissimilatory metal-reducing bacteria (DMRB) *Geobacter sulfurreducens* reduces soluble Pd(II) to Pd(0) nanoparticles (PdNPs) reducing the toxicity of metal ions and enables non-destructive Pd recovery due to extracellular nature of nanoparticle synthesis (Fig. 2.5B). PdNPs are distributed in the exopolysaccharide surrounding cells in H_2-fed cultures. Small quantities of PdNPs are also visible inside the cell. This microbial process is advantageous due to its reusability and effective Pd recovery (Yates et al. 2013). Microbes *Clostridium butyricum* (LMG 1217), *Citrobacter braakii* (ATCC 6750), *Klebsiella pneumoniae* (DSM 2026), *Enterococcus faecium* (PhIP-M1-a), *E. coli* (ATCC 25922), *Bacteroides vulgatus* (LMG 177672(2)), *Shewanella oneidensis* MR-1, *Bacillus benzeovorans* NCIMB 12555, *Cupriavidus necator* H16 (DSM 428), *Pseudomonas putida* (DSM 6125), *Paracoccus denitrificans* (DSM 413T) and *Calothrix pulvinata* strain ALCP 745A were reported to synthesize PdNPs, which provides a strong rationale for their use in Pd removal and recovery (Ghosh 2018).

Platinum nanostructures

Cupriavidus metallidurans CH34 is an aerobic, Gram-negative, facultative chemolithoautotrophic, rod-shaped *β*-Proteobacterium that exhibits resistance to a wide range of metallic cations. This resistance is attributed to metal-transporting ATPase efflux proteins in the cell envelope and cation reduction mechanisms in the cytoplasm, which in turn facilitate metal immobilization. *C. metallidurans* strains exhibit biofilm formation on the surfaces of Brazilian platinum (Pt) grains. They exhibit Pt immobilization that increases with time and Pt concentrations. These nanoscale Pt particles are distributed along the cell envelope where energy generation/electron transport occurs. Cells enriched in Pt shed outer membrane vesicles that are enriched in metallic, colloidal Pt, likely representing an important detoxification strategy. Formation of organo-Pt compounds and membrane-encapsulated nanophase Pt supports a role of bacteria in the formation and transport of Pt in natural systems that may help in removal and recovery of Pt (Campbell et al. 2018). Sulfate-reducing bacteria (SRB) *Desulfovibrio desulfuricans, Desulfovibrio fructosivorans* and *Desulfovibrio vulgaris* were also reported to show 90% biosorption of Pt within 5–15 min at pH 3 (Vargas et al. 2004). Mainly, hydrogenases of SRB contribute to the reduction and genetically modified yeasts displaying their hydrogenases on the cell membrane could reduce Pt ions to PtNPs under anaerobic conditions without any electron donors (Ito et al. 2016). Cyanobacteria *Plectonema boryanum* UTEX 485 precipitated amorphous spherical PtNPs in solutions and dispersed nanoparticles within bacterial cells as well that were connected into long beadlike chains by a continuous coating of organic material derived from the cyanobacterial cells. This process was facilitated by increase in temperature and reaction time (Lengke et al. 2006). *Escherichia coli* MC4100, *Shewanella algae* (Fig. 2.5C) and *Desulfovibrio alaskensis* G20 also synthesized PtNPs that can be used for removal of Pt toxicity from the environment and for metal recovery (Attard et al. 2012, Konishi et al. 2007, Capeness et al. 2015).

Zinc nanostructures

Mixed cultures of heavy metal-resistant and bioprecipitating *Ralstonia eutropha* CH34 (former name *Alcaligenes eutrophus*) together with the metal biosorbing *Pseudomonas mendocina* AS302 and *Arthrobacter* sp. BP7/26 could effectively remove Zn up to 95% (Diels et al. 2003). Fungi *Aspergillus fumigatus* and *A. terreus* can synthesize extracellular zinc oxide nanoparticles (ZnONPs), which is considered highly useful because of the potential for large-scale production, convenient downstream processing and economic viability. *A. fumigatus*-synthesized ZnONPs were in a size range from 1.2 to 6.8 nm, while those synthesized by *A. terreus* were between 54.8 and 82.6 nm. NPs synthesized using *Candida albicans* showed similar size range 15–25 nm as confirmed by SEM, TEM and XRD analysis (Agarwal et al. 2017). *Acinetobacter schindleri* SIZ7 produced ZnONPs at 37°C after an incubation period of 48 hours where predominantly the particles were spherical and polydispersed with diameters of 20–100 nm (Busi et al. 2016). Controllable biosynthesis of α-ZnS and β-ZnS quantum dots was achieved by incubating Na_2SO_4, EDTA-Zn and different doses of hydroxypropyl starch mixed with *Clostridiaceae* sp. cells in a neutral medium at mild temperature (25–32°C). Low dose of hydroxylpropyl starch yielded α-ZnS QDs with wurtzite structure and diameter of 8 ± 2 nm, while high dose yielded β-ZnS QDs sphalerite structure and diameter of 4 ± 1 nm (Yue et al. 2016). Fungi *Fusarium oxysporum* synthesized zinc sulfide nanoparticles (ZnS NPs) of size 42 nm (Mirzadeh et al. 2013). Metal ions get adsorbed on to the surface *Aspergillus flavus* biomass and create stress conditions leading to production of intracellular enzymes that produce ROS to regulate proliferation, differentiation, extracellular signal transduction, ion transport and immune response. Initially the sulfate ions are taken in (SO_4^{2-}) from the extracellular environment that are reduced to adenosine phosphosulfate by ATP sulfurylase. Phosphate gets added to adenosine phosphosulfate, to form 3′ phosphoadenosine phosphosulfate, which is further reduced to form sulfite ions (SO_3^{2-}) with the assistance of phosphoadenosine phosphosulfate reductase. These sulfite ions are reduced to sulfide ions (S^{2-}) with the help of sulfite reductase, which later gets coupled with Zn^{2+} to form ZnS NPs (Uddandarao and Mohan 2016).

Mechanisms of metal removal

Uptake capacities in microorganisms can broadly be classified as metabolism-independent binding of heavy metals to cell walls, extracellular polysaccharides, or other materials occurring in living and dead cells, which is generally a rapid process. Metabolism-dependent intracellular uptake or transport occurs in living cells and may be associated with toxic symptoms. Enhanced intracellular uptake may be due to higher degree of membrane permeability resulting from toxicity (Gadd 1990). Lead, uranium, thorium and several other metals are mostly accumulated by microbial biomass that is primarily surface-based without significant intracellular uptake unless by diffusion. As represented in Fig. 2.6, microbes were reported to remove metals by several possible mechanisms that can be summarized under four main phases (Brown and Lester 1979).

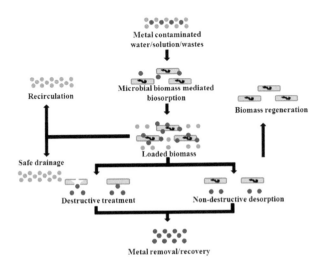

Figure 2.6. Schematic representation of metal removal/recovery from aqueous solutions by microbial biomass or derived products.

Physical trapping of precipitated metals

After growth, suspended bacterial cells exhibit aggregation. They also produce extracellular polymers that may play an important role in flocculation. At high metal concentrations, considerably large proportions of metals precipitate and may eventually be physically entrapped or entangled in the biological matrix (Ayangbenro and Babalola 2017).

Binding of soluble metal to extracellular polymers

Bacterial extracellular polymers may form a loose slime leading to increase in the viscosity of the medium, or capsules and microcapsules closely associated with the cell wall. The more soluble extracellular polymers detach from cells and remain in colloidal suspension while capsular polymers, present in higher concentrations, remain attached. Polymers comprising uronic acids or hexosamines confer properties of charge onto a polymer, unlike neutral polysaccharide. With the increase of extracellular polymer concentration, the surface charge becomes more negative. Hexuronic acid is one of the main components contributing to the overall negative value of the zeta potential. *Zoogloea ramigera* strain 115 was reported to absorb iron, cobalt, nickel, aluminum, calcium, potassium, magnesium, manganese, sodium, silver, copper, chromium, lithium, molybdenum, titanium, mercury and cadmium ions (Sag and Kutsal 1995).

Accumulation of soluble metal by the cell

Involvement of cell wall is also crucial in accumulation of metals, which may then be localized in the cytoplasm. Once inside cells, metal ions may be preferentially located within specific organelles and/or bound to proteins such as metallothionein. This deposition or precipitation renders the element non-toxic. Microbes generally metabolize metallic elements by the following transformations:

1) Chelate formation by the binding of metals to organic ligands,
2) Shifts in metal valencies,
3) Substitution of one metal for another and
4) Biomethylation of metals by microorganisms.

Substrate molecules chelate metals, which are further adsorbed, leading to accumulation of metals in the interior of a cell. Mercury, lead and selenium are methylated by certain micro-organisms, which may lead to detoxification. This metabolic transformation enables the bacteria to survive in an environment with high concentrations of toxic metals. Several mechanisms proposed for accumulation of heavy metals by microbes are discussed in detail in the following section (Jin et al. 2018, Gadd 1990).

Metabolism-independent mechanisms of metal accumulation

Both living and dead microbial biomass are important in metal accumulation. However, there may be variation of degrees of metal accumulation based on the mechanisms. Some metals, such as Cu, Zn, Mn or Co, are essential for microbial growth in trace amounts, while others, such as Au, As, Cd, Pb or Sn, are not. Elements such as uranium and thorium are rather highly toxic. One advantage of using dead biomass is overcoming the problems of toxicity and adverse conditions such as elevated temperature or nutrient limitation. On the other hand, live microbial cells are advantageous for morphological, physiological and genetic manipulation resulting in diverse groups of observable phenotypes that range from efficient accumulators to resistant strains. Rapid metal accumulation is contributed by metabolism-independent heavy metal binding with cell walls, extracellular polysaccharides, or other materials in live and dead cells. On the contrary, metabolism-dependent intracellular metal uptake in living microbes are occasionally associated with toxicity leading to further enhanced membrane permeability. Lead, uranium and thorium are accumulated by surface interactions, unlike an intracellular uptake. After entering the cells, metals either get located in specific compartments or get associated with proteins such as metallothionein. Thus, microbial cell factories accumulate metal by several mechanisms operating simultaneously (Etesami 2018).

In case of bacteria, metals are frequently associated with microbial biomass owing to non-directed physico-chemical interactions known as 'biosorption.' However, the mechanism of such interactions largely varies with the nature of the microbes. The carboxyl group of the glutamic acid of peptidoglycan in cell wall serves as metal chelator in Gram-positive bacteria like *Bacillus subtilis*. Similarly, teichoic and teichuronic acids are significant metal-binding sites in *Bacillus licheniformis*. Initial binding facilitated further inorganic deposition of increased amounts of metal leading to bio-accumulation up to levels as high as 50% of cellular dry weight (Dominguez-Benetton et al. 2018). However, in case of Gram-negative bacteria like *Escherichia coli* K-12 maximum metal deposition occurred at the polar head group regions of constituent membranes or along the peptidoglycan layer (Dominguez-Benetton et al. 2018). On the other hand, diversity in the chemical composition of fungal cell walls led to considerable variations in metal uptake and accumulation capacities between different species. Amine nitrogen of chitin in the cell wall plays a critical role in biosorption (Qiu et al. 2018). Temperature, pH, solubility of the metal/metal ions and presence of other anions/cations may also affect metal biosorption (Ting 2019). Copper, lead/zinc sulfides, zinc dust and ferric hydroxide were found to be efficiently removed in absorbed form using fungal biomass (Selvi et al. 2019).

Extracellular precipitation, complexation, and crystallization

Microbe-assisted metal removal may also take place using extracellular precipitation, complexation, and crystallization. Bacterial extracellular polysaccharides are efficient metal-binding agents that are exploited in activated sludge-mediated metal removal from solution (Jin et al. 2018). Metal crystallization on microbial surfaces warrants investigation if it is metabolism dependent (Ting 2019). Manganic oxides encrust a variety of bacteria, algae and fungi, leading to ferromanganese nodule formation on ocean floors (Selvi et al. 2019). Crystalline deposits of Au and U were also observed (Ting 2019). Precipitation of metal sulfides inside and on the cell surface may take place because of hydrogen sulfide production. Cadmium-grown *Klebsiella aerogenes* was associated with significant levels of cellular Cd and cell surface associated numerous electron-dense granules of CdS (Zou et al. 2019). High-affinity iron-binding molecules called siderophores, triggered by iron deficiency, bind Ga, Ni, U, Th and Cu.

Biosorption by derived microbial products

Industrial wastes comprising fungal biomass can be used as a source of compounds such as mannans, glucans, chitin, chitosan and melanin, acting as potential biosorptive agents. Phosphorylated derivatives of chitin and chitosan are more efficient biosorptive agents compared to their native non-phosphorylated forms (Ting 2019). Additionally, fungal phenolic polymers and melanins contain phenolic units, peptides, carbohydrates, aliphatic hydrocarbons and fatty acids (Selvi et al. 2019). Melanins are considered superior biosorptive agents in both purified as well as cell wall-associated form (Peng et al. 2019).

Metabolism-dependent intracellular accumulation

Metabolism-dependent uptake, although slower, may lead to higher metal accumulation in selected microbes like yeasts. Low temperature, metabolic inhibitors and the absence of an energy source reduces internalization of metal ions into microbial cells (Ting 2019). Metal uptake is also dependent upon metabolic state of cells and the external media composition (Ting 2019). Mostly, metal transport is mediated by electrochemical proton gradient across the cell membrane, which has two components, a chemical component called the pH gradient, and an electrical component called the membrane potential. Each of the components has the ability to facilitate transport of ionized solutes across membranes. Membrane potential is responsible for electrophoretic mono- and divalent cation transport in fungi (Selvi et al. 2019). Intracellular uptake may also occur by diffusion in case membrane permeability is altered by toxic effects. Metal resistance is a key feature associated with decreased uptake and/or impermeability (Ting 2019).

Intracellular localization and deposition

Compartmentalization or conversion of metals to more innocuous forms by precipitation and binding is the next important step after cellular uptake. Electron-dense bodies like polyphosphates are associated with intracellular metal accumulation in several bacteria, algae and fungi. Intracellular Co, Mn, Zn, Mg, and K are located in the vacuole, bound to low molecular weight polyphosphates in case of yeasts (Selvi et al. 2019). Synthesis of intracellular metal-binding proteins was also reported in bacteria, cyanobacteria, algae, fungi and yeasts. Metallothioneins are exploited for metal recovery owing to their property of binding with valuable metals, e.g., Au, Ag, Cu, Zn and Cd (Patankar et al. 2019).

Metal recovery from loaded biomass

The choice between living and dead cell systems is important because of the implications for designing metal recovery strategies. Simple, efficient and rapid non-destructive physical/chemical treatments may reverse metabolism-independent metal biosorption. On the other hand, metals subjected to intracellular accumulation can only be recovered by destructive processes like incineration or dissolution in acids or alkalis (Ting 2019). Waste biomass is considered most suitable when destructive recovery is employed as it is economical as well. Dilute mineral acids are considered as effective desorption agents. However, high concentrations may damage the biomass. Therefore, careful manipulation of pH of desorbing solutions may be an effective strategy for selective removal of metal ions (Selvi et al. 2019). It was observed that Cu and Zn were desorbed by lowering the eluant pH to 2, while Au, Ag and Hg were still strongly bound at this pH value. Gold is selectively eluted as Au^{3+} using mercaptoethanol. Carbonates and/or bicarbonates are used for facilitating non-destructive recovery (Salam 2019).

Immobilized cell systems

Small particle size, low mechanical strength and low density are some of the major limitations for separating the biomass from effluent. Pelleted biomass is more advantageous in packed-bed or fluidized-bed reactors as the benefits include easy separation of cells and effluent, high biomass loadings, minimal clogging and better capability for reuse due to provision of controlling particle size and high flow rates that can be exploited in a variety of reactor configurations (Sahmoune 2018). Polyacrylamide-immobilized *Streptomyces* could remove uranium, copper and cobalt that could be further desorbed for recovery using 0.1 M Na_2CO_3. This enables the recycle and reuse of the same biomass in repeated biosorption-desorption cycles. On supplementation with glycerol-2-phosphate, immobilized *Citrobacter* was able to remove a variety of metals. The underlying uptake mechanism involved phosphatase-mediated cleavage of glycerol-2-phosphate that released HPO_4^{2-}, which in turn precipitated metals extracellularly as insoluble phosphates, e.g., $CdHPO_4$ (Sahmoune 2018). This process is economical considering the feasibility of regenerating the biomass and reusing it over extended time periods in a wide temperature range from 2 to 45°C. Polyacrylamide-entrapped microbes can remove Au, Cu, Hg, and Zn, which can be selectively eluted. The use of biofilms grown on inert matrices is another popular strategy for exploiting immobilized microbes like *Citrobacter*. Simultaneous denitrification and metal removal can also be achieved using mixed bacterial cultures grown as a film on anthracite coal particles or immobilized on polyvinyl chloride or polypropylene webs (Gomaa and El-Meihy 2019). Even a large-scale commercial process using rotating-disc biological-contacting units aims at simultaneous cyanide, thiocyanate and ammonia degradation from gold mining and milling effluents, while removal of metals is accomplished by biosorption using microbial biofilm (Gomaa and El-Meihy 2019).

Volatilization of metal to the atmosphere

Mercury is lost from an activated sludge system indicating volatilization to the atmosphere as a probable mechanism of removal of mercury through microbe *K. aerogenes* (Yamada et al. 1959).

Applications

Nanostructured metals fabricated using microbes during removal from the environment may have immense utility in medicine, in improvement of the diagnosis and treatment of human diseases. Dispersed nanoparticles are usually employed in nanobiomedicine as fluorescent biological labels, as drug and gene delivery agents, and in applications such as biodetection of pathogens, tissue engineering, tumor destruction via heating (hyperthermia), MRI contrast enhancement, and phagokinetic studies. Biologically synthesized nanoparticles are being explored for their plausible applications in targeted drug delivery, cancer treatment, gene therapy and DNA analysis, antibacterial agents, biosensors, enhancing reaction rates, separation science, and MRI (Li et al. 2011). Some examples in the following sections illustrate these applications.

Drug delivery

Targeted drug delivery, controlled release and achievement of maximum therapeutic effect are the key issues in the design and development of novel drug delivery vehicles. Targeted nanocarriers must navigate through blood–tissue barriers to reach target cells. They must enter target cells to contact cytoplasmic targets via specific endocytotic and transcytotic transport mechanisms across cellular barriers. Because of their small size, nanosized drug carriers can bypass the blood–brain barrier and the tight epithelial junctions of the skin that normally impede delivery of drugs to the desired target site. Secondly, as a result of their high surface-to-volume ratio, nanocarriers show improved pharmacokinetics and biodistribution of therapeutic agents and thus minimize toxicity by their preferential accumulation at the target site. They improve the solubility of hydrophobic compounds and render them suitable for parenteral administration. Furthermore, they increase the stability of a variety of therapeutic agents like peptides and oligonucleotides (Ghosh 2018). Magnetic nanoparticles like Fe_3O_4 (magnetite) and Fe_2O_3 (maghemite) are known to be biocompatible. They have been actively investigated for targeted cancer treatment (magnetic hyperthermia), stem cell sorting and manipulation, guided drug delivery, gene therapy and DNA analysis, and MRI. Xiang and co-authors (2007) evaluated the toxicity of magnetosomes from *Magnetospirillum gryphiswaldense* towards mouse fibroblasts *in vitro* and found that the purified and sterilized magnetosomes were not toxic to such cells. Other authors studied the influence of native bacterial magnetic particles on mouse immune response (Meng et al. 2010). In their experiment, ovalbumin was used as an antigen, mixed with complete Freund's adjuvant, BacMps, and phosphate buffer solution, to immunize BALB/C mouse. After 14 days, the titers of the antiovalbumin (IgG) and subtype (IgG1, IgG2), the proliferation ability of T lymphocyte, and the expression of IL-2, IL-4, IL-10, and IFN-gamma were detected. The results showed that native bacterial magnetic particles did not have significant influence on mouse immune response and it was concluded that such magnetosomes have the potential to be used as novel drug or gene carriers for tumor therapy (Meng et al. 2010). In another study, Sun et al. (2007) loaded doxorubicin (DOX) onto bacterial magnetosomes (BMs) through covalent attachment and evaluated the ability of these particles to inhibit tumor growth. In this study performed on H22 tumor-bearing mice, these DOX-loaded BMs showed a comparable tumor suppression rate to DOX alone (86.8% versus 78.6%), but with much lower cardiac toxicity. Although, in this preliminary study, the particles were administrated subcutaneously into the solid tumor, the potential exists to magnetically manipulate these drug-loaded BMs, making them accumulate and execute therapeutic effects only at the disease sites (Sun et al. 2007). The biocompatibility and pharmacokinetics of BMs were also studied (Sun et al. 2009). The distribution of BMs in dejecta, urine, serum, and main organs were investigated when BMs were injected into the sublingual vena of Sprague-Dawley (SD) rats. Authors obtained BMs of high purity and narrow size-distribution using an effective method for purification and sterilization of BMs. Their results showed that BMs were only found in livers and there was no obvious evidence to indicate the existence of BMs in the dejecta and urine within 72 hours following the intravenous administration (Sun et al. 2009). Magnetotactic bacteria (MTB) MC-1 with magnetosomes was also used as drug delivery agent. Magnetotaxis was applied to change the direction of each MTB embedded with combination of nanoparticle magnetite and the flagella to steer in small-diameter blood vessels (Felfoul et al. 2007). However, in order to guide these MTBs towards a target, it is essential to be able to image these living bacteria *in vivo* using an existing medical imaging modality. It was shown that the magnetosomes embedded in each MTB can be used to track the displacement of these bacteria using an MRI system, since these magnetosomes disturb the local magnetic field affecting T_1 and T_2 relaxation times during MRI (Yan et al. 2017). Magnetic resonance, T_1-weighted and T_2-weighted images, as well as T_2 relaxivity of MTB were studied in order to validate the possibility of monitoring MTB drug delivery operations using a clinical MR scanner. It was found that MTB affect the T_2 relaxation rate much more than the T_1 relaxation rate and can be considered a negative contrast agent. As the signal decay in the T_2-weighted images was found to change proportionally to the bacterial concentration, a detection limit of cells/mL for bacterial concentration was achieved using a T_2-weighted image (Yan et al. 2017). Nanoparticle-mediated targeted delivery of drugs might significantly reduce the dosage of anticancer drugs with better specificity, enhanced efficacy, and low toxicities. Further, magnetic nanoparticles may be used for hyperthermia cancer treatment owing to local heating at specific sites using external magnetic field (Kitture et al. 2012).

Biosensors

Promising electronic and optical properties of biogenic nanoparticles are used in biosensor applications (Ghosh 2018). Spherical selenium nanoparticles formed by the *Bacillus subtilis* with diameters ranging from 50 to 400 nm with high surface-to-volume ratio, good adhesive ability, and biocompatibility were employed for building HRP (horseradish peroxidase) biosensor. These sensors exhibited good electrocatalytic activity towards the reduction of H_2O_2 due to the good adhesive ability and biocompatibility of Se nanomaterials. These biosensors have high sensitivity and affinity for H_2O_2, which is promising for a wide range of applications related to the detection of H_2O_2 in food, pharmaceutical, clinical, industrial and environmental analyses (Li et al. 2011).

Chemical catalysis

Microbe-synthesized nanoparticles are widely used in chemical reactions as reducing agents and catalysts due to their large surface areas and specific characteristics (Ghosh 2018). Magnetic nanoparticles are used to improve the microbiological reaction rates due to their catalytic functions and good ability to disperse. Microbial cells of *Pseudomonas delafieldii* coated with magnetic Fe_3O_4 nanoparticles helped in desulfurization of dibenzothiophene. The high surface energies of nanoparticles resulted in their strong adsorption on the cells. The application of an external magnetic field ensured that the cells were well diffused in the solution even without mixing and enhanced the potential to collect cells for reuse (Li et al. 2011).

Conclusions and future perspectives

This chapter provides the interrelationship and interdependence of metal resistance and metal removal by microbes, mainly bacteria from different environmental niches. Many bacteria, for example, *Bacillus cereus, Lysinibacillus* sp., *Bacillus* sp., *Kocuria rosea, Microbacterium oxydans, Serratia marcescens, Ochrobactrum* sp., sulfate-reducing bacteria *Desulfovibrio desulfuricans, Desulfovibrio fructosivorans* and *Desulfovibrio vulgaris*, and cyanobacteria *Plectonema boryanum* are used for effective and rapid removal of toxic metals. Metal-removing capabilities are sometimes surface associated and sometimes metabolism dependent. Conversion of metals to nanoscale, inert, less toxic forms seems to be the most vital microbe-assisted strategy. Among various proposed mechanisms, physical trapping, precipitation, complexation, crystallization and metallothioneins are accepted and being further explored for metal removal and recovery. Immobilized cell systems are being explored for large-scale commercial applications. Thus, microbial systems offer a low-cost, efficient and environment-friendly technology for the treatment of metal contaminated domestic or industrial wastewater.

Recombinant metal-binding proteins with higher reversible biosorbing capacity leading to metal removal and better desorption potential for simultaneous metal recovery may be designed in the near future. This will open a new attribute altogether towards a multidisciplinary research comprising material science, genetic engineering and metallurgy. There have been tremendous developments in the field of microbial synthesis of nanoparticles and their applications over the last decade. However, much work is needed to improve the synthesis efficiency and control of particle size and morphology by optimizing various parameters like inoculum density, age of culture, duration, concentration of precursors, pH of the reaction system, aeration and agitation. Reduction of synthesis time will make this biosynthesis route much more attractive. Several studies have shown that the nanoparticles formed by microorganisms lack stability and get decomposed after a certain period of time. Thus, the stability of nanoparticles produced by biological methods deserves further study and should be enhanced. Bioconversion of toxic metals to non-toxic nanoparticles is advantageous also because nanoparticles are sometimes coated with a lipid layer that confers biocompatibility and physiological solubility as well as stability, which is critical for biomedical applications and is the bottleneck of other synthetic methods. Research on manipulating cells at the genomic and proteomic levels is ongoing. With a better understanding of the synthesis mechanism on a cellular and molecular level, including isolation and identification of the compounds responsible for the reduction of nanoparticles, it is expected that heavy metal removal and recovery efficiency can be enhanced.

References

Abo-Amer, A.E., Ramadan, A.B., Abo-State, M., Abu-Ghabia, M.A., Ahmed H.E. 2012. Biosorption of aluminum, cobalt and copper ions by *Providencia rettgeri* isolated from wastewater. J. Basic. Microbiol. 52: 1–12.

Agarwal, H., Kumar, S.V., Rajeshkumar, S. 2017. A review on green synthesis of zinc oxide nanoparticles—An eco-friendly approach. Resour. Eff. Technol. 3: 406–413.

Aschner, M., Carvalho, C. 2019. The biochemistry of mercury toxicity. Biochim. Biophys. Acta Gen. Subj:129412.

Attard, G., Casadesús, M., Macaskie, L.E., Deplanche, K. 2012. Biosynthesis of platinum nanoparticles by *Escherichia coli* MC4100: Can such nanoparticles exhibit intrinsic surface enantioselectivity? Langmuir 28: 5267−5274.

Ayangbenro, A.S., Babalola, O.O. 2017. A new strategy for heavy metal polluted environments: A review of microbial biosorbents. Int. J. Environ. Res. Public Health. 14: 94.

Ayano, H., Miyake, M., Terasawa, K., Kuroda, M., Soda, S., Sakaguchi, T., Ike, M. 2014. Isolation of a selenite-reducing and cadmium-resistant bacterium *Pseudomonas* sp. strain RB for microbial synthesis of CdSe nanoparticles. J. Biosc. Bioen. 117: 576e581.

Bagai, I., Rensing, C., Blackburn, N., McEvoy, M. 2008. Direct metal transfer between periplasmic proteins identifies a bacterial copper chaperone. Biochemistry 47: 11408–11414.

Bai, H.J., Zhang, Z.M., Guo, Y., Yang, G.E. 2009. Biosynthesis of cadmium sulfide nanoparticles by photosynthetic bacteria *Rhodopseudomonas palustris*. Colloids Surf. B: Biointerfaces 70: 142–146.

Baruthio, F. 1992. Toxic effects of chromium and its compounds. Biol. Trace Elem. Res. 32: 145–153.

Bernhoft, R.A. 2012. Mercury toxicity and treatment: A review of the literature. J. Environ. Public Health 2012: Article ID 460508.

Bernhoft, R.A. 2013. Cadmium toxicity and treatment. Sci. World J. 2013: Article ID 394652.

Bestawy, E.E., Helmy, S., Hussien, H., Fahmy, M., Amer, R. 2013. Bioremediation of heavy metal-contaminated effluent using optimized activated sludge bacteria. Appl. Water Sci. 3: 181–192.

Bharde, A.A., Parikh, R.Y., Baidakova, M., Jouen, S., Hannoyer, B., Enoki, T., Prasad, B.L.V., Shouche, Y.S., Ogale, S., Sastry, M. 2008. Bacteria-mediated precursor-dependent biosynthesis of superparamagnetic iron oxide and iron sulfide nanoparticles. Langmuir 24: 5787–5794.

Britton, R.S., Leicester, K.L., Bacon, B.R. 2002. Iron toxicity and chelation therapy. Int. J. Hematol. 76: 219–228.

Brock, P.R., Knight, K.R., Freyer, D.R., Campbell, K.C.M., Steyger, P.S., Blakley, B.W., Rassekh, S.R., Chang, K.W., Fligor, B.J., Rajput, K., Sullivan, M., Neuwelt, E.A. 2012. Platinum-induced ototoxicity in children: A consensus review on mechanisms, predisposition, and protection, including a new international society of pediatric oncology Boston ototoxicity scale. J. Clin. Oncol. 30: 2408–2417.

Brown, M.J., Lester, J.N. 1979. Metal removal in activated sludge: the role of bacterial extracellular polymers. Water Res. 13: 817–837.

Burger, M.S., Krentz, C.A., Mercer, S.S., Gagnon, G.A. 2008. Manganese removal and occurrence of manganese oxidizing bacteria in full-scale biofilters. J. Water Supply: Res. Technol.—AQUA 57: 351–359.

Busi, S., Rajkumari, J., Pattnaik, S., Parasuraman, P., Hnamte, S. 2016. Extracellular synthesis of zinc oxide nanoparticles using *Acinetobacter schindleri* SIZ7 and its antimicrobial property against foodborne pathogens. J. Microbiol. Biotechnol. Food Sci. 5: 407–411.

Cai, L.M., Wang, Q.S., Luo, J., Chen, L.G., Zhu, R.L., Wang, S., Tang, C.H. 2019. Heavy metal contamination and health risk assessment for children near a large Cu-smelter in central China. Sci. Total Environ. 650: 725–733.

Campbell, G., MacLean, L., Reith, F., Brewe, D., Gordon, R.A., Southam, G. 2018. Immobilisation of platinum by *Cupriavidus metallidurans*. Minerals 8: 10.

Capeness, M.J., Edmundson, M.C., Horsfall, L.E. 2015. Nickel and platinum group metal nanoparticle production by *Desulfovibrio alaskensis* G20. New Biotechnol. 32: 727–731.

Carocci, A., Catalano, A., Sinicropi, M.S., Genchi, G. 2018. Oxidative stress and neurodegeneration: the involvement of iron. Biometals 31: 715–735.

Carrondo, M.A. 2003. Ferritins, iron uptake and storage from the bacterioferritin viewpoint. EMBO J. 22: 1959–1968.

Cassat, J.E., Skaar, E.P. 2013. Iron in infection and immunity. Cell Host Microbe. 13: 509–519.

Cayllahua, J.E.B., Carvalho, R.J., Torem, M.L. 2009. Evaluation of equilibrium, kinetic and thermodynamic parameters for biosorption of nickel(II) ions onto bacteria strain, *Rhodococcus opacus*. Minerals Eng. 22: 1318–1325.

Cempel, M., Nikel, G. 2006. Nickel: A review of its sources and environmental toxicology. Pol. J. Environ. Stud. 15: 375–382.

Chandrangsu, P., Rensing, C., Helmann, J.D. 2017. Metal homeostasis and resistance in bacteria. Nat. Rev. Microbiol. 15: 338–350.

Cheung, K.H., Lai, H.Y., Gu, J.D. 2006. Membrane-associated hexavalent chromium reductase of *Bacillus megaterium* TKW3 with induced expression. J. Microbiol. Biotechnol. 16: 855–862.

Colaco, H.G., Santo, P.E., Matias, P.M., Bandeiras, T.M., Vicente, J.B. 2016. Roles of *Escherichia coli* ZinT in cobalt, mercury and cadmium resistance and structural insights into the metal binding mechanism. Metallomics 8: 327–336.

Crossgrove, J., Zheng, W. 2004. Manganese toxicity upon overexposure. NMR Biomed. 17: 544–553.

Das, S. 2017. Handbook of Metal Microbe Interactions and Bioremediation. CRC Press Taylor & Francis Group, Boca Raton, FL.

Dekani, L., Johari, S.A., Joo, H.S. 2019. Comparative toxicity of organic, inorganic and nanoparticulate zinc following dietary exposure to common carp (*Cyprinus carpio*). Sci. Total Environ. 656: 1191–1198.

Desai, V., Kaler, S.G. 2008. Role of copper in human neurological disorders. Am. J. Clin. Nutr. 88: 855S–858S.

Diels, L., Spaans, P.H., Roy, S.V., Hooyberghs, L., Ryngaert, A., Wouters, H., Walter, E., Winters, J., Macaskie, L., Finlay, J., Pernfuss, B., Woebking, H., Pümpel, T., Tsezos, M. 2003. Heavy metals removal by sand filters inoculated with metal sorbing and precipitating bacteria. Hydrometallurgy 71: 235–241.

Dogan, N.M., Kantar, C., Gulcan, S., Dodge, C.J., Yilmaz, B.C., Mazmanci, M.A. 2011. Chromium(VI) bioremoval by *Pseudomonas* bacteria: Role of microbial exudates for natural attenuation and biotreatment of Cr(VI) contamination. Environ. Sci. Technol. 45: 2278–2285.

Dominguez-Benetton, X., Varia, J.C., Pozo, G., Modin, O., Heijne, A.T., Fransaer, J., Rabaey, K. 2018. Metal recovery by microbial electro-metallurgy. Prog. Mater. Sci. 94: 435–461.

Etesami, H. 2018. Bacterial mediated alleviation of heavy metal stress and decreased accumulation of metals in plant tissues: Mechanisms and future prospects. Ecotoxicol. Environ. Saf. 147: 175–191.

Felfoul, O., Mohammadi, M., Martel, S. 2007. Magnetic resonance imaging of Fe_3O_4 nanoparticles embedded in living magnetotactic bacteria for potential use as carriers for *in vivo* applications. Proceedings of the 29th Annual International Conference of the IEEE Engineering in Medicine and Biology Society (EMBS '07): 1463–1466.

Foresta, C., Garolla, A., Cosci, I., Menegazzo, M., Ferigo, M., Gandin, V., Toni, L.D. 2014. Role of zinc trafficking in male fertility: from germ to sperm. Hum. Reprod. 29: 1134–1145.

Fosmire, G.J. 1990. Zinc toxicity. Am. J. Clin. Nutr. 51: 225–227.

François, F., Lombard, C., Guigner, J.M., Soreau, P., Jaisson, F.B., Martino, G., Vandervennet, M., Garcia, D., Molinier, A.L., Pignol, D., Peduzzi, J., Zirah, S., Rebuffat, S. 2012. Isolation and characterization of environmental bacteria capable of extracellular biosorption of mercury. Appl. Environ. Microbiol. 78: 1097–1106.

Fredrickson, J.K., Kostandarithes, H.M., Li, S.W., Plymale, A.E., Daly, M.J. 2000. Reduction of Fe(III), Cr(VI), U(VI), and Tc(VII) by *Deinococcus radiodurans* R1. Appl. Environ. Microbiol. 66: 2006–2011.

Fukada, T., Yamasaki, S., Nishida, K., Murakami, M., Hirano, T. 2011. Zinc homeostasis and signaling in health and diseases: zinc signalling. J. Biol. Inorg. Chem. 16: 1123–1134.

Gadd, G.M. 1990. Heavy metal accumulation by bacteria and other microorganisms. Experientia 46: 834–840.

Gaetke, L.M., Chow, C.K. 2003. Copper toxicity, oxidative stress, and antioxidant nutrients. Toxicology 189: 147–163.

Gaetke, L.M., Chow-Johnson, H.S., Chow, C.K. 2014. Copper: toxicological relevance and mechanisms. Arch. Toxicol. 88: 1929–1938.

Gahlawat, G., Choudhury, A.R. 2019. A review on the biosynthesis of metal and metal salt nanoparticles by microbes. RSC Adv. 23 (in Press).

Gallardo, C., Monrás, J.P., Plaza, D.O., Collao, B., Saona, L.A., Durán-Toro, V., Venegas, F.A., Soto, C., Ulloa, G., Vásquez, C.C., Bravo, D., Pérez-Donoso, J.M. 2014. Low-temperature biosynthesis of fluorescent semiconductor nanoparticles (CdS) by oxidative stress resistant Antarctic bacteria. J. Biotechnol. 187: 108–15.

Ghosh, S. 2018. Copper and palladium nanostructures: a bacteriogenic approach. Appl. Microbiol. Biotechnol. 102: 7693–7701.

Gillet, S., Lawarée, E., Matroule, J.Y. 2019. Functional diversity of bacterial strategies to cope with metal toxicity. pp. 409–426. *In*: Das, S., Dash, H.R. (eds.). Microbial Diversity in the Genomic Era. Academic Press, Elsevier, Amsterdam.

Giri, S.S., Jun, J.W., Yun, S., Kim, H.J., Kim, S.G., Kang, J.W., Kim, S.W., Han, S.J., Park, S.C., Sukumaran, V. 2019. Characterisation of lactic acid bacteria isolated from the gut of *Cyprinus carpio* that may be effective against lead toxicity. Probiotics Antimicro. 11: 65–73.

Glade, M.J., Meguid, M.M. 2018. A glance at antioxidant and antiinflammatory properties of dietary cobalt. Nutrition 46: 62–66.

Gomaa, E.Z., El-Meihy, R.M. 2019. Bacterial biosurfactant from *Citrobacter freundii* MG812314.1 as a bioremoval tool of heavy metals from wastewater. Bull. Natl. Res. Cent. 43: 69.

González, J.F.C., Pérez, A.S.R., Morales, J.M.V., Martínez Juárez, V.M., Rodríguez, I.A., Cuello, C.M., Fonseca, G.G., Chávez, M.E.E., Morales, A.M. 2019. Bioremoval of Cobalt(II) from aqueous solution by three different and resistant fungal biomasses. Bioinorg. Chem. Appl. 2019: 8 pages.

Goyer, R.A. 1993. Lead toxicity: Current concerns. Environ. Health Perspect. 100: 177–187.

Gutiérrez, M.F., Alegría-Acevedo, L.F., Méndez-Bauer, L., Bermudez, J., Dávila-Sánchez, A., Buvinic, S., Hernández-Moya, N., Reis, A., Loguercio, A.D., Farago, P.V., Martin, J., Fernández, E. 2019. Biological, mechanical and adhesive properties of universal adhesives containing zinc and copper nanoparticles. J. Dent. 82: 45–55.

Halttunen, T., Salminen, S., Tahvonen, R. 2007. Rapid removal of lead and cadmium from water by specific lactic acid bacteria. Int. J. Food Microbiol. 114: 30–35.

Hansda, A., Kurnar, V., Anshumali. 2015. Biosorption of copper by bacterial adsorbents: A review. Res. J. Environ. Toxicol. 9: 45–58.

Hosseinkhani, B., Emtiazi, G. 2011. Synthesis and characterization of a novel extracellular biogenic manganese oxide (Bixbyite-like Mn_2O_3) nanoparticle by isolated *Acinetobacter* sp. Curr. Microbiol. 63: 300–305.

Hu, Y., Cheng, H. 2013. Application of stochastic models in identification and apportionment of heavy metal pollution sources in the surface soils of a large-scale Region. Environ. Sci. Technol. 47: 3752–3760.

Ito, R., Kuroda, K., Hashimoto, H., Ueda, M. 2016. Recovery of platinum(0) through the reduction of platinum ions by hydrogenase-displaying yeast. AMB Expr. 6: 88.

Jang, E., Shim, H.W., Ryu, B.H., An, D.R., Yoo, W.K., Kim, K.K., Kim, D.W., Kim, T.D. 2015a. Preparation of cobalt nanoparticles from polymorphic bacterial templates: A novel platform for biocatalysis. Int. J. Biol. Macromol. 81: 747–753.

Jang, G.G., Jacobs, C.B., Gresback, R.G., Ivanov, I.N., Meyer, H.M., Kidder, M., Joshi, P.C., Jellison, G.E., Phelps, T.J., Graham, D.E., Moon, J.W. 2015b. Size tunable elemental copper nanoparticles: extracellular synthesis by thermoanaerobic bacteria and capping molecules. J. Mater. Chem. C 3: 644–650.

Jin, Y., Luan, Y., Ning, Y., Wang, L. 2018. Effects and mechanisms of microbial remediation of heavy metals in soil: A critical review. Appl. Sci. 8: 1336.

Kapusta, P., Sobczyk, L. 2015. Effects of heavy metal pollution from mining and smelting on enchytraeid communities under different land management and soil conditions. Sci. Total Environ. 536: 517–526.

Kashefi, K., Lovley, D.R. 2000. Reduction of Fe(III), Mn(IV), and toxic metals at 100 degrees C by *Pyrobaculum islandicum*. Appl. Environ. Microbiol. 66: 1050–1056.

Khanna, K., Jamwal, V.L., Gandhi, S.G., Ohri, P. and Bhardwaj, R. 2019. Metal resistant PGPR lowered Cd uptake and expression of metal transporter genes with improved growth and photosynthetic pigments in *Lycopersicon esculentum* under metal toxicity. Sci. Rep. 9: 5855.

Kielhorn, J., Melber, C., Keller, D., Mangelsdorf, I. 2002. Palladium: a review of exposure and effects to human health. Int. J. Hyg. Environ. Health 205: 417–432.

Kim, J.J., Kim, U.S., Kumar, V. 2019. Heavy metal toxicity: An update of chelating therapeutic strategies. J. Trace. Elem. Med. Biol. 54: 226–231.

Kishore, D., Chandra, S.N., Kumar, P.A., Kumar, M.A. 2019. Assessment of mercury toxicity on sensitivity and behavioural response of freshwater tropical worm, *Branchiura sowerbyi* Beddard, 1892. J. Aquacult. Trop. 34: 27–35.

Kitture, R., Ghosh, S., Kulkarni, P., Liu, X.L., Maity, D., Patil, S.I., Jun, D., Dushing, Y., Laware, S.L., Chopade, B.A., Kale, S.N. 2012. Fe$_3$O$_4$-citrate-curcumin: Promising conjugates for superoxide scavenging, tumor suppression and cancer hyperthermia. J. Appl. Phys. 111: 064702–064707.

Kiyono, M., Hou, H.P. 2006. Genetic engineering of bacteria for environmental remediation of mercury. J. Health Sci. 52: 199–204.

Komori, H., Higuchi, Y. 2015. Structural insights into the O$_2$ reduction mechanism of multicopper oxidase. J. Biochem. 158: 293–298.

Konishi, Y., Ohno, K., Saitoh, N., Nomura, T., Nagamine, S., Hishida, H., Takahashi, Y., Uruga, T. 2007. Bioreductive deposition of platinum nanoparticles on the bacterium *Shewanella algae*. J. Biotechnol. 128: 648–653.

Kozlowski, H., Janicka-Klos, A., Brasun, J., Gaggelli, E., Valensin, D., Valensin, G. 2009. Copper, iron, and zinc ions homeostasis and their role in neurodegenerative disorders (metal uptake, transport, distribution and regulation). Coord. Chem. Rev. 253: 2665–2685.

Krishnan, K.P., Fernandes, C.E.G., Fernandes, S.O., Bharathi, P.A.L. 2006. Tolerance and immobilization of cobalt by some bacteria from ferromanganese crusts of the Afanasiy Nikitin Seamounts. Geomicrobiol. J. 23: 31–36.

Krumholz, L.R., Elias, D.A., Suflita, J.M. 2003. Immobilization of cobalt by sulfate-reducing bacteria in subsurface sediments. Geomicrobiol. J. 20: 61–72.

Kumar, S.A., Ansary, A.A., Ahmad, A., Khan, M.I. 2007. Extracellular biosynthesis of CdSe quantum dots by the fungus, *Fusarium oxysporum*. J. Biomed. Nanotechnol. 3: 190–194.

Kuyucak, N., Volesky, B. 1989. Accumulation of cobalt by marine alga. Biotechnol. Bioeng. 33: 809–814.

Lengke, M.F., Fleet, M.E., Southam, G. 2006. Synthesis of platinum nanoparticles by reaction of filamentous cyanobacteria with platinum(IV)-chloride complex. Langmuir 22: 7318–7323.

Leyssens, L., Vinck, B., Van Der Straeten, C., Wuyts, F., Maes, L. 2017. Cobalt toxicity in humans—A review of the potential sources and systemic health effects. Toxicology 387: 43–56.

Li, X., Xu, H., Chen, Z.S., Chen, G. 2011. Biosynthesis of nanoparticles by microorganisms and their applications. J. Nanomater. 2011: 270974.

Li, Z., Ma, Z., van der Kuijp, T.J., Yuan, Z., Huang, L. 2014. A review of soil heavy metal pollution from mines in China: pollution and health risk assessment. Sci. Total Environ. 468-469: 843–853.

Lindsay, D., Kerr, W. 2011. Cobalt close-up. Nat. Chem. 3: 494.

Martins, M., Assuncao, A., Martins, H., Matos, A.P., Costa, M.C. 2013. Palladium recovery as nanoparticles by an anaerobic bacterial community. J. Chem. Technol. Biotechnol. 88: 2039–2045.

Marusak, K.E., Feng, Y., Eben, C.F., Payne, S.T., Cao, Y., You, L., Zauscher, S. 2016. Cadmium sulphide quantum dots with tunable electronic properties by bacterial precipitation. RSC Adv. 6: 76158–76166.

Meng, C., Tian, J., Li, Y., Zheng, S. 2010. Influence of native bacterial magnetic particles on mouse immune response. Wei Sheng Wu Xue Bao 50: 817–821.

Mergeay, M., Monchy, S., Vallaeys, T., Auquier, V., Benotmane, A., Bertin, P., Taghavi, S., Dunn, J., van der Lelie, D., Wattiez, R. 2003. *Ralstonia metallidurans*, a bacterium specifically adapted to toxic metals: towards a catalogue of metal-responsive genes. FEMS Microbiol. Rev. 27: 385–410.

Mirzadeh, S., Darezereshki, E., Bakhtiari, F., Fazaelipoor, M.H., Hosseini, M.R. 2013. Characterization of zinc sulfide (ZnS) nanoparticles biosynthesized by *Fusarium oxysporum*. Mat. Sci. Semicon. Proc. 16: 374–378.

Monrás, J.P., Collao, B., Molina-Quiroz, R.C., Pradenas, G.A., Saona, L.A., Durán-Toro, V., Ordenes-Aenishanslins, N., Venegas, F.A., Loyola, D.E., Bravo, D., Calderón, P.F., Calderón, I.L., Vásquez, C.C., Chasteen, T.G., Lopez, D.A., Pérez-Donoso, J.M. 2012. Enhanced glutathione content allows the *in vivo* synthesis of fluorescent CdTe nanoparticles by *Escherichia coli*. PLoS ONE 7: e48657.

Monsieurs, P., Moors, H., Houdt, R.V., Janssen, P.J., Janssen, A., Coninx, I., Mergeay, M., Leys, N. 2011. Heavy metal resistance in *Cupriavidus metallidurans* CH34 is governed by an intricate transcriptional network. Biometals 24: 1133–1151.

Moore, M.J., Distefano, M.D., Zydowsky, L.D., Cummings, R.T., Walsh, C.T. 1990. Organomercurial lyase and mercuric ion reductase: nature's mercury detoxification catalysts. Acc. Chem. Res. 23: 301–308.

Naskar, A., Bera, D. 2017. Mechanistic exploration of Ni(II) removal by immobilized bacterial biomass and interactive influence of coexisting surfactants. Environ. Prog. Sustain. Energy 1–13.

Nealson, K.H., Ford, J. 1980. Surface enhancement of bacterial manganese oxidation: Implications for aquatic environments. Geomicrobiol. J. 2: 21–37.

Patankar, H.V., Al-Harrasi, I., Kharusi, L.A., Jana, G.A., Al-Yahyai, R., Sunkar, R., Yaish, M.W. 2019. Overexpression of a metallothionein 2A gene from date palm confers abiotic stress tolerance to yeast and *Arabidopsis thaliana*. Int. J. Mol. Sci. 20: 2871.

Pavani, K.V., Kumar, N.S., Sangameswaran, B.B. 2012. Synthesis of lead nanoparticles by *Aspergillus* species. P. J. Microbiol. 61: 61–63.

Peng, H., Li, D., Ye, J., Xu, H., Xie, W., Zhang, Y., Wu, M., Xu, L., Liang, Y., Liu, W. 2019. Biosorption behavior of the *Ochrobactrum* MT180101 on ionic copper and chelate copper. J. Environ. Manage. 235: 224–230.

Piazza, A., Ciancio Casalini, L., Pacini, V.A., Sanguinetti, G., Ottado, J., Gottig, N. 2019. Environmental bacteria involved in manganese(II) oxidation and removal from groundwater. Front Microbiol. 10: 119.

Qiu, X., Yao, Y., Wang, H., Duan, Y. 2018. Live microbial cells adsorb Mg^{2+} more effectively than lifeless organic matter. Front Earth Sci. 12: 160–169.

Qu, Y., Zhang, X., Xu, J., Zhang, W., Guo, Y. 2014. Removal of hexavalent chromium from wastewater using magnetotactic bacteria. Sep. Purif. Technol. 136: 10–17.

Raghu, G., Balaji, V., Venkateswaran, G., Rodrigue, A., Mohan, P.M. 2008. Bioremediation of trace cobalt from simulated spent decontamination solutions of nuclear power reactors using *E. coli* expressing NiCoT genes. Appl. Microbiol. Biotechnol. 81: 571–578.

Sağ, Y., Kutsal, T. 1995. Biosorption of heavy metals by *Zoogloea ramigera*: use of adsorption isotherms and a comparison of biosorption characteristics. Chem. Eng. J. Biochem. Eng. J. 60: 181–188.

Sahmoune, M.N. 2018. Performance of *Streptomyces rimosus* biomass in biosorption of heavy metals from aqueous solutions. Microchem. J. 141: 87–95.

Salam, K.A. 2019. Towards sustainable development of microalgal biosorption for treating effluents containing heavy metals. Biofuel. Res. J. 22: 948–961.

Salunke, B.K., Sawant, S.S., Lee, S.I., Kim, B.S. 2015. Comparative study of MnO_2 nanoparticle synthesis by marine bacterium *Saccharophagus degradans* and yeast *Saccharomyces cerevisiae*. Appl. Microbiol. Biotechnol. 99: 5419–5427.

Selvi, A., Rajasekar, A., Theerthagiri, J., Ananthaselvam, A., Sathishkumar, K., Madhavan, J., Rahman, P.K.S.M. 2019. Integrated remediation processes toward heavy metal removal/recovery from various environments—A review. Front Environ. Sci. 7: Article 66.

Seshadri, S., Saranya, K., Kowshik, M. 2011. Green synthesis of lead sulfide nanoparticles by the lead resistant marine yeast, *Rhodosporidium diobovatum*. Biotechnol. Prog. 27: 1464–1469.

Sharma, S.K., Petrusevski, B., Schippers, J.C. 2005. Biological iron removal from groundwater: A review. J. Water Supply Res. T—AQUA 54: 239–246.

Shen, H., Wang, Y.T. 1993. Characterization of enzymatic reduction of hexavalent chromium by *Escherichia coli* ATCC 33456. Appl. Environ. Microbiol. 59: 3771–3777.

Shim, H.W., Lim, A.H., Kim, J.C., Jang, E., Seo, S.D., Lee, G.H., Kim, T.D., Kim, D.W. 2013. Scalable one-pot bacteria-templating synthesis route toward hierarchical, porous-Co_3O_4 superstructures for supercapacitor electrodes. Sci. Rep. 3: 2325.

Shim, H.W., Park, S., Song, H.J., Kim, J.C., Jang, E., Hong, K.S., Kim, T.D., Kim, D.W. 2015. Biomineralized multifunctional magnetite/carbon microspheres for applications in Li-ion batteries and water treatment. Chemistry 21: 4655–4663.

Shokunbi, O.S., Adepoju, O.T., Mojapelo, P.E.L., Ramaite, I.D.I., Akinyele, I.O. 2019. Copper, manganese, iron and zinc contents of Nigerian foods and estimates of adult dietary intakes. J. Food Compos. Anal. 82: 103245.

Soliman, H.A.M., Hamed, M., Lee, J.S., Sayed, A.E.H. 2019. Protective effects of a novel pyrazolecarboxamide derivative against lead nitrate induced oxidative stress and DNA damage in *Clarias gariepinus*. Environ. Pollut. 247: 678–684.

Srivastava, S., Thakur, I.S. 2007. Evaluation of biosorption potency of *Acinetobacter* sp. for removal of hexavalent chromium from tannery effluent. Biodegradation 18: 637–646.

Srivastava, S.K., Constanti, M. 2012. Room temperature biogenic synthesis of multiple nanoparticles (Ag, Pd, Fe, Rh, Ni, Ru, Pt, Co, and Li) by *Pseudomonas aeruginosa* SM1. J. Nanoparticle Res. 14: 831.

Sun, J.B., Duan, J.H., Dai, S.L., Ren, J., Zhang, Y.D., Tian, J.S., Li, Y. 2007. *In vitro* and *in vivo* antitumor effects of doxorubicin loaded with bacterial magnetosomes (DBMs) on H22 cells: the magnetic bionanoparticles as drug carriers. Cancer Lett. 258: 109–117.

Sun, J.B., Wang, Z.L., Duan, J.H., Ren, J., Yang, X.D., Dai, S.L., Li, Y. 2009. Targeted distribution of bacterial magnetosomes isolated from *Magnetospirillum gryphiswaldense* MSR-1 in healthy sprague-dawley rats. J. Nanosci. Nanotechnol. 9: 1881–1885.

Tajer-Mohammad-Ghazvini, P., Kasra-Kermanshahi, R., Nozad-Golikand, A., Sadeghizadeh, M., Ghorbanzadeh-Mashkani, S., Dabbagh, R. 2016. Cobalt separation by Alphaproteobacterium MTB-KTN90: Magnetotactic bacteria in bioremediation. Bioprocess Biosyst. Eng. 39: 1899–1911.

Thatheyus, A.J., Ramya, D. 2016. Biosorption of chromium using bacteria: An overview. Sci. Int. 4: 74–79.

Ting, A.S.Y. 2019. Green approach: Microbes for removal of dyes and metals via ion binding. pp. 1–23. *In*: Inamuddin, Ahamed, M., Asiri, A. (eds.). Applications of Ion Exchange Materials in the Environment. Springer, Cham.

Uddandarao, P., Mohan, B.R. 2016. ZnS semiconductor quantum dots production by an endophytic fungus *Aspergillus flavus*. Mater. Sci. Eng. B 207: 26–32.

Vargas, I., Macaskie, L.E., Guibal, E. 2004. Biosorption of palladium and platinum by sulfate-reducing bacteria. J. Chem. Technol. Biotechnol. 79: 49–56.

Wong, M.F., Chua, H., Lo, W., Leung, C.K., Yu, P.H. 2001. Removal and recovery of copper(II) ions by bacterial biosorption. Appl. Biochem. Biotechnol. 91: 447–457.

Wu, Q., Leung, J.Y., Geng, X., Chen, S., Huang, X., Li, H., Huang, Z., Zhu, L., Chen, J., Lu, Y. 2015a. Heavy metal contamination of soil and water in the vicinity of an abandoned e-waste recycling site: implications for dissemination of heavy metals. Sci. Total Environ. 506-507: 217–225.

Wu, S.M., Su, Y., Liang, R.R., Ai, X.X., Qian, J., Wang, C., Chen, J.Q., Yan, Z.Y. 2015b. Crucial factors in biosynthesis of fluorescent CdSe quantum dots in *Saccharomyces cerevisiae*. RSC Adv. 5: 79184–79191.

Xiang, L., Wei, J., Jianbo, S., Guili, W., Feng, G., Ying, L. 2007. Purified and sterilized magnetosomes from *Magnetospirillum gryphiswaldense* MSR-1 were not toxic to mouse fibroblasts *in vitro*. Lett. Appl. Microbiol. 45: 75–81.

Yamada, M., Dazai, M., Tonomura, K. 1959. Change of mercurial compounds in activated sludge. J. Ferment. Technol. 47: 155–160.

Yan, L., Da, H., Zhang, S., López, V.M., Wang, W. 2017. Bacterial magnetosome and its potential application. Microbiol. Res. 203: 19–28.

Yates, M.D., Cusick, R.D., Logan, B.E. 2013. Extracellular palladium nanoparticle production using *Geobacter sulfurreducens*. ACS Sustain. Chem. Eng. 1: 1165−1171.

Yi, Y.J., Lim, J.M., Gu, S., Lee, W.K., Oh, E., Lee, S.M., Oh, B.T. 2017. Potential use of lactic acid bacteria *Leuconostoc mesenteroides* as a probiotic for the removal of Pb(II) toxicity. J. Microbiol. 55: 296–303.

Yue, L., Qi, S., Wang, J., Cai, J., Xin, B. 2016. Controllable biosynthesis and characterization of α-ZnS and β-ZnS quantum dots: Comparing their optical properties. Mat. Sci. Semicon. Proc. 56: 115–118.

Zou, D., Li, Y., Kao, S.J., Liu, H., Li, M. 2019. Genomic adaptation to eutrophication of ammonia-oxidizing archaea in the Pearl River estuary. Environ. Microbiol. 21: 2320–2332.

Chapter 3

Biomineralization of Mine Sediments for Synthesis of Iron-Based Nanomaterials

Angela Banerjee,[1] *Abhilash,*[1,*] *B.D. Pandey*[1] and *Y.L. Gurevich*[2]

Introduction

The interaction between iron minerals and their reduction by bacteria is an important phenomenon that has become known in the 20th century, and much evidence has been provided of Fe(III) reduction as the earliest form of respiration (Abhilash et al. 2011). Bacterial Fe(III) reduction plays a crucial role in the biogeochemical cycling of iron where Fe(III) minerals act as the terminal electron acceptor. Not only the redox cycling of Fe but also dissimilatory iron reducers have been studied for their bioreduction of toxic metals, remediation by utilization of aromatic compounds and metal corrosion (Lovley 2000). The iron reducers range from archaea to bacteria (Bomberg et al. 2019). Most of them, though they are facultative aerobes, reduce iron under anaerobic conditions. The Fe reducers mainly reduce Fe(III) by completely oxidizing multicarbon compounds to CO_2 or by incompletely oxidizing the multicarbon compounds. Whatever is the electron acceptor, almost all the Fe-reducing bacteria (FeRB) have been isolated or known to survive under a neutral condition. While FeRB growing at low pH have always been highly desirable, only a few (Postgate 1959, Lovley 1997) have been able to survive acidic pH and a rare few have been able to reduce Fe(III) at such pH. Also, most of the reducers have their reduction ability inhibited at high Fe(III) concentration, and the reduction takes place in anaerobic conditions (Berg et al. 2019). Until recently it was believed that iron reduction in the environment was non-enzymatic, led by the belief that anoxic environment was created by the microbial

[1] CSIR National Metallurgical Laboratory, Jamshedpur, India.
[2] ISCEOR, Krasnoyarsk Science Center, Krasnoyarsk, Russia.
* Corresponding author: abhibios@gmail.com

consumption of oxygen and by the production of reduced metabolites in the sediments that finally ensured Fe(III) reduction (Leitholf et al. 2019). This belief has now been rejected on the basis that a mere redox environment is not enough for Fe(III) reduction to Fe(II) (Vargas et al. 1998).

It has been proved that the extent of abiotic reduction of Fe(III) was far less than in the presence of iron-reducing microorganisms (Semerena et al. 1980). Dissimilatory iron-reducing bacteria while performing the iron reduction also crystallize iron nanoparticles (Raiswell and Canfield 1998). This nanoparticle formation is similar to an autogenic mineral formation with a difference of bacteria acting as a nucleation center surrounding which the crystals form (Frankel and Bazylinski 2003). The iron-reducing bacteria use insoluble Fe(III) hydroxides in the environment as the terminal electron acceptor. Their anionic cell wall surfaces react with the iron, and sorption of iron occurs. Finally, reduction of the Fe(III) to Fe(II) makes Fe more reactive, and it then reacts with other elements such as sulfur, silicates, and carbonates to form sulfides, silicates, and carbonates of iron (Mclean et al. 1992, Konhauser 1997, Phoenix et al. 2003). If the reduced Fe(II) present on the bacterial cell wall reacts with the excessive Fe(III) oxides present in nature, iron oxide nanoparticles such as magnetite and haematite are formed (Bazylinksi and Frankel 2000, Kirschvink and Hagadorn 2000). The nanoparticles of magnetite formed on the reduction of Fe(III) have been studied for their magnetic ability, which has many commercial purposes. Iron oxide nanoparticles have been applied in the field of molecular imaging, for example in magnetic resonance imaging (MRI). Moreover, iron oxide nanoparticles have been used for drug delivery, hyperthermia treatment of cancerous tissue and tissue labeling purposes. Magnetite nanoparticles have been observed not only on earth but also in Martian meteorites, suggesting the importance of Fe(III) reduction by microbes (Kucera and Wolfe 1957, Ishaq et al. 1964, Lovley et al. 1993, Rickard et al. 2001). Here, we have also highlighted the potential of iron-reducing bacteria to biomineralize iron from mine sediments.

Collection of samples and microbial isolation vis-à-vis screening

Two different samples, copper mine sediment sample (Fig. 3.1a) and iron oxide filing sample (Fig. 3.1b), were used to carry out experimental studies. The sediment sample was obtained from a depth of 3 km at the Musabani Copper Mines (IRL), East Singhbhum, India. The samples were collected with the help of a hand auger into a sterilized bag and were immediately placed in an ice-cooled portable glove bag before isolation of the microbes in the laboratory. At the time of sample collection, the measured pH of the sample was 3.66 and its redox potential was 368 mV. Based on SEM-EDS analysis reports, it can be concluded that the copper sediment sample was a chalcopyrite-rich matrix with iron and silica associations. Iron oxide filings were also obtained from the mine. The filings were prepared by crushing the sample with mortar and pestle and drying it at 50°C. SEM-EDS analysis of the iron oxide sample revealed that it was a pyrite-rich matrix.

The isolates from copper mine sediment sample were designated as 236-S1 and 236-S2, and those from the iron oxide sample were designated as 236-M1. A defined medium as stated by Roh et al. (2006) was used initially for bacterial isolation of strain 236-S1 (Table 3.1).

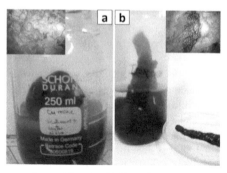

Figure 3.1. Copper mine samples used for microbial isolation: (a) copper-rich sediment; (b) iron-rich oxide sediment.

Table 3.1. Major components (g L^{-1}) of defined medium (Roh et al. 2006).

Component	Concentration (g L^{-1})
NaHCO$_3$	2.5
CaCl$_2$ × 2H$_2$O	0.08
NH$_4$Cl	1
MgCl$_2$ × 6H$_2$O	0.2
NaCl	10
HEPES	7.2
Resazurin	0.1
Yeast extract	0.5

Additionally supplemented with vitamin solution (1 ml) and trace minerals (10 ml).

A 1 ml copper mine sample was added to 99 ml of the broth. The pH of the broth was maintained at 7.0 and the culture was incubated in an orbital shaker (ORBITEK-LEH™) at 100 rpm and 35°C. The pH of the broth was adjusted daily with 10N H$_2$SO$_4$ and 2N NaOH. Repeated subculture resulted in the isolation of strain 236-S1. Plates of defined medium were prepared using 2% agar. The bacterial growth was determined by the change in color of an indicator, resazurin, which color changed from blue to pink with the growth of the culture. Sterile vitamin elixir was added after autoclaving the medium and mineral elixir (both detailed in Table 3.2).

The culture though isolated on defined medium further showed poor growth on it, and another medium known as Basic Salt Solution (BSS, detailed in Table 3.3) was used, supplemented with trace minerals and vitamins (see Table 3.4). Vitamin elixir was added after autoclaving the media and mineral elixir. The growth of strain 236-S1 in BSS media was indicated by precipitation of ferric citrate (FeC$_6$H$_5$O$_7$).

Strains 236-S2 and 236-M1 were isolated in Casitone Glycerol Yeast medium (CGY) (HiMedia). Copper sediment sample and iron oxide sediment were initially diluted and spread on the plates with CGY agar. Isolates 236-S2 and 236-M1 were then transferred to CGY broth, incubated at 37°C, on an orbital shaker (100 rpm) for 7 days.

Growth studies of each of the three strains were done by culturing isolate 236-S1 in BSS medium and strains 236-S2 and 236-M1 in CGY medium. Hourly

Table 3.2. Composition of vitamin elixir and mineral elixir (Bazylinksi and Frankel 2000).

Vitamin elixir		Mineral elixir	
Component	g L^{-1}	Component	g L^{-1}
Biotin	0.02	Nitrilotriacetic acid	15
Folic acid	0.02	$FeCL_3 \times 6H_2O$	2
Pyridoxine (B6)	0.1	$MgCL_2 \times 6H_2O$	1
Thiamine (B1)	0.05	Sodium molybdate	0.3
Riboflavin (B2)	0.05	$MnCl_2 \times 4H_2O$	1
Niacin	0.05	$CoCl_2 \times 6H_2O$	1
Pantothenic acid	0.05	$CaCl_2 \times 2H_2O$	50
Cyanocobalamin (B12)	0.001	$ZnCl_2$	0.5
p-Aminobenzoic acid	0.05	$CuCl_2 \times 2H_2O$	0.02
Lipoic acid	0.05	H_3BO_3	0.5
		NaCl	10
		$Na_2Se_2O_3$	0.17
		$NiCl_2 \times 6H_2O$	0.24

Table 3.3. Major components of BSS medium (Bazylinksi and Frankel 2000) (g L^{-1}).

Component	
$C_6H_5FeO_7$	14.694
$NaHCO_3$	2.5
NH_4Cl	1.5
K_2HPO_4	0.6
KCl	0.1
$CH_3COONa \times 3H_2O$	1.36

Table 3.4. Trace vitamins and minerals of the BSS culture medium (Bazylinksi and Frankel 2000) (mg L^{-1}).

Trace vitamins		Trace minerals	
Biotin	20	Nitrilotriacetic acid	15
Folic acid	20	$MgSO_4 \times 7H_2O$	30
Pyridoxine hydrochloride	100	$MnSO_4 \times 2H_2O$	5
Thiamine hydrochloride	50	NaCL	10
Riboflavin	50	$FeSO_4 \times 7H_2O$	1
Nicotinic acid	50	$COSO_4 \times CoCL_2$	1
DL-Calcium pantothenate	50	$CaCL_2 \times 2H_2O$	1
Vitamin B12	1	$CuSO_4 \times 5H_2O$	0.1
p-Aminobenzoic acid	50	$ZnSO_4$	1
Lipoic acid	50	H_3BO_3	0.1
		$AlK(SO_4)_2$	0.1
		$Na_2MoO_4 \times 2H_2O$	0.1

measurements of turbidity using a spectrophotometer and determination of the cell number using a Petroff Hauser chamber were carried out. Redox potential and pH were also measured to assess cell growth. Hourly staining of the strains was done to understand the growth pattern of strains 236-S2 and 236-M1.

To determine bioreduction ability, experiments were performed by inoculating 1 ml of the bacterial culture of strain 236-S1 in 8 conical flasks of BSS broth, which were divided into two sets, A and B, of four flasks each. Set A contained all the components of BSS broth with the microbial inoculum. Set B contained BSS broth without $FeSO_4.7H_2O$ as a trace mineral and the microbial inoculum. The conical flasks were numbered 1–4 for each set. Initially, flasks 1 were purged with $N_2:O_2$ in the ratio of 80:20 for 10 minutes and sealed with thick rubber stoppers and parafilm crimps. Likewise, flasks 2, flasks 3 and flasks 4 were treated with an $N_2:O_2$ in ratios of 70:30, 60:40 and 50:50, respectively and were checked for the reduction ability. Isolates 236-S2 and 236-M1 were cultured in CGY broth. The reduction of Fe(III) by the test strain was studied by monitoring the redox potential and repeated microscopic observation at an hourly interval to check cell viability. As the cells in all of the flasks showed good reduction ability and viability, the ratio of $N_2:O_2$ was reversed to $O_2:N_2$ in ratios of 80:30, 70:30 and 60:40 for flasks 1, 2, 3, respectively in both sets. Flasks 4 of both sets remained the same. The reduction in ferric citrate was determined hourly by titration using 0.05 N potassium dichromate and against barium diphenylamine sulfonate (BDAS) as an indicator. The reduction potential and pH for both the used broths were measured hourly. Strains 236-S1, 236-S2 and 236-M1 were also cultivated in ferric citrate and ferrous sulfate solution devoid of any vitamin or mineral supplementation for 24 hours. Although the growth of the strains was observed in all of the cultures, the growth rate was slow and the marked rise in growth was observed only after 24 hours. After the initial 24 hours of growth the nitriloacetic acid (NTA), which is known to increase the bioavailability of iron, was added to determine its effect on microbial growth.

To determine the metal tolerance capability of strains they were cultured in a medium having three different ferric salts. Flask 1 for a culture of each bacterial strain, namely 236-S1, 236-S2 and 236-M1, consisted of ferric citrate ($FeC_6H_5O_7$) as the Fe(III) ion, ferrous sulfate ($Fe_2SO_4.7H_2O$) as the Fe(II) ion, and nitriloacetic acid (0.0015 g per 100 ml). Flask 2 and flask 3 for a culture of strains 236-S1, 236-S2 and 236-M1 consisted of the same components except for a different Fe(III) salt. Ferric chloride ($FeCl_3.6H_2O$) and ferric sulfate ($Fe_2(SO_4)_3$) were used in flask 2 and flask 3, respectively. No other vitamin or trace mineral was added to the medium. Cell viability was studied by hourly determination of OD at 600 nm and microscopic observation after 24 hours of incubation.

Microbial characteristics

Strains 236-S1, 236-S2 and 236-M1 showed varied colony morphology as expected. DM broth turned pink with the growth of bacteria for 2 days (Fig. 3.2).

Colonies of the bacteria grown on DM were pink (due to resazurin), domed and circular (Fig. 3.3).

In BSS medium supplemented with mineral and vitamin components, colonies were white and about 1 mm in diameter (Fig. 3.4).

Figure 3.2. Color change from blue to pink during the growth of strain 236-S1 in defined medium (DM).

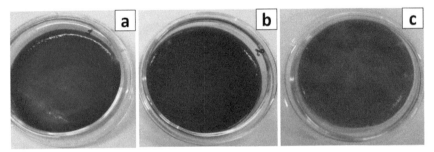

Figure 3.3. Gradual growth of aerobic iron-reducing bacteria observed on the agar plates.

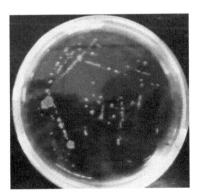

Figure 3.4. Growth of strain 236-S1 isolated from copper mine sample on BSS medium.

When strain 236-S1 grew in BSS broth the complete growth was observed after 2 days of incubation by a distinct change in the appearance of the broth. The color of the medium provided by the presence of ferric citrate ($FeC_6H_5O_7$) lightened and, after a certain period, for unknown reasons, precipitation of ferric citrate ($FeC_6H_5O_7$) occurred, forming a white layer at the top. Cells of the strain 236-S1 were found to be motile, non-capsulated, curved Gram-positive rods (Fig. 3.5a). Flagella detected by using flagellar staining suggests the ability of 236-S1 to form direct contact with the Fe(III) mineral (Fig. 3.5c) (Childers et al. 2002).

Strain 236-S2 showed unique colony morphology in CGY medium (Fig. 3.6).

Various staining procedures showed that isolate 236-S2 was Gram-positive bacilli and could not form endospore or capsule, but showed motility during motility staining. Negative staining showed clear cytoplasm. Another peculiar feature found during the microscopic observation was the presence of crystalline inorganic material surrounded by sheath-shaped structure, though no inorganic matter had been added to the broth cultures (Fig. 3.7a).

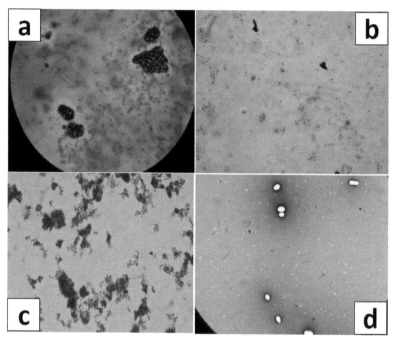

Figure 3.5. Microscopic features of strain 236-S1: (a) Gram staining; (b) endospore staining showing negative results; (c) flagellar staining showing the presence of flagella; (d) negative staining of 236-S1.

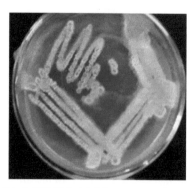

Figure 3.6. Growth of strain 236-S2 isolated from copper mine sample in CGY medium.

Precipitation of formed material was not observed in broth culture.

Strain 236-M1 showed circular, pale yellow colonies on CGY medium (Fig. 3.8). Culture of isolate 236-M1 in CGY broth showed no precipitated material. Strain 236-M1 was a Gram-positive bacillus, though the size of the cells was comparatively larger than those of strain 236-S2. Results of endospore and capsule staining were negative for the strain 236-M1. Negative staining showed clear cytoplasm (Fig. 3.9).

Over the growth period of isolate 236-S1 in BSS medium, the pH increased from 7.0 to 8.47 after 24 hours of growth. Conversely, the Eh increased from 28.4 to 89.8

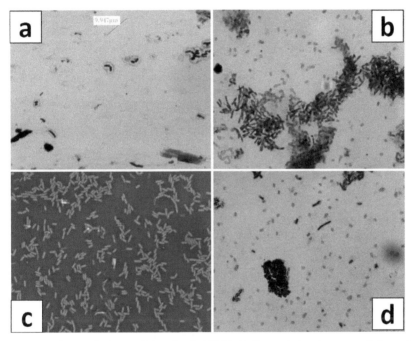

Figure 3.7. Microscopic characteristics of strain 236-S2: (a) Gram staining, the arrow pointing to the inorganic structure formed; (b) endospore staining giving negative results; (c) negative staining; (d) capsule staining showing an absence of capsule.

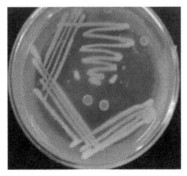

Figure 3.8. Growth of strain 236-M1 isolated from iron oxide filings on CGY medium.

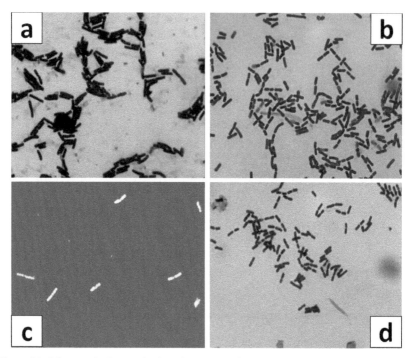

Figure 3.9. Microscopic characterization of 236-M1 strain: (a) Gram staining of 236-M1 showing positive result; (b) absence of endospore in endospore staining; (c) negative staining of 236-M1; (d) capsule staining showing the absence of capsule.

after 8 hours and then reduced to −102.5 after 24 hours of growth. The isolate had an exponential rise in cell number as depicted in Fig. 3.10, where it reached a maximum cell number of 1.17×10^9 cells ml^{-1} after 24 hours of incubation.

The optical density also showed a similar pattern of growth as observed by the cell count. Figure 3.11 denotes a first-order plot having ln (cell number (t)/initial cell number) as the x-axis and time as the y-axis. A straight line was drawn to obtain the specific growth rate μ. An R^2 value of 0.926 indicates a very good fit and μ calculated from the slope of the line was found to be 0.1319 h^{-1}; the generation time of strain 236-S1 at pH 7 incubated at 37°C was 6.43 hours. The growth in the defined medium was very slow and not much change in OD was observed at the end of 24-hour incubation. It was concluded that the low growth was affected by the low concentration of Fe(III) in the medium and further experiments were carried out with the use of BSS medium.

In contrast, during the 24-hour growth of strains 236-S2 and 236-M1, the pH of the cultures decreased from 7.20 to 5.73 and 7.0 to 5.87, respectively. Conversely, the Eh decreased from −9.4 to −309 mV, and from 15.3 to −247 mV after 6 hours, and then increased to −164.6 and 89.1 mV after 24 hours of growth, respectively. Both of the isolates had an exponential rise in cell number as depicted in Figs. 3.12 and 3.14, where they reached a maximum cell number of 121.75×10^7 and 378.75×10^6 cells ml^{-1} after 24 hours of incubation, respectively.

Figures 3.13 and 3.15 denote the first-order plot having ln (cell number (t)/initial cell number) as the x-axis and time as the y-axis. A straight line was drawn to obtain

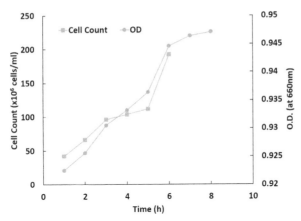

Figure 3.10. Changes in cell number of strain 236-S1 to the optical density (OD) at 600 nm with growth time.

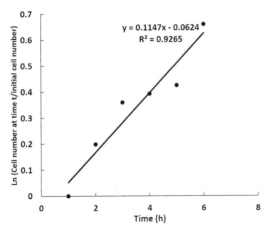

Figure 3.11. Variation in cell number with time for estimation of generation time of strain 236-S1.

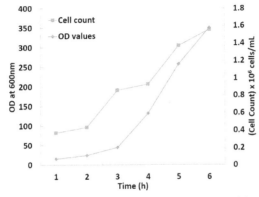

Figure 3.12. Changes in cell number of strain 236-S2 with time to an optical density (OD) at 600 nm.

the specific growth rate μ. An R^2 value of 0.947 and 0.824 was gained and μ calculated from the slope of the line was found to be 0.305 h^{-1} and 0.191 h^{-1}; the generation time of strains 236-S2 and 236-M1 at pH 7 incubated at 37°C was 2.27 and 3.62 hours, respectively.

Growth pattern study of strains 236-S2 and 236-M1 in CGY medium showed negligible changes for the first 2 hours of incubation. From the third hour of growth, there was a prolific increase in cell number and size of strain 236-S2, whereas small cell size and minimal growth were observed for strain 236-M1. At 26 hours of incubation, full-grown cells along with inorganic metal crystallization were observed for strain 236-S2 (Fig. 3.16).

Similarly, microscopic study of strain 236-M1 revealed fully grown cells after 26 hours of growth, whereas, after 27 and 28 hours of incubation of strain 236-S2 in broth, sheath-like structures enclosing inorganic material were observed (Fig. 3.17a, b).

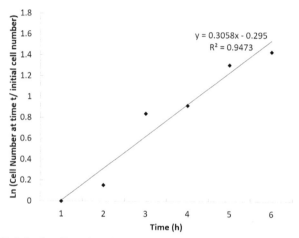

Figure 3.13. Variation in cell number with time for estimation of generation time for strain 236-S2.

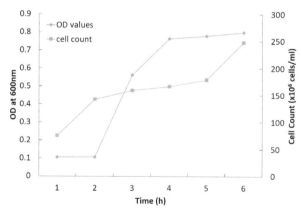

Figure 3.14. Changes in cell number of strain 236-M1 with time to an optical density (OD) at 600 nm.

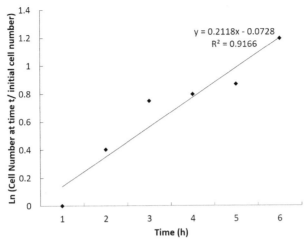

Figure 3.15. Variation in cell number with time for estimation of generation time for strain 236-M1.

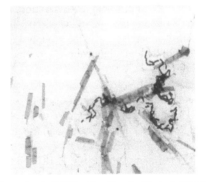

Figure 3.16. Crystalline formation in CGY broth cultured with strain 236-S2 after 26 hours of growth.

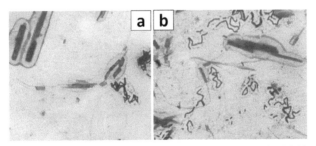

Figure 3.17. Encapsulated inorganic material observed in the growth of strain 236-S2 after 27 hours (a) and 28 hours (b) in broth medium.

Bio-reduction abilities of the isolates

Most of the iron-reducing bacteria isolated to date have shown the capability of reducing Fe(III) under anaerobic conditions as being thermodynamically more suitable (Lovely 2000). Strain 236-S1 showed the rare capability to reduce iron even when a high concentration of oxygen was present. Fe(III) reduction took place in both the media, which was concluded from the observation of the reduction in the Eh values. pH and ORP (Eh) served as good indicators for Fe(III) reduction, with a significant decrease in the Eh with reduction. Also, where acetate served as the electron acceptor, a decrease in the pH was observed till reduction was achieved, after which the pH started increasing. The reduction of iron leads to the synthesis of nanoparticles (Bazylinksi and Frankel 2000), the reasons for which are still unknown, and SEM studies are currently being pursued to detect its nature.

Microscopic observations also showed the viability of the bacterial culture in such a high concentration of O_2. Moreover, spores were formed when the amount of O_2 and N_2 was in the ratio above 60:40. It was also noticed that after 24 hours of growth the cell number in flasks with $FeSO_4 \times 7H_2O$ as a trace mineral was higher than in flasks without such a trace mineral; the cells grown in the absence of $FeSO_4 \times 7H_2O$ were comparatively larger. The isolates 236-S2 and 236-M1 showed reduction ability. This can be concluded from the decrease in redox potential during the time of growth. The redox potential of strain 236-S2 decreased from –9.4 to –309 during 6 hours of growth in CGY media. Similarly, isolate 236-M1 decreased the reduction potential of the broth from –15.3 to 247 in a 6-hour time duration. A decreased growth rate was observed for the first 24 hours, though all the strains showed adaptability to the harsh condition. As expected, a marked rise in the growth was observed on addition of nitriloacetic acid at the end of 24 hours incubation. Also, there was an increase in the cell size on the addition of NTA.

All the three strains 236-S1, 236-S2 and 236-M1 showed an extraordinary capability to tolerate metal without the presence of other trace mineral and vitamins to support their growth. However, a certain decrease in the absorbance was seen after 24 hours of growth depicting the activity of strains in the presence of three different ferric salts as a substrate. Under microscopic view, hexagonal crystals were observed in broths of ferric sulfate of strains 236-S2 (Fig. 3.18) and 236-M1, whereas isolate 236-S1 showed cubic crystals of varied size (Fig. 3.19). In broth with ferric chloride, concentric colonization of cells was observed for all the cultures, whereas in broth with ferric citrate, though maximum enumeration of all the strains was observed, no crystal formation was seen.

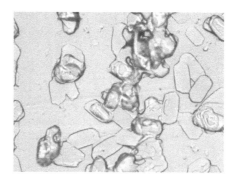

Figure 3.18. Microscopic view (1000x) of hexagonal crystals observed in the culture of strain 236-S2 growing in presence of ferric sulfate.

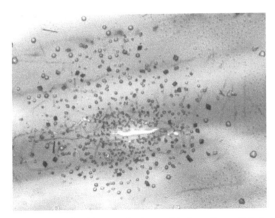

Figure 3.19. Microscopic view of cubic crystals observed in the culture of strain 236-S1 growing in the presence of ferric sulfate.

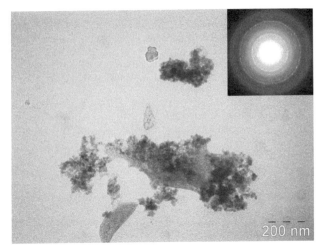

Figure 3.20. TEM-SAED of the precipitate formed during the growth of strain 236-S1 with ferric chloride as a substrate.

The iron particles formed in the culture of strain 236-S1 grown in the presence of ferric chloride was analyzed by TEM and is reported in Fig. 3.20. The SAED pattern depicts the crystallinity of the iron particles synthesized with size below < 50 nm.

Conclusions

In summary, the method for remediation of old mine sites by biologically induced mineralization is presented with a view to utilization of raw materials and conversion to valuable products. The iron reducers, strains 236-S1, 236-S2 and 236-M1, showed good reduction capability as well as metal tolerance ability. The characteristic study performed on these isolates proves that they are good models for further work using other copper and iron leanings. As observed microscopically, the inorganic crystal formation ability of all the cultures with ferric sulfate as an iron source further supports their use for nanoparticle synthesis. Further tests need to be carried out with increased concentration of iron and more specific conditions to elucidate the mechanism of biosynthesis and trials for bulk synthesis.

Acknowledgments

This investigation was carried out as part of the DST-ILTP (Indo-Russian) cooperation project with SB-RAS (A.2.6.1).

References

Abhilash, Revati, K., Pandey, B.D. 2011. Microbial synthesis of iron-based nanomaterials—A review. Bull. Mater. Sci. 34(2): 191–198.

Bazylinski, D.A., Frankel, R.B. 2000. Magnetic iron oxide and iron sulfide minerals within microorganisms. pp. 25–47. *In*: Edmund Baeuerlein (ed.). Biomineralisation from Biology to Biotechnology and Medical Application. Wiley–VCH, Michigan, US.

Berg, J.S., Jézéquel, D., Duverger, A., Lamy, D., Laberty-Robert, C., Miot, J. 2019. Microbial diversity involved in iron and cryptic sulfur cycling in the ferruginous, low-sulfate waters of Lake Pavin. PLoS ONE 14(2): e0212787.

Bomberg, M., Mäkinen, J., Salo, M., Kinnunen, P. 2019. High diversity in iron cycling microbial communities in acidic, iron-rich water of the Pyhäsalmi Mine, Finland. Geofluids, Article ID 7401304, 17 pages.

Childers, S.E., Ciufo, S., Lovley, D.R. 2002. *Geobacter metallireducens* accesses insoluble Fe(III) oxide by chemotaxis. Nature. 416: 767–769.

Frankel, R.B., Bazylinski, D.A. 2003. Biologically induced mineralization by bacteria. Rev. Mineral. Geochem. 54: 95–114.

Ishaq, C.M., Wilcom, M.J., Reid, G. 1964. Isolation and cultivation of iron and sulfur bacteria from domestic sewage. Proceedings for the Oklahoma Society of Sciences 1: 229–233.

Kirschvink, J.L., Hagadorn, J.W. 2000. A grand unified theory of biomineralisation. pp. 139–150. *In*: Bäuerlein, E. (ed.). The Biomineralisation of Nano- and Macro-Structures. Wiley-VCH Verlag GmbH Weinheim, Germany.

Konhauser, K.O. 1997. Bacterial iron biomineralization in nature. FEMS Microbiol. Rev. 20: 315–326.

Kucera, S., Wolfe, R.S. 1957. A selective enrichment method for Gallionellaferruginia. J. Bacteriol. 74: 344–349.

Leitholf, A.M., Fretz, C.E., Mahanke, R., Santangelo, Z., Senko, J.M. 2019. An integrated microbiological and electrochemical approach to determine distributions of Fe metabolism in acid mine drainage-induced "iron mound" sediments. PLoS ONE 14(3): e0213807.

Lovley, D.R., Giovanonni, S.J., White, D.C., Champine, J.E., Philips, E.J.P, Gorby, Y.A., Godwin, S. 1993. *Geobacter metallireducens* gen. nov. sp. nov., a microorganism capable of coupling the complete oxidation of organic compounds to the reduction of iron and other metals. Arch. Microbiol. 159: 336–344.

Lovley, D.R. 1997. Microbial Fe(III) reduction in subsurface environments. FEMS Microbiol. Rev. 20: 305–313.

Lovley, D.R. 2000. Fe(III) and Mn(IV) reduction. *In*: Environmental Microbe—Metal Interaction. ASM Press 3–30.

Mclean, R.J.C., Beauchemin, D., Beveridge, T.J. 1992. Influence of oxidation state on iron binding by *Bacillus lincheniformis* capsule. Appl. Environ. Microbiol. 58: 405–408.

Phoenix, V.R., Konhauser, K.O., Ferris, F.G. 2003. Experimental study of iron and silica immobilization by bacteria in mixed Fe-Si systems: implications for microbial silicification in hot springs. Cand. J. Earth Sci. 40: 1669–1678.

Postgate, J.R. 1959. Sulphate reduction by bacteria. Ann. Rev. Microbiol. 13: 505–520.

Raiswell, R., Canfield, D.E. 1998. Sources of iron for pyrite formation in marine sediments. Amer. J. Sci. 298: 219–245.

Rickard, D., Butler, I.B., Oldroyd, A. 2001. A novel iron sulfide mineral switch and its implication for the earth and planetary sciences. Earth Planet. Sci. Letters 189: 85–91.

Roh, Y., Gao, H., Vali, H., Kennedy, D.W., Yang, Z.K., Gao, W., Dohnalkova, A.C., Stapleton, R.D., Moon, J.W., Phelps, T.J., Fredrickson, J.K., Zhou, J. 2006. Metal reduction and iron biomineralization by a psychrotolerant Fe(III)-reducing bacterium, *Shewanella* sp. strain PV-4. Appl. Environ. Microbiol. 72: 3236–3244.

Semerena, J.C.E, Blakemore, R.P., Wolfe, R.S. 1980. Nitrate dissimilation under microaerophilic conditions by a magnetic spirillum. Appl. Environ. Microbiol. 40: 429–430.

Vargas, M., Kashefi, K., Blunt-Harris, E.L., Lovley, D.R. 1998. Microbiological evidence for Fe(III) reduction on the early earth. Nature 39: 65–67.

Chapter 4

Green Synthesis of Metal and Metal Oxide Nanomaterials Using Seaweed Bioresources

*Govindaraju Kasivelu** and *Tamilselvan Selvaraj*

Introduction

Nanoscience and nanotechnology generally refer to materials with at least one structural dimension ranging from 1 to 100 nanometres. Various nanomaterials such as liposomes (Kisak et al. 2004), polymers (Xu et al. 2007), dentrimers (Paleos et al. 2004), and carbon nanotubes (Wu et al. 2005) have been used for various delivery applications. Among nanomaterials, inorganic nanomaterials play important roles in numerous applications in various fields, such as optical, electronics, catalysis, environmental and biomedical particularly biomedicine, due to their unique chemical, electrical magnetic and physical properties (De et al. 2008, Shipway et al. 2000, Guo and Wang 2011). The unique properties of metal and metal oxide nanomaterials can be finely tuned through the engineering of their compositions and dimensions such as diameter, length, ratio and shape (e.g., spherical, rods, triangles, cubes, plates, wires) (Lu et al. 2007, Xia et al. 2013, Polavarapu et al. 2015). In general, there are two approaches in the preparation of metal and metal oxide nanomaterials, either from "top to bottom" or "bottom to top" (Fig. 4.1). In the top to bottom approach, suitable bulk materials are split into fine particles by thermal/laser ablation, milling, grinding, sputtering and other techniques. In the bottom to top approach, nanomaterials can be synthesized using chemical reduction and biological methods via self-assembly of atoms to new nuclei that grow into a nanoscale particle (Niu and Xu 2011).

Various classical physico-chemical methods that include greener and more environment-friendly approaches have been adopted for synthesis of nanomaterials (Brayner et al. 2012). For the last two decades, various metal and metal oxide

Centre for Ocean Research, Sathyabama Institute of Science and Technology, Chennai 600119, India.
* Corresponding author: govindtu@gmail.com

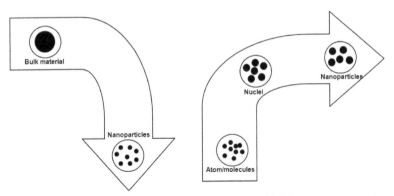

Figure 4.1. Nanomaterials synthesis (a) top down approach; (b) bottom up approach.

nanoparticles synthesized by biological methods have been proved to be useful in various applications. Besides, these biological methods meet the criteria of green chemistry, defined as "the design of chemical products and processes that eliminate the use and generation of toxic/hazardous chemicals". Biological synthesis of nanomaterials using various bioresources has gained much attention in the area of nanotechnology in last few decades because it is cost-effective, non-toxic, and eco-friendly. Researchers have turned their attention towards the synthesis of nanoparticles using various biological resources such as actinomycetes, bacteria, fungi, plants, seaweeds, yeasts, and even viruses (Pramila and Sushil Kumar 2016, Singh et al. 2016). Algae have been used for reduction of metal ions and subsequently for the synthesis of nanomaterials (Hassaan and Hosny 2018). They are also being used as a "bio-factory" for synthesis of metallic nanoparticles. Among different genre of bio-reductants, seaweeds (marine macroalgae) have distinct advantages due to their high metal reduction and uptake capacity, low cost and macroscopic structure. In this context, a green chemistry approach for the synthesis of metal and metal oxide nanomaterials using parts of seaweeds is now under active investigation.

Seaweed-based green synthesis of gold nanoparticles and their applications

During the last few decades, nanoscale gold has been used in a wide variety of applications such as biosensors, cosmetics, catalysis, drug delivery, and biomedical and environmental applications. Because of its chemical stability and unique properties, gold core is fundamentally inert, biocompatible and low in toxicity, and thus an ideal starting point for carrier construction. Nanoscale gold with a wide range of core sizes, i.e., 1–150 nm, can be fabricated easily with controlled dispersity (Ghosh et al. 2008). Both size and shape dispersity are key aspects for diagnosis and drug delivery systems. Various chemical and physical procedures have been developed for synthesis of nanoscale gold, but because of use of toxic chemicals and elevated temperature in the preparation, these techniques were proved to be toxic and harmful to biological and environmental applications (Khan et al. 2017). Biological preparation of nanoscale gold is becoming more common because of easy reduction and because it is simple, cost-effective, less toxic and environment-friendly.

Gold nanoparticles play a vital role in nano-biotechnology as biomedicine because of convenient surface bio-conjugation with bio-molecular probes and remarkable surface plasmon resonance and optical properties (Daniel and Astruc 2004, Wu and Chen 2010). Gold nanoparticles have an important function in the delivery of biological materials such as proteins, nucleic acids, gene therapy, *in vivo* delivery, and targeting (Tiwari et al. 2011). In most cases, physical and chemical methods are routinely used to synthesize nanomaterials on a large scale for a wide range of industrial applications (Chirea et al. 2011). Besides, biological techniques adopt green chemistry principles that avoid harsh reaction conditions and employ eco-friendly technologies (Akhtar et al. 2013). Biological synthesis of gold nanoparticles using various macroalgal species such as *Sargassum wightii* (Singaravelu et al. 2007), *Laminaria japonica* (Ghodake and Lee 2011), *Sargassum swartzii* (Stalin Dhas et al. 2014), *Turbinaria conoides* (Vijayaraghavan et al. 2011, Rajeshkumar et al. 2013), *Stoechospermum marginatum* (Rajathi et al. 2012), *Turbinaria ornata* (Ashokkumar and Vijayaraghavan 2016), *Egregia* sp. (Angel Colin et al. 2018), and *Gracilaria verrucosa* (Chellamuthu et al. 2019) has been reported. Several methodologies developed for the biosynthesis of gold nanoparticles resulted in the formation of nanostructures of various shapes and sizes that can be used in various applications for catalytic, anticancer or antimicrobial activity (Table 4.1).

Seaweed-based green synthesis of silver nanoparticles and their applications

Nanoscale silver particles receive considerable attention for their distinguishing catalytic, electrical, electronic, magnetic, thermal, optical, sensing, conductivity, chemical stability and antimicrobial functionalities compared to bulk metal, which depend on physico-chemical properties such as shape, size, inter-particle spacing and surface functional molecules. The antimicrobial properties of nanoscale silver particles have more applications in catheters, ceramic water filters, air filters, wound dressing materials, biodegradable poly(l-lactide) fibres, biosensing, cosmetics, and electronic appliances. The advantage of nanoscale silver particles compared to bulk material is the slow and synchronized release of silver from nanoparticles, which provides long-term protection against bacterial infection. Since the development of the concept of green nanoparticle preparation, there has been increasing demand for eco-friendly processes of silver nanoparticle synthesis that do not employ toxic chemicals (Chen et al. 2008, Mubarak Ali et al. 2011, Padalia et al. 2015, Thovhogi et al. 2015).

Marine polysaccharides are isolated from various marine natural bioresources such as marine algae, seaweeds (*Sargassum wightii*) (Shanmugam et al. 2014), and *Caulerpa racemose* (Kathiraven et al. 2015). Several researchers demonstrated that silver nanoparticles can be synthesized using macroalgal species. The work carried out by de Aragao et al. (2016) revealed that green synthesis of silver nanoparticles can be achieved using polysaccharide extracted from red algae *Gracilaria birdiae*. Studies have also demonstrated the pesticidal and antimicrobial properties of silver nanoparticles synthesized using extracts of *Sargassum wightii* (Govindaraju et al. 2009), *Gracilaria corticata*, *G. edulis*, *Hypnea musciformis* and *Spyridia hypnoides*

Table 4.1. Seaweed bio-resources based green synthesis of gold nanoparticles and their applications.

Phylum	Class	Order	Species	Collection site	Shape	Size [nm]	Application	References
Ochrophyta	Phaeophyceae	Fucales	*Sargassum wightii*	Mandapam Camp, South East Coast of Tamil Nadu, India	Spherical	8–12	-	Singaravelu et al. 2007
			Sargassum swartzii	Mandapam camp, Rameswaram, Tamil Nadu, India	Spherical	35	Anticancer studies	Stalin Dhas et al. 2014
			Sargassum crassifolium	Seashore of Biga, Lugait, Misamis Oriental in the Philippines	Triangles, Pentagons and Hexagons	25–200		Ouano et al. 2018
Ochrophyta	Phaeophyceae	Fucales	*Turbinaria conoides*	Mandapam Camp, South East Coast of Tamil Nadu, India	-	20–80	-	Vijayaraghavan et al. 2011
			Turbinaria conoides	-	Triangle, Rectangle and Square	60	Antibacterial activity	Rajeshkumar et al. 2013
			Turbinaria conoides	Mandapam coastal region and Gulf of Mannar, Southeast Coast of India	Triangular	2–19	Antimicrofouling activity	Vijayan et al. 2014
			Turbinaria ornata	Beaches of Mandapam region of Tamil Nadu, India	Spherical	7–11	-	Ashokkumar and Vijayaraghavan 2016
			Turbinaria conoides	Mandapam Camp, South East Coast of Tamil Nadu, India	Spherical	27–35	Catalytic activity	Ramakrishna et al. 2016
Ochrophyta	Phaeophyceae	Fucales	*Sargassum tenerrimum*					
Ochrophyta	Phaeophyceae	Laminariales	*Laminaria japonica*	Seafood Market, South Korea	-	15–20	-	Ghodake and Lee 2011

Table 4.1 contd. ...

...Table 4.1 contd.

Phylum	Class	Order	Species	Collection site	Shape	Size [nm]	Application	References
			Gracilaria verrucosa	South East Coast of Tamil Nadu, India	Spherical, Triangular, Oval, Octahedral, Rhombus and Pentagonal	73	Biocompatibility	Chellamuthu et al. 2019
Ochrophyta	*Phaeophyceae*	*Dictyotales*	*Stoechospermum marginatum*	Tuticorin Coast, Tamil Nadu, India	Spherical, Hexagonal and Triangle	18–93	Antibacterial activity	Rajathi et al. 2012
Ochrophyta	*Phaeophyceae*	*Laminariales*	*Egregia* sp.	Mexico company Baja-Kelp	Spherical	8	-	Angel Colin et al. 2018
Ochrophyta	*Phaeophyceae*	*Dictyotales*	*Padina tetrastromatica* and *Turbinaria ornata*	Mandapam Camp, South East Coast of Tamil Nadu, India	Cubics	18–90	Antimicrobial activity	Kayalvizhi et al. 2014

(Roseline et al. 2019), *Padina gymnospora* (Rajaboopathi and Thambidurai 2018), *Sargassum plagiophyllum* (Stalin Dhas et al. 2014), *Jania rubens* and *Sargassum dentifolium* (Saber et al. 2017), *Sargassum tenerrimum* (Kumar et al. 2012), *Padina tetrastromatica* and *Turbinaria ornata* (Kayalvizhi et al. 2014), *Padina pavonica* (Sahayaraj et al. 2012), *Sargassum longifolium* (Rajeshkumar et al. 2014), *Turbinaria conoides* (Vijayan et al. 2014), *Gelidiella acerosa* (Vivek et al. 2011), *Gracilaria corticata* (Kumar et al. 2013). Silver nanoparticles synthesized using *Sargassum siliquosum*-based sulfated polysaccharide were an effective hepatoprotective against paracetamol-induced liver toxicity (Vasquez et al. 2016).

Biocompatible silver nanoparticles are synthesized from aqueous extracts such as *Sargassum polycystum* (Palanisamy et al. 2017), *Enteromorpha compressa* (Ramkumar et al. 2017), *Turbinaria conoides* (Ali et al. 2019), *Sargassum ilicifolium* (Kumar et al. 2012) and *Turbinaria ornata* (Deepak et al. 2017) as potential antioxidant, anticancer activity, cardioprotective, liver protective and mosquito larvicidal activity agent. The synthesis and characterization of silver nanoparticles using variable macroalgae, including *Cystophora moniliformis* (Prasad et al. 2013), *Padina tetrastromatica* (Jegadeeswaran et al. 2012), *Padian boeregeseni* (Hashemi et al. 2015), *Gracilaria edulis* (Murugesan et al. 2011), *Sargassum plagiophyllum*, *Ulva reticulata* and *Enteromorpha compressa* (Dhanalakshmi et al. 2012) have been presented in Table 4.2.

Seaweed based green synthesis of Copper oxide nanoparticles and their applications

Copper oxide (CuO) nanoparticles a p-type semiconductor with a band gap of 1.2 eV, large surface area and reactive morphology is particularly interesting due to its good chemical and thermal stability, low toxicity, ease of handling, low cost and high catalytic reusability (Dipankar and Murugan 2012). CuO nanoparticles are stable and have longer shelf life compared to organic and microbial agents (Das et al. 2013). CuO nanoparticles have attracted huge attention due to antibacterial, catalytic, optical, photonic and electric properties, which depend on their size and shape (Padil and Cernik 2013). Application of micro algal extracts as capping and reducing agents in nanoparticle synthesis makes the process non-toxic and environmentally benign. Biosynthesized CuO nanoparticles by using *Sargassum polycystum* (Naika et al. 2015, Ramaswamy et al. 2016), *Gracilaria edulis* (Abraham et al. 2018) had significant antimicrobial and anticancer activities and photo catalytic activity for wastewater treatment. Furthermore, Cu nanoparticles were synthesized using aqueous extracts of *Corallina officinalis* Linnaeus and *Corallina mediterranea* Areschoug. Algae based biomolecules such as proteins, complex carbohydrates, carboxylic acids act as encapsulating and reducing agents led to formation of copper nanoparticles from copper sulphate. Cu nanoparticles have been reported to have toxic effect on *Lyngbya majuscula* in a dose dependent manner (El-Kassas and Okbah 2017) (Table 4.3).

Table 4.2. Seaweed bio-resources based green synthesis of silver nanoparticles and their applications

Phylum	Class	Order	Species	Collection site	Shape	Size [nm]	Application	References
Ochrophyta	Phaeophyceae	Fucales	Sargassum wightii	Mandapam camp, South East Coast of Tamil Nadu, India	Spherical shape	8 to 27	Antibacterial activity	Govindaraju et al. 2009
			Sargassum tenerrimum	Gulf of Mannar, South East Coast of Tamil Nadu, India	Spherical	20	Antimicrobial activity	Kumar et al. 2012a
			Sargassum ilicifolium	Puthumadam, Gulf of Mannar Marine Biosphere Reserve Trust	Spherical morphology	33–44	Antibacterial activity and Cytotoxicity	Kumar et al. 2012b
			Sargassum plagiophyllum, Ulva reticulata and Enteromorpha compressa	Kovalam and Muttukadu, Chennai, India	Spherical shape	20–50	-	Dhanalakshmi et al. 2012
			Sargassum plagiophyllum	Mandapam Camp, South East Coast of Tamil Nadu, India	Spherical	18–42	Antimicrobial activity	Stalin Dhas et al. 2014
			Sargassum wightii	Mandapam Costal Region, Gulf of Manner, Southeast Coast of India	Spherical shape	15 to 20	Antibacterial activity	Shanmugam et al. 2014
			Sargassum longifolium	Tuticorin coastal area, Tamil Nadu, India	Spherical, Truncated and Ellipsoidal	-	Antifungal activity	Rajeshkumar et al. 2014
			Sargassum siliquosum	Bolo Beach, Alaminos Pangasinan, Philippines, Diliman	Spherical shape	20–480	Heptoprotective studies	Vasquez et al. 2016
			Sargassum polycystum	Gulf of Mannar region, Mandapam, South East Coast of Tamil Nadu, India	Spherical	28	Antioxidant and Anticancer activity	Palanisamy et al. 2017

Ochrophyta Rhodophyta	Phaeophyceae Florideophyceae	Fucales Corallinales	Sargassum dentifolium and Jania rubens	Red sea in Hurghada, Egypt	Spherical shape	113–155	Antimicrobial activity	Saber et al. 2017
Rhodophyta	Florideophyceae	Gracilariales	Gracilaria birdiae	Northeast Brazil	Spherical	20–94	Antimicrobial activity	de Aragao et al. 2016
Rhodophyta	Florideophyceae Gracilariales Gigartinales Ceramiales	Gracilariales	Gracilaria corticata Gracilaria edulis Hypnea musciformis Spyridia hypnoides	Puthumadam, Pamban, Gulf of Mannar, Tamil Nadu India	Spherical	37 54 53 49	Antibacterial and antifungal activity	Roseline et al. 2019
Ochrophyta	Phaeophyceae	Fucales	Turbinaria conoides	Mandapam coastal region Gulf of Mannar, Southeast coast of India	Spherical	2–17	Antimicrofouling Activity	Vijayan et al. 2014
			Turbinaria ornata	Mandapam, South East Coast of Tamil Nadu, India	Spherical	20–32	Mosquito larvicidal activity	Deepak et al. 2017
			Turbinaria conoides	-	-	50–100	Cardioprotective and Heptoprotective	Ali et al. 2019
Ochrophyta	Phaeophyceae	Dictyotales	Padina pavonica	Tuticorin Coast, Tamil Nadu	Spherical	10–72	Antimicrobial activity	Sahayaraj et al. 2012
			Padina tetrastromatica	Mandapam coastal region, Gulf of Mannar, Tamil Nadu, South India	Spherical	20	-	Jegadeeswaran et al. 2012
Ochrophyta	Phaeophyceae	Dictyotales	P. tetrastromatica and T. ornata	South East Coast of Tamil Nadu, India	Cubics	20–90	Antimicrobial activity	Kayalvizhi et al. 2014
Ochrophyta	Phaeophyceae	Fucales						

Table 4.2 contd. ...

... Table 4.2 contd.

Phylum	Class	Order	Species	Collection site	Shape	Size [nm]	Application	References
Ochrophyta	*Phaeophyceae*	*Dictyotales*	*Padian boeregeseni*	Iran's Southern coast, Omen	Spherical	43.3	-	Hashemi et al. 2015
			Padina gymnospora	Mandapam, Coastal region of Tamil Nadu, India	Spherical	2–20	Antimicrobial activity against *S. aureus* and *E. coli*	Rajaboopathi and Thambidurai 2018
Rhodophyta	*Florideophyceae*	*Gelidiales*	*Gelidiella acerosa*	Mandapam coastal region, Gulf of Mannar, Tamil Nadu, South India	Spherical	22	Antifungal activity	Vivek et al. 2011
Rhodophyta	*Florideophyceae*	*Gracilariales*	*Gracilaria edulis*	Mandapam, South East Coast of Tamil Nadu, India	Spherical	12.5–100	-	Murugesan et al. 2011
			Gracilaria corticata	Mandapam, South East Coast of Tamil Nadu, India	Spherical	18–46	Antifungal activity	Kumar et al. 2013
Ochrophyta	*Phaeophyceae*	*Fucales*	*Cystophora moniliformis*	Glenelg Beach Adelaide, South Australia	Monodisperse	75	-	Prasad et al. 2013
Chlorophyta	*Ulvophyceae*	*Bryopsidales*	*Caulerpa racemosa*	Gulf of Mannar, Southeast Coast of India	Spherical, Triangular	5–25	Antibacterial activity	Kathiraven et al. 2015
Chlorophyta	*Ulvophyceae*	*Ulvales*	*Enteromorpha compressa*	Pudumadam Coastal Region, Southeast coast of India	Spherical	4–24	Cytotoxicity	Ramkumar et al. 2017

Table 4.3. Seaweed bio-resources based green synthesis of CuO nanoparticles and their applications.

Phylum	Class	Order	Species	Collection site	Shape	Size [nm]	Application	References
Ochrophyta	*Phaeophyceae*	*Fucales*	*Sargassum polycystum*	Gulf of Mannar (Rameshwaram), Tamil Nadu, India	-	-	Antimicrobial and Anticancer activities	Ramaswamy et al. 2016
Rhodophyta	*Florideophyceae*	*Gigartinales*	*Kappaphycus alvarezii*	Sabah, Malaysia	Core-shell structure	60	-	Khanehzaei et al. 2015
Rhodophyta	*Florideophyceae*	*Gracilariales*	*Gracilaria edulis*	Intertidal rocky shore, Mandapam, Gulf of Mannar region, India	-	60–100	-	Abraham et al. 2018
Rhodophyta	*Florideophyceae*	*Corallinales*	*Corallina officinalis* and *Corallina mediterranea*	-	Spherical	12.7–13.6	Lyngbya majuscula	El-Kassas and Okbah 2017

Seaweed based green synthesis of FeO nanoparticles and their applications

Iron oxide (FeO) based nanomaterials play a vital role in several applications as gas sensor (Chen et al. 2005), catalyst (Shekhah et al. 2003) and electrode material (Poizot et al. 2000). Magnetic iron oxide (hematite) nanomaterials (Fe_3O_4 and Fe_2O_3) have been studied for various biological and bio-medical applications including magnetic resonance imaging (MRI) as contrast agents, drug-delivery vehicles and the magnetic separation of biological molecules (Rana et al. 2017). Fe nanoparticles were also used in fields of analytical chemistry, pathogen, enzyme or antigen detection (Ruan et al. 2018). Green synthesis of FeO nanoparticles was performed by using *Chaetomorpha antennina* and *Turbinaria turbinata* extract. These studies showed that nanoparticles synthesized at pH 8 showed higher ferromagnetism, while the nanoparticles synthesized at pH 6 showed the least magnetism (Siji et al. 2018). Seaweed extracts of *Kappaphycus alvarezii* was used for synthesis Fe_3O_4 nanoparticles by using a hot plat combustion method. Fe_3O_4 nanoparticles were studied in various fields such as catalytic degradation of textile dyeing waste water and antibacterial analysis (Arularasu et al. 2018). Seaweed derived Fe_3O_4 nanoparticles were successfully synthesized from *K. alvarezii* which was employed as a green reducing and stabilizing agents (Yew et al. 2016).

Fe_3O_4 nanoparticles were also synthesized by reduction of ferric chloride using a completely green biosynthetic method with brown seaweed *Padina pavonica*, *Sargassum acinarium* and *Sargassum muticum* aqueous extract containing sulphated polysaccharides as a key factor which acted as reducing and efficient stabilizing agent. Studies revealed that efficacy of the bioremediation of heavy metal pollution increased while using the biosynthesized Fe_3O_4 nanoparticles. Fe_3O_4 nanaoparticles particularly increased the lead removing efficiency (El-Kassas et al. 2016, Mahdavi et al. 2013) (Table 4.4).

Seaweed based green synthesis of ZnO nanoparticles and their applications

Environment friendly technology for the synthesis of ZnO nanoparticles has been applied in recent studies due to its unique chemical and physical properties *viz*, high chemical stability, high electrochemical coupling coefficient, II–VI semiconductor with wide band gap energy, which is 3.3 eV and high excitation binding energy, that is, 60 eV. Therefore, ZnO can tolerate large electric fields, high temperature and high power operations and have potential applications as they exhibit non-toxic, bio-safe, anti-bacterial, anti-fungal, anti-diabetic, anti-inflammatory, wound healing, antioxidant, drug carriers, cosmetics and optic properties (Nagarajan and Kuppusamy 2013, Agarwal et al. 2017). Highly stable and spherical ZnO nanoparticles were produced by zinc acetate utilizing by the biocomponents of crude extract of *Turbinaria conoides* (Raajshree and Durairaj 2017). Seaweeds *Caulerpa peltata*, *Hypnea valencia* and *Sargassum myriocystum* (Nagarajan and Kuppusamy 2013), *Padina tetrastromatica* (Pandimurugan and Thambidurai 2017), *Ulva lactuca* (Ishwarya et al. 2018) extracts

Table 4.4. Seaweed bio-resources based green synthesis of FeO nanoparticles and their applications.

Phylum	Class	Order	Species	Collection site	Shape	Size [nm]	Application	References
Chlorophyta	Ulvophyceae	Cladophorales	*Chaetomorpha antennina*	-	-	-	-	Siji et al. 2018
Ochrophyta	Phaeophyceae	Fucales	*Turbinaria turbinata*					
Rhodophyta	Florideophyceae	Gigartinales	*Kappaphycus alvarezii*	-	Spherical	14.7	-	Yew et al. 2016
			Kappaphycus alvarezii	-	Hexagonal	10–30	Antimicrobial activity	Arularasu et al. 2018
Ochrophyta	Phaeophyceae	Dictyotales	*Padina pavonica*	Egyptian Mediterranean Coast, especially in Alexandria coast	Spherical	10–19.5; 21.6–27.4	Lead Bioremediation	El-Kassas et al. 2016
Ochrophyta	Phaeophyceae	Fucales	*Sargassum acinarium*					
Ochrophyta	Phaeophyceae	Fucales	*Sargassum muticum*	Persian Gulf waters	Cubic	18	-	Mahdavi et al. 2013

were found to have the potential for the synthesis of ZnO nanoparticles which act as antibacterial agent against Gram-positive (*Staphylococcus aureus* and *Streptococcus mutans*) than the Gram-negative bacteria (*E. coli, Neisseria gonorrhoeae, Klebsiella pneumoniae* and *Vibrio cholerae*) and they can be used in effluent treatment process for reducing microbial load. ZnO nanoparticles have been also synthesized using seaweed polysaccharides from *Ulva fasciata* and *Sargassum muticum* (Azizi et al. 2014) as well as *Padina gymnospora* (Rajaboopathi and Thambidurai 2017). It was claimed that macroalgae synthesized ZnO-MgO nanoparticles could be employed as a proficient catalyst and photocatalytic activity for the pollutants removal applications (Vijai Anand et al. 2019).

Dumbrava et al. (2018) demonstrated that the ZnO particles onto calcium carbonate was synthesized using extract of green seaweeds *Ulva lactuca* and tested as a wound dressings for the treatment of thermal burns, to promote the skin healing processes. Preparation of ZnO nanoparticles was accomplished through new natural green chemistry approach, using the *U. lactuca* extract as a reducing and stabilizing agent. ZnO nanoparticles treated methylene blue dye was efficiently degraded under sunlight. Similarly, seaweed-ZnO-polyaniline hybrid composite could be employed as an efficient adsorbent, much more than the parent material adsorption of methylene blue dye (Pandimurugan and Thambidurai 2016). Promising larvicidal activity was shown by the *Ulva lactuca* derived ZnO nanoparticles on Zika virus vectors, namely *Aedes aegypti*, while *S. wightii* showed larvicidal activity against malaria vector *Anopheles stephensi* and pupicidal activity against cotton bollworm *Helicoverpa armigera* (Ishwarya et al. 2018, Murugan et al. 2018). Seaweed bioresources of ZnO nanoparticles, their properties and activities are presented in Table 4.5.

Seaweed based green synthesis of other oxide nanomaterials and their applications

A facile and green combustion technique has been optimized for the preparation of zirconia (ZrO_2) nanoparticles using seaweed *Sargassum wightii*. The synthesized ZrO_2 nanoparticles showed significant anti-bacterial activity against gram-positive and gram-negative bacterial species due to their large surface area resulting from nanosize (Kumaresan et al. 2018). The green synthesis of palladium oxide nanoparticles has been carried out using *Dictyota indica* seaweed extract. Palladium oxide nanoparticles can be used as a safe and cost effective method in water treatment. The removal of cadmium by the palladium oxide nanoparticles was investigated (Shargh et al. 2018). Magnesium oxide (MgO) nanoparticles were synthesized using the brown alga *Sargassum wightii* as the reducing and capping agent. The synthesized MgO nanoparticles were physico-chemical characterized by spectroscopic, microscopic techniques and showed potent antimicrobial activities against both human pathogenic bacterial and fungal species in a dose dependent manner. In addition, the MgO nanoparticles showed potent photocatalytic activity in the organic dye methylene blue degradation under both sunlight and UV irradiation. MgO nanoparticles also showed strong cytotoxic activity against lung cancer cell lines A549 (Pugazhendhi et al. 2019) (Table 4.6).

Table 4.5. Seaweed bio-resources based green synthesis of ZnO nanoparticles and their applications.

Phylum	Class	Order	Species	Collection site	Shape	Size [nm]	Application	References
Chlorophyta	*Ulvophyceae*	*Bryopsidales*	*Caulerpa peltata*	Intertidal rocky shore regions in Mandapam, Pudhumadam and Kilakarai coast of the Gulf of Mannar region, India	Triangle, Radial, Hexagonal, Rod and Rectangle	76–186	Antibacterial	Nagarajan and Kuppusamy 2013
Rhodophyta	*Florideophyceae*	*Gigartinales*	*Hypnea valencia*					
Ochrophyta	*Phaeophyceae*	*Fucales*	*Sargassum myriocystum*					
Ochrophyta	*Phaeophyceae*	*Fucales*	*Sargassum muticum*	-	Hexagonal	30–57	-	Azizi et al. 2014
			Sargassum wightii	Gulf of Mannar Biosphere Reserve, Southeast Coast of India	Spherical	20–62	Larvicidal and Pupicidal	Murugan et al. 2018
Chlorophyta	*Ulvophyceae*	*Ulvales*	*Ulva lactuca*	Black Sea coastline, in Constanta city	Spherical	-	Skin healing	Dumbrava et al. 2018
			Ulva lactuca	Intertidal stony coast areas in Tuticorin, Tamil Nadu, India	Triangles, Hexagons, Rods and Rectangles	10–50	Photocatalytic Activity, Antibacterial and Antibiofilm Studies	Ishwarya et al. 2018
Ochrophyta	*Phaeophyceae*	*Dictyotales*	*Padina tetrastromatica*	Intertidal region of Mandapam located in the South Coast of India	-	-	Adsorption of methylene blue dye	Pandimurugan and Thambidurai 2016
			Padina tetrastromatica	Gulf of Mannar, Tamil Nadu, India	Hexagonal wurtzite	28–24	Antibacterial activity	Pandimurugan and Thambidurai 2017
			Padina gymnospora	Mandapam Coastal area, South Coast of India	Hexagonal, Irregular	20–50	Photocatalytic activity	Rajaboopathi and Thambidurai 2017
Ochrophyta	*Phaeophyceae*	*Fucales*	*Turbinaria conoides*	Mandapam Coastal region, Gulf of Mannar, Southeast Coast of India	Spherical	80–130	-	Raajshree and Durairaj 2017

Table 4.6. Seaweed bio-resources based green synthesis of ZrO_2 and MgO nanoparticles and their applications.

Phylum	Class	Order	Species	Collection site	Shape	Size [nm]	Application	References
Ochrophyta	Phaeophyceae	Fucales	Sargassum wightii	Mandapam Camp, South East Coast of Tamil Nadu, India	Spherical	5	Antibacterial activity	Kumaresan et al. 2018
			Sargassum wighitii	South Indian Coastal area, Tamil Nadu	Flower shaped structure	68.06	Anticancer, Antimicrobial and Photocatalytic activities	Pugazhendhi et al. 2019
Ochrophyta	Phaeophyceae	Dictyotales	Dictyota indica	Oman Sea Coasts in Chabahar, Iran	Spherical	8–43	Removal of cadmium	Shargh et al. 2018

Conclusion

In this review, we sum up a decade of results from research and exploration in the field of biological synthesis of nanomaterials using marine bioresources particularly macro algal (seaweeds) species. The macroalgae currently known to be involved in the reduction and formation of various nanomaterials belong to different classes. To date, macroalgae have been studied for synthesis of gold, silver, zinc oxide, iron oxide, copper oxide, zirconia and magnesium oxide nanoparticles. Depending upon the type of seaweed resource, concentration and other experimental parameters (temperature, pH, etc.), nanoparticles of various sizes and shapes can be obtained for various biological applications. Biological synthesis of nanoparticles using algal bioresources is in its infancy compared to the other bioresource such as plants. Almost, few decades over number of papers have been published and dealing with plant-mediated biosynthesis of nanoparticles published per year. In this review, significant boom in the biological synthesis of various nanoparticles using macro algae (seaweed) and their extracts by increasing studies to include a wider range of various algae and synthesis processes of different nanoparticles that will be a strong potential in various commercial, environmental and various biological applications.

Acknowledgement

We thank Indian Council of Agricultural Research (ICAR), Ministry of Agriculture, Government of India for their financial support. Also thank the management of Sathyabama Institute of Science and Technology, Chennai for stanch support in research activities.

References

Abraham, N., John, V.S., Rajini, P.S.P. 2018. Green synthesis and characterization of copper oxide nanoparticles using a red seaweed *Gracilaria edulis*. Int. J. Eng. Sci. Inven. 7(10)Ver III: 9–13.
Agarwal, H., Venkat Kumar, S., Rajeshkumar, S. 2017. A review on green synthesis of zinc oxide nanoparticles—an eco-friendly approach. Resour. Eff. Technol. 3: 406–413.
Akhtar, M.S., Panwar, J., Yun, Y.S. 2013. Biogenic synthesis of metallic nanoparticles by plant extracts. ACS Sus. Chem. Eng. 1(6): 591–602.
Ali, M.S., Anuradha, V., Yogananth, N., Krishnakumar, N. 2019. Heart and liver regeneration in zebrafish using silver nanoparticle synthesised from *Turbinaria conoides—in vivo*. Biocatal. Agricul. Biotechnol. 17: 104–109.
Angel Colin, J., Pech-Pech, I.E., Oviedo, M., Aguila, S.A., Romo-Herrera, J.M., Contreras, O.E. 2018. Gold nanoparticles synthesis assisted by marine algae extract: Biomolecules shells from a green chemistry approach. Chem. Phys. Lett. 708(16): 210–215.
Aragao, A.P., Oliveira, T.M., Quelemes, P.V., Perfeito, M.L.G., Araújo, M.C., Santiago, J.A.S., Cardoso, V.S., Quaresma, P., de Almeida Leite, J.R.S., da Silva, D.A. 2016. Green synthesis of silver nanoparticles using the seaweed *Gracilaria birdiae* and their antibacterial activity. Arab. J. Chem. https://doi.org/10.1016/j.arabjc.2016.04.014.
Arularasu, M.V., Devakumar, J., Rajendran, T.V. 2018. An innovative approach for green synthesis of iron oxide nanoparticles: Characterization and its photocatalytic activity. Polyhed. 156: 279–290.
Ashokkumar, T., Vijayaraghavan, K. 2016. Brown seaweed-mediated biosynthesis of gold nanoparticles. J. Environ. Biotechnol. Res. 2(1): 45–50.

Azizi, S., Ahmad, M.B., Namvar, F., Mohamad, R. 2014. Green biosynthesis and characterization of zinc oxide nanoparticles using brown marine macro alga *Sargassum muticum* aqueous extract. Mater Lett. 116: 275–277.

Brayner, R., Coradin, T., Beaunierc, P., Grenèche, J.M., Djediat, C., Yepremian, C., Coute, A., Fievet, F. 2012. Intracellular biosynthesis of superparamagnetic 2-lines ferri-hydrite nanoparticles using *Euglena gracilis* microalgae. Colloids Surf. B: Biointerf. 93: 20–23.

Chellamuthu, C., Balakrishnan, R., Patel, P., Shanmuganathan, R., Pugazhendhi, A., Ponnuchamy, K. 2019. Gold nanoparticles using red seaweed *Gracilaria verrucosa*: Green synthesis, characterization and biocompatibility studies. Process Biochem. 80: 58–63.

Chen, J., Wang, J., Zhang, X., Jin, Y. 2008. Microwave-assisted green synthesis of silver nanoparticles by carboxymethyl cellulose sodium and silver nitrate. Mater Chem. Phys. 108: 421–424.

Chen, J., Xu, L., Li, W., Gou, X.C. 2005. α-Fe$_2$O$_3$ nanotubes in gas sensor and lithium-ion battery applications. Adv. Mat. 17: 582–586.

Chirea, M., Freitas, A., Vasile, B.S., Ghitulica, C., Pereira, C.M., Silva, F. 2011. Gold nanowire networks: synthesis, characterization and catalytic activity. Langmuir. 27(7): 3906–3913.

Daniel, M.C., Astruc, D. 2004. Gold nanoparticles: assembly, supramolecular chemistry, quantum-size-related properties, and applications toward biology, catalysis and nanotechnology. Chem. Rev. 104: 293–346.

Das, D., Nath, B.C., Phukon, P., Dolui, S.K. 2013. Synthesis and evaluation of antioxidant and antibacterial behavior of CuO nanoparticles, Colloids Surf. B: Biointerf. 101: 430–433.

De, M., Ghosh, P.S., Rotello, V.M. 2008. Applications of nanoparticles in biology. Adv. Mat. 20(22): 4225–4241.

Deepak, P., Rajamani, S., Rajendiran, R., Govindasamy, B., Dilipkumar, A., Pachiappan, P. 2017. Structural characterization and evaluation of mosquito-larvicidal property of silver nanoparticles synthesized from the seaweed, *Turbinaria ornata* (Turner) J. Agardh 1848. J. Artif. Cells Nanomed. Biotechnol. 45(5): 990–998.

Dhanalakshmi, P.K., Riyazulla, A., Rekha, R., Poonkodi, S., Thangaraju, N. 2012. Synthesis of silver nanoparticles using green and brown seaweeds. Phykos. 42(2): 39–45.

Dipankar, C., Murugan, S. 2012. Green synthesis of CuO nanoparticles by aqueous extract *Gundelia tournefortii* and evaluation of their catalytic activity for the synthesis of N-monosubstituted ureas and reduction of 4-nitrophenol. Colloids Surf. B: Biointerf. 98: 112–119.

Dumbrava, A., Berger, D., Matei, C., Radu, M.D., Gheorghe, E. 2018. Characterization and applications of a new composite material obtained by green synthesis, through deposition of zinc oxide onto calcium carbonate precipitated in green seaweeds extract. Ceram Int. 44(5): 4931–4936.

El-Kassas, H.Y., Okbah, M.A.E.A. 2017. Phytotoxic effects of seaweed mediated copper nanoparticles against the harmful alga: *Lyngbya majuscule*. J. Gen. Engg. Biotechnol. 15(1): 41–48.

El-Kassas, H.Y., Aly-Eldeen, M.A., Gharib, S.M. 2016. Green synthesis of iron oxide (Fe$_3$O$_4$) nanoparticles using two selected brown seaweeds: Characterization and application for lead bioremediation. Acta Oceanol Sin. 35(8): 89–98.

Ghodake, G., Lee, D.S. 2011. Biological synthesis of gold nanoparticles using the aqueous extract of the brown algae *Laminaria japonica*. J. Nanoelec. Optoelec. 6: 268–271.

Ghosh, P., Han, G., De, M., Kim, C.K., Rotello, V.M. 2008. Gold nanoparticles in delivery applications. Adv. Drug Del. Rev. 60: 1307–1315.

Govindaraju, K., Kiruthiga, V., Kumar, V.G., Singaravelu, G. 2009. Extracellular synthesis of silver nanoparticles by a marine alga, *Sargassum wightii* Grevilli and their antibacterial effects. J. Nanosci. Nanotechnol. 9(9): 5497–501.

Guo, S., Wang, E. 2011. Noble metal nanomaterials: Controllable synthesis and application in fuel cells and analytical sensors. Nano Today. 6: 240–264.

Hashemi, S., Givianrad, M.H., Moradi, A.M., Larijani, K. 2015. Biosynthesis of silver nanoparticles using brown marine seaweed *Padian boeregeseni* and evaluation of physic chemical factors. Indian J. Geo-Marine Sci. 44(9): 1415–1421.

Hassaan, M.A., Hosny, S. 2018. Green synthesis of Ag and Au nanoparticles from micro and macro algae—review. Int. J. Atmos. Ocec. Sci. 2(1): 10–22.

Ishwarya, R., Vaseeharan, B., Kalyani, S., Banumathi, B., Govindarajan, M., Alharbi, N.S., Kadaikunnan, S., Al-anbr, M.N., Khaled, J.M., Benelli, G. 2018. Facile green synthesis of zinc oxide nanoparticles using *Ulva lactuca* seaweed extract and evaluation of their photocatalytic, antibiofilm and insecticidal activity. J Photochem Photobiol B: Biol. 178: 249–258.

Jegadeeswaran, P., Shivaraj, R.,Venckatesh, R. 2012. Green synthesis of silver nanoparticles from extract of *Padina tetrastromatica* leaf. Digest J. Nanomater Biostruct. 7(3): 991–998.

Kathiraven, T., Sundaramanickam, A., Shanmugam, N., Balasubramanian, T. 2015. Green synthesis of silver nanoparticles using marine algae *Caulerpa racemosa* and their antibacterial activity against some human pathogens. Appl. Nanosci. 5: 499–504.

Kayalvizhi, K., Nabikhan, A., Vasuki, S., Kandasamy, K. 2014. Purification of silver and gold nanoparticles from two species of brown seaweeds (*Padina tetrastromatica* and *Turbinaria ornata*). J. Med. Plan Stuc. 2(4): 32–37.

Khan, I., Saeed, K., Khan, I. 2017. Nanoparticles: Properties, applications and toxicities. Arab. J. Chem. DOI: 10.1016/j.arabjc.2017.05.011

Khanehzaei, H., Ahmad, M.B., Kamyar Shameli, K. Ajdari, Z. 2014. Synthesis and characterization of Cu@Cu2O core shell nanoparticles prepared in seaweed *kappaphycus alvarezii* media. Int. J. Electrochem. Sci. 9: 8189–819.

Kisak, E.T., Coldren, B., Evans, C.A., Boyer, C., Zasadzinski, J.A. 2004. The vesosome—a multicompartment drug delivery vehicle. Curr. Med. Chem. 11: 199–219.

Kumar, P., Senthamil Selvi, S., Govindaraju, M. 2013. Seaweed-mediated biosynthesis of silver nanoparticles using *Gracilaria corticata* for its antifungal activity against *Candida* spp. Appl. Nanosci. 3: 495–500.

Kumar, P., Senthamil Selvi, S., Lakshmi Prabha, A., Prem Kumar, K., Ganeshkumar, S., Govindaraju, M. 2012. Synthesis of silver nanoparticles from *Sargassum tenerrimum* and screening phytochemicals for its antibacterial activity. Nano Biomed. Eng. 4: 12–16.

Kumar, P., Senthamil Selvi, S., Lakshmi Prabha, A., Selvaraj, M., Macklin Rani, L., Suganthi, P., Sarojini Devi, B., Govindaraju, M. 2012. Antibacterial activity and invitro cytotoxicity assay against brine shrimp using silver nanoparticles synthesized from *Sargassum ilicifolium*. Diges. J. Nano Biostruc. 7(4): 1447–1455.

Kumaresan, M., Vijai Anand, K., Govindaraju, K., Tamilselvan, S., Ganesh Kumar, V. 2018. Seaweed *Sargassum wightii* mediated preparation of zirconia (ZrO_2) nanoparticles and their antibacterial activity against gram positive and gram negative bacteria. Microb. Pathogen. 124: 311–315.

Lu, A.H., Salabas, E.L., Schuth, F. 2007. Magnetic nanoparticles: synthesis, protection, functionalization and application. Angew. Chem. Int. Ed. 46: 1222–1244.

Mahdavi, M., Namvar, F., Ahmad, M.B., Mohamad, R. 2013. Green biosynthesis and characterization of magnetic iron oxide (Fe_3O_4) nanoparticles using seaweed (*Sargassum muticum*) aqueous extract. Molecules 18(5): 5954–5964.

Mubarak Ali, A., Thajuddin, N., Jeganathan, K., Gunasekaran, M. 2011. Plant extract mediated synthesis of silver and gold nanoparticles and its antibacterial activity against clinically isolated pathogens. Colloids Surf. B: Biointerf. 85: 360–365.

Murugan, K., Roni, M., Panneerselvam, C., Aziz, A.T., Suresh, U., Rajaganesh, R., Aruliah, R., Mahyoub, J.A., Trivedi, S., Rehman, H., Naji Al-Aoh, H.A., Kumar, S., Higuchi, A., Vaseeharan, B., Wei, H., Senthil-Nathan, S., Canale, A., Benelli, G. 2018. *Sargassum wightii*-synthesized ZnO nanoparticles reduce the fitness and reproduction of the malaria vector *Anopheles stephensi* and cotton bollworm *Helicoverpa armigera*. Physiol. Mol. Plant Pathol. 101: 202–213.

Murugesan, S., Elumalai, M., Dhamotharan, R. 2011. Green synthesis of silver nanoparticles from marine algae *Gracillaria edulis*. Bosci. Biotechnol. Res. Comm. 4(1): 105–110.

Nagarajan, S., Kuppusamy, K.A. 2013. Extracellular synthesis of zinc oxide nanoparticles using seaweeds of gulf of Mannar, India. J. Nanobiotecnol. 11: 39.

Naika, H.R., Lingaraju, K., Manjunath, K., Kumar, K., Nagaraju, G., Suresh, D., Nagabhushana, H. 2015. Green synthesis of CuO nanoparticles using *Gloriosa superba* L. extract and their antibacterial activity. J. Taibah. Uni. Sci. 9: 41–49.

Niu, W., Xu, G. 2011. Crystallographic control of noble metal nanocrystals. Nano Today. 6: 265–285.

Ouano, Johnny Jim S., Que, Mar Christian O., Basilia, Blessie A., Alguno, Arnold C. 2018. Controlling the absorption spectra of gold nanoparticles synthesized via green synthesis using brown seaweed (*Sargassum crassifolium*) Extract. Key Engineering Materials 772: 78–82.

Padalia, H., Moteriya, P., Chanda, S. 2015. Green synthesis of silver nanoparticles from marigold flower and its synergistic antimicrobial potential. Arab. J. Chem. 8: 732–741.

Padil, V.V.T., Cernik, M. 2013. Green synthesis of copper oxide nanoparticles using gum karaya as a biotemplate and their antibacterial application. Int. J. Nanomed. 8: 889–898.

Palanisamy, S., Rajasekar, P., Vijayaprasath, G., Ravi, G., Manikandan, R., Prabhu, N., Marimuthu. 2017. A green route to synthesis silver nanoparticles using *Sargassum polycystum* and its antioxidant and cytotoxic effects: an *in vitro* analysis. Mater Let. 189: 196–200.

Paleos, C.M., Tsiourvas, D., Sideratou, Z., Tziveleka, L. 2004. Acid and salt-triggered multifunctional poly(propylene imine) dendrimer as a prospective drug delivery system. Biomacromol. 5: 524–529.

Pandimurugan, R., Thambidurai, S. 2016. Synthesis of seaweed-ZnO PANI hybrid composite for adsorption of methylene blue dye. J. Environ. Chem. Eng. 4(1): 1332–1347.

Pandimurugan, R., Thambidurai, S. 2017. UV protection and antibacterial properties of seaweed capped ZnO nanoparticles coated cotton fabrics. Int. J. Biol. Macromol. 105(1): 788–795.

Poizot, P., Laruelle, S., Grugeon, S., Dupont, L., Tarascon, J.M. 2000. Nano-sized transition-metal oxides as negative-electrode materials for lithium-ion batteries. Nature 407: 496–499.

Polavarapu, L., Mourdikoudis, S., Pastoriza-Santos, I., Pérez-Juste, J. 2015. Nanocrystal engineering of noble metals and metal chalcogenides: controlling the morphology, composition and crystallinity. Cryst. Engg. Comm. 17: 3727–3762.

Pramila, K., Sushil Kumar, S. 2016. Microbes mediated synthesis of metal nanoparticles: current status and future prospects. Int. J. Nanomat. Biostruc. 6(1): 1–24.

Prasad, T.N., Kambala, V.S.R., Naidu, R. 2013. Phyconanotechnology: synthesis of silver nanoparticles using brown marine algae *Cystophora moniliformis* and their characterisation. J. Appl. Phycol. 25: 177–182.

Pugazhendhi, A., Prabhu, R., Muruganantham, K., Shanmuganathan, R., Natarajan, S. 2019. Anticancer, antimicrobial and photocatalytic activities of green synthesized magnesium oxide nanoparticles (MgONPs) using aqueous extract of *Sargassum wightii*. J. Photochem. Photobiol. B: Biol. 190: 86–97.

Raajshree, R.K., Durairaj, B. 2017. Biosynthesis of zinc oxide nanoparticles using *Turbinaria conoides*—a green approach. Int. J. Cur. Res. 9(07): 54461–54464.

Rajaboopathi, S., Thambidurai, S. 2017. Green synthesis of seaweed surfactant based CdO-ZnO nanoparticles for better thermal and photocatalytic activity. Cur. App. Phys. 17(12): 1622–1638.

Rajaboopathi, S., Thambidurai, S. 2018. Evaluation of UPF and antibacterial activity of cotton fabric coated with colloidal seaweed extract functionalized silver nanoparticles. J. Photochem. Photobiol. B: Biol. 183: 75–87.

Rajathi, F.A.A., Parthiban, C., Ganesh Kumar, V., Anantharaman, P. 2012. Biosynthesis of antibacterial gold nanoparticles using brown alga, *Stoechospermum marginatum* (kutzing). Spectroch Acta Part A: Mol. Biomol. Spec. 99: 166–173.

Rajeshkumar, S., Malarkodi, C., Paulkumar, K., Vanaja, M., Gnanajobitha, G., Annadurai, G. 2014. Algae mediated green fabrication of silver nanoparticles and examination of its antifungal activity against clinical pathogens. Int. J. Metals. 692643.

Rajeshkumar, S., Malarkodi, C., Gnanajobitha, G., Paulkumar, K., Vanaja, Kannan, M., Annadurai, G. 2013. Seaweed-mediated synthesis of gold nanoparticles using *Turbinaria conoides* and its characterization. J. Nanostruc. Chem. 3: 44.

Ramakrishna, M., Rajesh Babu, D., Gengan, R.M., Chandra, S., Nageswara Rao, G. 2016. Green synthesis of gold nanoparticles using marine algae and evaluation of their catalytic activity. J. Nanostruct. Chem. 6: 1–13 https://doi.org/10.1007/s40097-015-0173-y.

Ramaswamy, S.V.P., Narendhran, S., Sivaraj, R. 2016. Potentiating effect of ecofriendly synthesis of copper oxide nanoparticles using brown alga: antimicrobial and anticancer activities. Bull. Mater. Sci. 39: 361–364.

Ramkumar, V.S., Arivalagan, P., Kumar, G., Periyasamy, S., Ganesh, D.S. Thi Ngoc, B.D., Ethiraj, K. 2017. Biofabrication and characterization of silver nanoparticles using aqueous extract of seaweed *Enteromorpha compressa* and its biomedical properties. Biotechnol. Rep. 14: 1–7.

Rana, V.K., Kissner, R., Jauregui-Haza, U., Gaspard, S., Levalois-Grutzmacher, J. 2017. Enhanced chlordecone (Kepone) removal by FeO-nanoparticles loaded on activated carbon. J. Environ. Chem. Eng. 5(2): 1608–1617.

Roseline, T.A., Murugan, M., Sudhakar, M.P., Arunkumar, K. 2019. Nanopesticidal potential of silver nanocomposites synthesized from the aqueous extracts of red seaweeds. Environ. Technol. Innov. 13: 82–93.

Ruan, H., Wu, X., Yang, C., Li, Z., Xia, Y., Xue, T., Shen, Z., Wu, A. 2018. A supersensitive CTC analysis system based on triangular silver nanoprisms and SPION with function of capture, enrichment, detection and release. ACS Biomatt. Sci. Eng. 4(3): 1073–1082.

Saber, H., Alwaleed, E.A., Ebnalwaled, K.A., Sayed, A., Wesam, S. 2017. Efficacy of silver nanoparticles mediated by *Jania rubens* and *Sargassum dentifolium* macroalgae; Characterization and biomedical applications. Egyp. J. Basic. Appl. Sci. 4(4): 249–255.

Sahayaraj, K., Rajesh, S., Rathi, J.M. 2012. Silver nanoparticles biosynthesis using marine alga *Padina pavonica* (linn.) and its microbicidal activity. Digest J. Nanomat. Biostruc. 7(4): 1557–1567.

Shanmugam, N., Rajkamal, P., Cholan, S., Kannadasan, N., Sathishkumar, K., Viruthagiri, G., Sundaramanickam, A. 2014. Biosynthesis of silver nanoparticles from the marine seaweed *Sargassum wightii* and their antibacterial activity against some human pathogens. Appl. Nanosci. 4: 13204-013-0271.

Shargh, A.Y., Sayadi, M.H., Heidari, A. 2018. Green biosynthesis of palladium oxide nanoparticles using *Dictyota indica* seaweed and its application for adsorption. J. Water Environ. Nanotechnol. 3(4): 337–347.

Shekhah, O., Ranke, W., Schüle, A., Kolios. G., Schlogl, R. 2003. Styrene synthesis: High conversion over unpromoted iron oxide catalysts under practical working conditions. Angew. Chem. Int. Ed. 42(46): 5760–5763.

Shipway, A.N., Katz, E., Willner, I. 2000. Nanoparticle arrays on surfaces for electronic, optical and sensor applications. Chem. Phys. Chem. 1(1): 18–52.

Siji, S., Njana, J., Amrita, P.J., Dalia, V. 2018. Biogenic synthesis of iron oxide nanoparticles from marine algae. TKM Arts College Int. J. Multidis Res. 1: 1–7.

Singaravelu, G., Arockiamary, J.S., Ganesh Kumar, V., Govindaraju, K. 2007. A novel extracellular synthesis of monodisperse gold nanoparticles using marine alga, *Sargassum wightii* Greville. Colloids Surf B: Biointerf. 57: 97–101.

Singh, P., Kim, Y.J., Zhang, D., Yang, D.C. 2016. Biological synthesis of nanoparticles from plants and microorganisms. Trends Biotechnol. 34(7): 588–599.

Stalin Dhas, T., Ganesh Kumar , V., Karthick, V., Jini Angel, K., Govindaraju, K. 2014. Facile synthesis of silver chloride nanoparticles using marine alga and its antibacterial efficacy. Spectrochim Acta Part A: Mol. Biomol. Spec. 120: 416–420.

Thovhogi, N., Diallo, A., Gurib-Fakim, A., Maaza, M. 2015. Nanoparticles green synthesis by *Hibiscus sabdariffa* flower extract: main physical properties. J. Alloys Compd. 647: 392–396.

Tiwari, P.M., Vig, K., Dennis, V.A., Singh, S.R. 2011. Functionalized gold nanoparticles and their biomedical applications. Nanomaterials 1: 31–63.

Vasquez, R.D., Apostol, J.G., deLeon, J.D., Mariano, J.D., Mirhan, C.M.C., Pangan, S.S., Reyes, A.G.M., Zamora, E.T. 2016. Polysaccharide-mediated green synthesis of silver nanoparticles from *Sargassum siliquosum* J.G. Agardh: Assessment of toxicity and hepatoprotective activity. Open Nano. 1: 16–24.

Vijai Anand, K., Aravind Kumar, J., Keerthana, K., Paribrita Deb, Tamilselvan, S., Theerthagiri, J., Rajeswari, V., Muthamil Selvan, S., Govindaraju, K. 2019. Photocatalytic degradation of rhodamine B dye using biogenic hybrid ZnO-MgO nanocomposites under visible light. ChemistrySelect 4: 5178–5184.

Vijayan, S.R., Santhiyagu, P., Singamuthu, M., Kumari Ahila, N., Jayaraman, R., Ethiraj, K. 2014. Synthesis and characterization of silver and gold nanoparticles using aqueous extract of seaweed, *Turbinaria conoides* and their antimicrofouling activity. Sci. World J. 938272.

Vijayaraghavan, K., Mahadevan, A., Sathishkumar, M., Pavagadhi, S., Balasubramanian, R. 2011. Biosynthesis of Au(0) from Au(III) via biosorption and bioreduction using brown marine alga *Turbinaria conoides*. Chem. Eng. J. 167: 223–227.

Vivek, M., Kumar, P.S., Steffi, S., Sudha, S. 2011. Biogenic silver nanoparticles by *Gelidiella acerosa* extract and their antifungal effects. Avicenna J. Med. Biotechnol. 3: 143–148.

Wu, C.C., Chen, D.H. 2010. Facile green synthesis of gold nanoparticles with gum arabic as a stabilizing agent and reducing agent. Gold Bull. 43: 234–239.

Wu, W., Wieckowski, S., Pastorin, G., Benincasa, M., Klumpp, C., Briand, J.P., Gennaro, R., Prato, M., Bianco, A. 2005. Targeted delivery of amphotericin B to cells by using functionalized carbon nanotubes. Angew. Chem. Int. Ed. 44: 6358–6362.

Xia, Y., Xia, X., Wang, Y., Xie, S. 2013. Shape-controlled synthesis of metal nanocrystals. MRS Bull. 38: 335–344.

Xu, P.S., Van Kirk, E.A., Zhan, Y.H., Murdoch, W.J., Radosz, M., Shen, Y.Q. 2007. Targeted charge-reversal nanoparticles for nuclear drug delivery. Angew. Chem. Int. Ed. 46: 4999–5002.

Yew, Y.P., Kamyar, S., Mikio, M., Noriyuki, K., Nurul, B.B.A.K., Shaza, E.B.M., Kar, X.L. 2016. Green synthesis of magnetite (Fe_3O_4) nanoparticles using seaweed (*Kappaphycus alvarezii*) extract. Nanoscal. Res. Let. 11: 276.

Chapter 5

Biosynthesis of Nanoparticles by Green Algae

Mechanism and Applications

Agnieszka Pawłowska and *Zygmunt Sadowski**

Introduction

Algae are among the most widespread types of aquatic photosynthetic organisms on Earth, and they contain unique features. Algal cells are able to produce molecular oxygen by way of photosynthesis and are classified by: (1) pigmentation and photosynthetic mechanism, (2) nature of cell wall components, (3) nuclear organization, or (4) nature of food reserve. They are also classified by color: green, red, brown, and blue-green. In the present study, green algae were considered as a substrate for nanoparticle synthesis.

Though microorganisms are increasingly applied in the biosynthesis of nanoparticles, the fabrication of nanoparticles can lead to enzymatic lysis of the cells. Surfactants might be good candidates for extraction and stabilization of particles on the nanoscale (Korbekandi et al. 2009).

The biosynthesis of metal and metal oxide nanoparticles using algal species has systematically increased in this decade (Sharma et al. 2016) and falls into four categories (Dahoumane et al. 2016). The first method makes use of algal extract as a substrate; in the second method, the supernatant after the centrifugation of algal biomass is used; in the third method, whole algal cells in distilled water suspension are applied; the last method requires the application of algae under their normal culturing conditions (see Fig. 5.1).

The nanoparticles formed in the biosynthesis process may have a toxic effect on algae. The toxicity mechanism is manifold and depends on physicochemical properties

Department of Chemical Engineering, Wroclaw University of Science and Technology, Wybrzeze Wyspianskiego 27, 50-370 Wroclaw, Poland.
* Corresponding author: zygmunt.sadowski@pwr.edu.pl

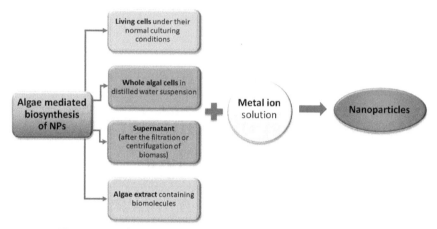

Figure 5.1. Methodologies for the biosynthesis of nanoparticles using algae.

of nanoparticles such as size, shape, charge, and crystal structure. Toxicity studies demonstrated that anatase TiO_2 was more toxic than rutile (Siddiqi and Husen 2016).

The importance of extracellular polymeric substance (EPS) addressing the harmful effect of NPs to *Chlorella vulgaris* green algae was demonstrated by Zheng and colleagues (2019). Algal-synthesized nanoparticles show antibacterial, antifungal, and anticancer activity. For these reasons, they have potential applications in medicine (Oscar et al. 2016). Nanoparticles of microalgal origin (*Botryococcus braunii*) were also applied as a catalyst for the synthesis of biologically important benzimidazoles (Arya et al. 2019).

The biogenic synthesis of nanoparticles is a viable alternative to both the chemical and physical method. This chapter presents various methods for the biosynthesis of nanoparticles using algae and discusses the toxic effect of nanoparticles on live microalgae.

Green algae and their characteristics

Seven thousand species identified as green algae (Slade and Bauen 2013) can be found in salt or fresh water, soil or rock surfaces. They may be unicellular, multicellular, or colonial and are divided into *Embryophyta, Charophyta, Chlorophyta, Rhodophyta,* and *Heterokontophyta*. They owe their bright green color to chloroplasts that contain chlorophyll *a* and *b* (α-1,4 and α-1,6 polymer of glucose) and their surfaces contain various functional groups with different metal affinities. Kiefer and co-authors (1997) measured the adsorption of Cu at pH 6.9 for the *Cyclotella cryptica* diatom and the green alga *Chlamydomonas reinhardtii* (family *Chlamydomonadaceae*). The bonding of copper with the green algae surface was interpreted with the Langmuir equation at constant pH, which assumes that the adsorption of Cu occurs in the single type of surface group (Kiefer et al. 1997).

Algae are a rich source of organic compounds such as pigments, proteins, carbohydrates, fats, nucleic acids, and secondary metabolites, including alkaloids, macrolides, peptides, and terpenes. These compounds can play an important role as reducing agents in nanoparticle formation (Siddiqi and Husen 2016). A rich range of biochemicals and enzymes present in algal cells can participate in the biosynthesis of nanoparticles, which may be performed via extracellular or intracellular pathways.

Extracellular pathway to biosynthesis nanoparticles

Aziz and co-workers (2015) have suggested that *Chlorella pyrenoidosa* (*Chlorellaceae*) can be used for efficient extracellular biosynthesis of silver nanoparticles. Biosynthesis took place when the following procedure was used. A 50 ml portion of deionized water was added to 1.5 g of fresh algal biomass, kept at 100°C for 5 min and cooled down. The cooled suspension was then sonicated for 5 min and separated by centrifugation. The supernatant, like an algal extract, was used for AgNP synthesis. Biosynthesis was carried out by mixing 90 ml of 1 mM $AgNO_3$ with 10 ml of algal cell extract. After 72 h of incubation, the reduction of silver ions was complete. The FTIR measurements were taken to understand the stabilization mechanism of AgNPs (Fig. 5.2).

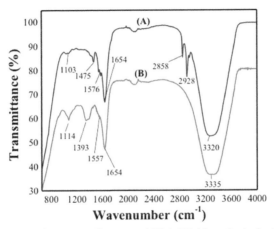

Figure 5.2. FTIR spectra of (A) water cell extract and (B) AgNPs biosynthesized using *C. pyrenoidosa* alga. Reprinted with permission from Langmuir (Aziz et al. 2015).

The FTIR spectra analysis of green extract showed absorption bands at 1654 cm^{-1} corresponding to C=O vibration frequencies of amides and at 3320 cm^{-1} to O-H stretching of aromatic compounds. The peaks at 1475 and 1103 cm^{-1} indicate the C-H banding of alkanes and alkenes respectively. The vibrational frequencies at 1576 cm^{-1} are characteristic for C-H bonding associated with hydrocarbons (Aziz et al. 2015). The results confirmed that proteins act as capping and stabilizing agents.

The water extract of powdered *Spirogyra varians* (*Zygnemataceae*) was used in silver nanoparticle synthesis (Salari et al. 2016). About 10 ml of algae extract was added to 90 ml of an aqueous solution of silver nitrate. This mixture was kept for 20 min in the dark. The red color of the solution indicated that silver nanoparticles were formed. The UV-vis spectra of the silver NPs showed the absorption band between 420 and 430 nm similar to those reported in the literature (Sadowski et al. 2008). Dry powder of silver NPs were characterized by X-ray diffraction analysis. Four peaks were observed at 2Θ of 38.1°, 44.3°, 64.5°, and 76.4° that corresponded to (111), (200), (220), and (311) planes, respectively.

The preparation of algal water extract using *Pithophora oedogonia* (*Pithophoraceae*) was similar. The 10 g of algal powder was mixed with 100 ml of water and incubated at 60°C for 15 min. The extract was combined with aqueous 1 mM $AgNO_3$ solution and the brown-yellow color of the reaction solution indicated the presence of AgNPs. Additionally, the chemical analyses of *P. oedogonia* extract demonstrated the presence of carbohydrate, saponins, steroid, tannins, terpenoids, and proteins (Sinha et al. 2015).

Caulerpa racemosa is seaweed of the family *Caulerpaceae. C. racemosa* is a type of green algae commonly known as "sea grapes". Experiments have shown that the extract of marine seaweed *C. racemosa* is capable of producing AgNPs extracellularly (Kathiraven et al. 2015). The procedure of AgNP biosynthesis was similar to the one using *Pithophora oedogonia*.

The water extract obtained from *Enteromorpha flexuosa* (*Ulvaceae*) was used for silver ion reduction and stabilization of silver nanoparticles (Yousefzadi et al. 2014). The algal extract was prepared using 5 g of fresh algal biomass mixed with 100 ml of deionized water and boiled for 10 minutes. The cold extract was filtrated and stored at 4°C. The synthesis was carried out by mixing 10 ml of the algae extract with 90 ml of 1 mM $AgNO_3$ aqueous solution. The reaction mixture was kept in the dark. The reduction of silver ions was monitored by the measurements of the UV-Vis spectra. The absorption peak at 430 nm confirmed the presence of AgNPs (Yousefzadi et al. 2014).

A different synthesis procedure was proposed by Sharma and co-workers (2014) for gold nanoparticle fabrication using marine green alga *Prasiola crispa* (*Prasiolaceae*). One gram of dry algal powder was mixed with 100 ml of aqueous solution of chloroauric acid (concentration 10^{-3} M) and stirred for 12 h. The green color of the algal biomass did not change even when bioreduction of gold ions was complete, which suggested that the biosynthesis of gold nanoparticles occurred extracellularly. The synthesized AuNPs were characterized by UV-Vis, XRD, and FTIR analysis. The XRD pattern of green algal biomass after gold nanoparticle synthesis showed five different peaks at 2Θ equal to 38.62°, 44.72°, 65.02°, 72.84°, and 81.90°, which corresponded to 111, 200, 220, 311 and 222 planes. The average size of gold crystallite size estimated by the Debye-Scherrer formula was found to be 9.8 nm (Sharma et al. 2014).

The biosynthesis of gold nanoparticles from algae was also reported by Castro and co-workers (2013). The gold NPs were synthesized using *Spirogyra insignis*, a green alga of the *Zygnemataceae* family. Algal biomass was directly added to the $HAuCl_4$ solution. When the pH of the nanoparticles solution was changed within the range from 2 to 10, the color changed from pale yellow to dark blue or pink. This color change corresponds to the size and shape of nanoparticles. TEM observations showed that in strongly acidic conditions, polygonal (mainly triangular and hexagonal) gold NPs were formed. When pH was increased to 4, spherical nanoparticles ($\sim$ 30 nm) were formed. Further increase in solution pH also favored the synthesis of spherical nanoparticles with diameter 30–50 nm. The highest concentration of AuNPs was reached after 2 h of incubation. The gold ion reduction was accompanied by the oxidation of hydroxyl to carbonyl groups present in the algal biomass. The biosynthesis of AuNPs occurs according to the following reaction:

$$AuCl_4^- + 3\ R\text{-}OH = Au^\circ + 3\ R{=}O + 3H^+ + 4\ Cl^-$$

The water extract from a dry powder of green alga *Pithophora oedogonia* (*Pithophoraceae*) was used for gold NP synthesis (Li and Zhang 2016). A 20 ml of algal extract was added to 20 ml of HAuCl₄ solution and sonicated for 1 h. Scanning electron microscopy (SEM) determined the morphology of AuNPs. The biosynthesized gold nanoparticles showed a spherical shape with small agglomeration. The size distribution of AuNPs was determined by the DLS technique. The mean size of AuNPs was 32.06 nm. Moreover, the electroactivity of biosynthesized AuNPs was used for the detection of trace amounts of carbendazim in soil samples (Li and Zhang 2016).

Xie and co-workers (2007) tried to explain the role of algal biomolecules in the nucleation and growth of gold clusters of specific shapes and sizes. They used the extract obtained from *Chlorella vulgaris* (*Chlorellaceae*) for gold nanoparticle preparation. The algal extract was separated into low fractions (< 12 kDa) and high fractions (> 12 kDa). The high-molecular-weight fraction was further fractionated into eight additional fractions. The series of gold synthesis experiments showed that only one fraction containing biomolecules with the molecular weight of 28 kDa gave a high yield of nanoparticles (≈ 90%). This protein fraction was named GSP (gold shape-directing protein) and used in further studies. Subsequent experiments showed that the GSP concentration can affect the shape and size of AuNPs in three different ways: leading to a decrease in the average particle size, leading to a decrease in the yield of synthesized AuNPs, or strongly influencing the morphology of biosynthesized AuNPs.

Intracellular pathway of biosynthesis

Eukaryotic green alga *Rhizoclonium fontinale* (*Cladophoraceae*) was used as a bioreagent for gold nanoparticle synthesis (Parial and Pal 2015). Pure algal biomass was exposed to gold chloride solution for 24 h and then to 7.5 mM sodium citrate solution, following the treatment with ultrasonic vibration for 30 min. The suspension was then centrifuged for 5 min at 3000 rpm. The supernatant was examined for the presence of gold NPs. The absorption band at 530 nm indicated the occurrence of gold NPs. Dynamic light scattering study showed that gold NPs obtained at pH 5, 7 and 9 had different sizes. The average sizes were found to be 156.6, 148 and 145 nm, respectively. Microphotographs of Au³⁺-treated algae cells showed the degradation of chlorophyll and gold nanoparticle deposition after 48 h (Parial and Pal 2015). The optical properties of surface plasmon resonance are directly dependent on the size of gold nanoparticles. The shape of the prepared AuNPs was dependent on both the reduction rate and pH value of the reaction mixture (Parial et al. 2012). Unlike the AuNPs fabricated from *Ulva intestinalis* (*Ulvaceae*), which were mainly spherical in shape, the AuNPs synthesized from *Rhizoclonium fontinale* (*Cladophoraceae*) had variable shapes and sizes (Parial et al. 2012). These experiments suggested that a high reduction rate at low pH leads to nano-rod formation. Although the process of intracellular formation of AuNPs by algal biomass is not fully understood, it has been explained that the reduction of gold ions is associated with cellular metabolism and reducing enzymes are involved in this process (Parial and Pal 2015).

Surface enhanced Raman spectroscopy (SERS) was used to show the cellular images with intracellularly biosynthesized gold nanoparticles (Lahr and Vikesland 2014). SERS is a vibration spectroscopic method that provides information about the position of molecules in the vicinity of noble metal. The SERS cellular images were collected to follow intercellular growth of AuNPs. These observations enhanced the understanding of nanoparticle biosynthesis.

Commercial application of algal synthesized nanoparticles

The silver nanoparticles synthesized using green alga *Chlorella pyrenoidosa* exhibited a strong antibacterial activity (Aziz et al. 2015). This activity was tested against three Gram-negative bacteria, namely *Klebsiella pneumoniae*, *Aeromonas hydrophila,* and *Acinetobacter* sp., and one Gram-positive bacterium, *Staphylococcus aureus*. Chemically synthesized AgNPs were used in these studies as the control sample. The strong inhibitory effect of AgNPs was observed against *A. hydrophila* and *Acinetobacter* sp., whereas *K. pneumoniae* was less sensitive. Inhibition effect in the case of *S. aureus* was observed only for the lowest concentration of biogenic NPs. These experiments have shown that biogenic AgNPs act as efficient antibacterial agents in comparison to chemically synthesized nanoparticles (Aziz et al. 2015).

The antimicrobial activity of biosynthesized AgNPs using the extract from green algae *E. flexuosa* was tested against Gram-negative and Gram-positive bacteria as well as fungi (Yousefzadi et al. 2014), as shown in Table 5.1.

The antibacterial and antifungal mechanism of silver and gold nanoparticles obtained with the use of algae extracts was extensively discussed by El-Sheekh and

Table 5.1. Antimicrobial activity of AgNPs synthesized using *E. flexuosa*. Reprinted with permission from Materials Letters (Yousefzadi et al. 2014).

Microorganisms	Algae extract			Synthesized silver nanoparticle			Ampicillin[c]	Nystatine[c]	Ag NP[d]
	IZ[a]	MIC[b]	MBC[b]	IZ	MIC	MBC	IZ	IZ	IZ
Bacillus subtilis	15±0.9	25	50	18±0.8	12.5	50	14±0.4	Nt	16±0.6
Bacillus pumulis	16±1.0	25	50	19±1.2	6.25	25	15±0.3	Nt	17±0.9
Enterococcus faecalis	11±0.7	50	100	12±0.9	50	100	11±0.3	Nt	10±0.6
Staphylococcus aureus	12±0.6	50	100	14±0.7	25	50	13±0.3	Nt	11±0.9
Staphylococcus epidermidis	18±1.2	25	50	20±1.5	6.25	12.5	19±0.5	Nt	17±0.8
Escherichia coli	11±0.9	50	100	13±0.9	50	100	12±0.2	Nt	12±0.6
Klebsiella pneumoniae	10±0.5	100	200	10±0.4	50	100	0	Nt	9±0.5
Pseudomonas aeruginosa	0	Nt	Nt	0	Nt	Nt	10±0.3	Nt	0
Aspergillus niger	0	Nt	Nt	0	Nt	Nt	–	16±0.4	0
Candida albicans	12±0.8	50	100	14±0.8	25	100	–	18±0.5	13±0.5
Saccharomyces cerevisiae	13±0.9	25	100	16±0.6	25	50	–	18±0.2	15±0.7

[a] Inhibition Zone (IZ) includes a diameter of the disc (6 mm).
[b] Minimum inhibitory concentration (MIC) and minimum bactericidal concentration (MBC) values in μg/ml.
[c] Tested at 10 μg ampicillin/disc and 30 μg nystatin/disc.
[d] AgNP: chemically synthesized silver nanoparticles.

El-Kassas (2016). Those authors assumed that Ag and Au nanoparticles could play an important role in medicine as they could control the drug effect on bacteria. It was reported that biosynthesized AgNPs at low concentrations were non-toxic to the human body, prevented bacterial colonization on different surfaces, e.g., on human skin, and displayed effective antifungal properties. Once they are adsorbed into the fungal cell membrane and penetrate the cell, they interact with DNA, which destroys fungal reproduction. The authors also studied the cytotoxic effect of green-synthesized AuNPs on cancer cells (El-Sheekh and El-Kassas 2016). It was shown that gold nanoparticles derived from algae uniquely interacted with breast cancer cells, which suggested that they might be used in anticancer therapy.

Toxic effect of nanoparticles on algae

The use of nanoparticles has been systematically increasing and their potential toxic activity poses a considerable risk. The examples presented below will illustrate how nanoparticles can be dangerous for algae. Three techniques were used to demonstrate the interactions of AgNPs with the green alga *Raphidocelis subcapitata* (Selenastraceae) (Sekine et al. 2017).

Metallic nanoparticles affect algal growth in three specific ways: (1) metal ion diffusion across bulk media to the algal surface leading to toxic effect after internalization, (2) attachment of nanoparticles to the algal cell surface and dissolution, and (3) penetration of metallic nanoparticles into algal cells and liberation of ions (Quigg et al. 2013).

Direct toxic effect of NPs depends on both their chemical composition and surface reactivity. Toxicological effect of NPs may include oxidative stress, protein denaturation and reduction of photosynthetic activity in algal cells (Lei et al. 2016).

The AgNPs released to the aquatic environment might be a source of ionic silver, having a toxic effect on green algae, even at low concentrations. Although the toxic effect of AgNPs and silver ions (Ag^+) on *Chlamydomonas reinhardtii*, a single-cell green alga (about 10 µm in diameter), was investigated (Navarro et al. 2008), the mechanism of its toxicity has not been satisfactorily explained. It was assumed that the toxicity of Ag^+ ions toward *Chlamydomonas reinhardtii* cells depends on intracellular accumulation of silver ions. Cysteine, which is a component of the algal cell surface, is a strong silver ion ligand. To examine the role of silver ions in the toxicity of AgNPs to algal photosynthesis, the following experiments were carried out, namely algae samples containing 5 or 10 µM AgNPs and different cysteine concentrations (10, 50, 100 and 500 mM) were exposed to light for 1 h. The obtained results provide indirect evidence that the toxicity of AgNPs is mediated by Ag^+ ions. The exclusion of toxicity to photosynthesis at an equimolar concentration of cysteine indicated that the protective effect of cysteine was due to the complexation of Ag^+ ions (Navarro et al. 2008).

Nam and colleagues (2018) analyzed the toxic effect of AgNPs on the soil alga *Chlamydomonas reinhardtii* (*Chlamydomonadaceae*) by measurement of change in chlorophyll levels in algal biomass within 6 days. In comparison to the control sample, both the algal biomass and photosynthetic activity of *C. reinhardtii* decreased when the concentration of silver nanoparticles in soil increased.

The pH effect on copper toxicity to green microalgae was investigated by De Schamphelaere and team (2005). The number of the protonated binding sites at the algal surface increased with pH increase, because more copper ions could bind to an algal surface. These binding sites were biotic ligands and determined the toxic effect of nanoparticles. This phenomenon was demonstrated for two green algal species, *Pseudokirchneriella subcapitata* (*Chlorophyceae*) and *Chlorella* sp. (*Trebouxiophyceae*).

The toxic effects induced by 50 nm AgNPs were investigated using a freshwater microalga *Chlorella vulgaris* (*Trebouxiophyceae*) and a marine alga *Dunaliella tertiolecta* (*Chlorophyceae*) (Oukarroum et al. 2012). When the algal cultures were treated with 0.1 mg/L and 10 mg/L of AgNPs for 24 h, the cell aggregates were formed and the amount of chlorophyll inside the cells decreased significantly. In the case of *C. vulgaris*, 34% and 51% decrease of chlorophyll amount was observed compared to control sample, respectively at 1 and 10 mg/l AgNPs. In *D. tertiolecta* the decrease was 44% and 75% respectively. These results suggest that the effect of AgNPs on *C. vulgaris* was less hazardous than on *D. tertiolecta*. These differences could be due to the cell wall structure. It was also demonstrated that AgNPs induced the formation of reactive oxygen species (ROS) in cells of both alga species. It is known that a high level of ROS can lead to damage of cell structure resulting from lipid peroxidation (Oukarroum et al. 2012).

Leonardo and colleagues (2016) examined the toxic effects of silver NPs and silver ions on unicellular green microalga *Coccomyxa actinabiotis* (*Trebouxiophyceae*) to understand the fundamental mechanisms of silver accumulation and toxicity actions. *C. actinabiotis* was isolated from an environment contaminated with radioactive silver, so it exhibited an extreme radiotolerance. Most of the silver was accumulated inside the algal cells. A variety of spectroscopic methods were used to study the speciation of silver, namely UV-visible, infrared, or Raman spectroscopy, X-ray absorption spectroscopy (XAS), laser spectroscopy, mass spectroscopy, and nuclear magnetic resonance. The localization of silver NPs in the microalga cell was shown in Fig. 5.3.

The results suggest that Ag$^+$ ions accumulated by *C. actinabiotis* at a low silver concentration (10^{-5} M) react with cellular molecules and create a complex with

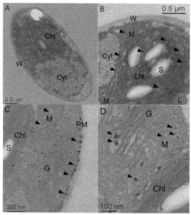

Figure 5.3. Localization of silver nanoparticles inside microalgal cells. Reprinted with permission from Environmental Science & Technology (Leonardo et al. 2016).

sulfur-containing groups. The thiol-containing species such as cysteine, glutathione, metallothioneins or phytochelating groups are responsible for adsorption mechanisms. When the silver concentration increases (10^{-4} and 10^{-3} M), silver nanoparticles appear in almost all cellular compartments. It was shown that silver nanoparticles were mainly localized in the chloroplasts and in mitochondria (Fig. 5.3).

Quigg and co-authors (2013) found that the toxicity of TiO_2 nanoparticles toward algal cells has been found to be light-dependent. At a concentration of 100 mg/L of TiO_2 NPs, these nanoparticles were accumulated in algal cells. The photon absorption by TiO_2 nanoparticles leads to the reaction of highly excited electron-hole pairs, which act as strong oxidizing and reducing agents (Quigg et al. 2013).

Inhibition of algal growth and prevention of photosynthesis by shading effect caused by nano-TiO_2 were reported by Li et al. (2015). They claimed that the biosynthesis of nanoparticles on the surface of algal cells might reduce light availability for photosynthesis. However, the shading effect was not significant to nano-TiO_2 toxicity. The presence of TiO_2 nanoparticles was attributed to ROS production.

The inhibitory effect of TiO_2 nanoparticles was also tested with *Karenia brevis* (*Kareniaceae*) red-tide algae and *Skeletonema costatum* (*Skeletonemataceae*) diatom. TiO_2 nanoparticles were mainly agglomerated on the algal surface (Fig. 5.4).

Algal cells were treated with nano-TiO_2 for 12 h to check the effect of the produced ROS. TiO_2 nanoparticles were able to enter via algal cell wall and adhere

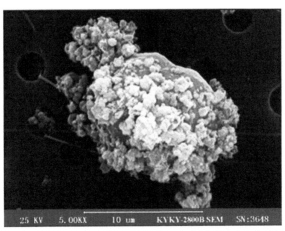

Figure 5.4. The cell of *K. brevis* treated by 100 mg/L nano-TiO_2. Reprinted with permission from Aquatic Toxicology (Li et al. 2015).

to the chloroplast. The production of ROS caused lipid peroxidation in algal cells. The content of the three radicals (O^{2-}, OH, and HO_2) increased in both *K. brevis* and *S. costatum*. The presence of ROS caused oxidative stress and broke the balance between anti- and pro-oxidation enzymes in the algal cells. Finally, algal growth was inhibited (Li et al. 2015).

The interaction of P25 TiO_2 NPs on algal cells of *Chlorella* and *Scenedesmus* (*Scenedesmaceae*) was investigated by Roy and co-workers (2016). Cytotoxic effect of TiO_2 NPs was connected with the generation of ROS as an effect of irradiation. These investigations showed that there were significant differences between *Chlorella* and *Scenedesmus*, which can be explained by the specific mechanism underpinning the stress condition associated with the cellular composition.

The impact of TiO_2 nanoparticles on algal growth was mainly negative. However, their effect varied with the type of algae group: *Scenedemus quadricauda* (*Scenedesmaceae*), *Chlamydomonas moewusii* (*Chlamydomonadaceae*), and *Chlorella vulgaris* (*Chlorellaceae*) (Cardinale et al. 2012).

The toxic effect of ZnO and TiO_2 nanoparticles to the green algae *Pseudokirchneriella subcapitata* (*Selenastraceae*) was investigated using visible, UVA and UVB irradiation (Lee and An 2013). The results of this study showed that algal growth was completely inhibited at the concentration of Zn^{2+} equal to 0.1 mg/L of $ZnCl_2$. The used nanoparticles had a commercial origin (Sigma-Aldrich, USA, and Evonik Degussa, Germany). The hydrodynamic sizes of ZnO and TiO_2 nanoparticles were 264 ± 5 and 486 ± 7 nm, respectively. The toxic effect of ZnO NPs was similar to that of Zn^{2+} and was associated with algal cell destruction resulting from cell membrane destabilization. A different behavior was observed for TiO_2 NPs. Surprisingly, the shading effect of TiO_2 NPs was not observed. The toxicity differences were correlated with the nanoparticle size, crystal structure, and method of preparation.

Quantum dots such as CdS, CdSe and CdTe are small assemblies of semiconductor materials within the range of 2–10 nm and can induce algae cell death by lipid peroxidation (Quigg et al. 2013).

The copper oxide nanoparticles are frequently used as an antifouling component of marine paint, which mainly consists of a polymer film. Presence of Cu_2O nanoparticles in water can have a toxic effect on green algae. *Chlamydomonas reinhardtii* (*Chlamydomonadaceae*) was used to test the toxicity of copper oxide nanoparticles (Saison et al. 2010). It was observed that exposure of *C. reinhardtii* cells to Cu_2O-polystyrene core-shell nanoparticles caused an increase in cellular aggregation. This effect was dependent on nanoparticle concentration. The presence of Cu_2O core-shell nanoparticles induced the formation of ROS, which is responsible for chlorophyll deterioration. In consequence, the photochemical process tapered off (Saison et al. 2010).

The toxic effect of Al_2O_3 nanoparticles on algae was investigated using micro-alga *Dunaliella salina* (*Dunaliellaceae*) (Ayatallahzadeh Shirazi et al. 2015). *D. salina* is a unicellular green alga widely distributed in seawater. Aluminium nanoparticles used for the toxic experiments were of commercial origin (Pishgaman Nanomavad Iran Company). The toxic effect of Al_2O_3 NPs was monitored by observation of algal kinetic growth and the concentrations of both chlorophyll and carotenoid in algal cells. The results showed that the aluminium oxide nanoparticles had a toxic effect and reduced the specific growth rate of *D. salina*. Simultaneously, Al_2O_3 nanoparticles had a significant effect on the chlorophyll and carotenoid concentrations inside the algal cell. The concentration changes of chlorophyll and carotenoid are presented in Table 5.2.

The micro-alga *Dunaliella salina* (*Dunaliellaceae*) was also used to investigate the toxicity of SiO_2 nanoparticles (Ayatallahzadeh Shirazi et al. 2016). The growth cycle of algae was observed when the different quantities of SiO_2 nanoparticles were added to the algal suspension. The experimental data showed that SiO_2 NPs significantly inhibited the growth of *D. salina*. As the concentration of SiO_2 NPs increased, algal cell density decreased. These results overlap with those obtained for

Table 5.2. Toxic effect of Al_2O_3 NPs on the concentrations of chlorophyll and carotenoid inside the cells of *Dunaliella salina* (own elaboration based on Ayatallahzadeh Shirazi et al. 2015).

Al_2O_3 NPs concentration [mg/L]	Chlorophyll concentration [mg/L]	Carotenoid concentration [mg/L]
0 (control)	1887.41	792.67
0.2	342.73	483.00
0.4	114.63	245.67
2.0	416.68	66.67

Chlorella kessleri and *Pseudokirchneriella subcapitata* (Fujiwara et al. 2008, van Hoecke et al. 2009).

Conclusions

In recent years, biosynthesis of nanoparticles has made significant progress. Many studies have been carried out to better understand the biosynthesis process. Nanoparticles originate extra- or intracellularly. These two synthesis pathways are strongly dependent on the procedure used. Bioreduction of metal ions is caused by complex photosynthetic enzymatic machinery. However, it has not yet been fully explained. Since nanoparticles synthesized using green algae show a strong antibacterial effect, they have the potential to be applied in medicine. There is a downside, though: they have also a toxic effect on algae. The presence of inorganic nanoparticles destabilized the growth of microalgae by causing cell destruction.

Acknowledgments

This work was financed by a statutory activity subsidy from the Polish Ministry of Science and Higher Education for the Faculty of Chemistry of Wroclaw University of Science and Technology.

References

Arya, A., Mishra, V., Chundwat, S.T. 2019. Green synthesis of silver nanoparticles from green algae (*Botryococcus braunii*) and its catalytic behavior for the synthesis of benzimidazoles. Chemical Data Collections. 20:100190.

Ayatallahzadeh Shirazi, M., Shariati, F., Keshavarz, A.K., Ramezanpour, Z. 2015. Toxic effect of aluminium oxide nanoparticles on green micro-algae *Dunaliella salina*. Int. J. Environ. Res. 9(2): 585–594.

Ayatallahzadeh Shirazi, M., Shariati, F., Ramezanpour, Z. 2016. Toxicity effect of SiO_2 nanoparticles on green micro-algae *Dunaliella salina*. Int. J. Nanosci. Nanotechnol. 12(4): 269–275.

Aziz, N., Faraz, M., Pandey, R., Shakir, M., Fatma, T., Varma, A., Barman, I., Prasad, R. 2015. Facile algae-derived route to biogenic silver nanoparticles: synthesis, antibacterial, and photocatalytic properties. Langmuir. 31(42): 11605–11612.

Cardinale, B.J., Bier, R., Kwan, C. 2012. Effects of TiO_2 nanoparticles on the growth and metabolism of three species of freshwater algae. J. Nanopart. Res. 14: 913–921.

Castro, L., Blazquez, L.M., Munoz, A.J., Gonzalez, F., Ballester, A. 2013. Biological synthesis of metallic nanoparticles using algae. IET Nanobiotechnol. 7(3): 109–116.

Dahoumane, A.S., Mechouet, M., Alvarez, J.F., Agathos, N.S., Jeffryes, C. 2016. Microalgae: An outstanding tool in nanotechnology. Bionatura. 1(4): 196–201.

De Schamphelaere, K.A., Stauber, J.L., Wilde, K.L., Markich, S.J., Brown, P.L., Franklin, N.M., Creighton, N.M., Janssen, C.R. 2005. Toward a biotic ligand model for freshwater green algae: Surface-bound and internal copper are better predictors of toxicity than free Cu^{2+}-ion activity when pH is varied. Environ. Sci. Technol. 39(7): 2067–2072.

El-Sheekh, M.M., El-Kassas, H.Y. 2016. Algal production of nano-silver and gold: Their antimicrobial and cytotoxic activities: A review. J. Genetic Eng. Biotechnol. 14(2): 299–310.

Fujiwara, K., Suematsu, H., Kiyomira, E., Aoki, M., Moritoki, N. 2008. Size-dependent toxicity of silica nanoparticles to *Chlorella kessleri*. J. Environ. Sci. Health A Tox. Hazard. Subst. Environ. Eng. 43(10): 1167–1173.

Kathiraven, T., Sundaramanickam, A., Shanmugam, N., Balasubramanian, T. 2015. Green synthesis of silver nanoparticles using marine algae *Caulerpa racemosa* and their antibacterial activity against some human pathogens. Appl. Nanosci. 5: 499–504.

Kiefer, E., Sigg, L., Schosseler, P. 1997. Chemical and spectroscopic characterization of algae surfaces. Environ. Sci. Technol. 31(3): 759–764.

Korbekandi, H., Iravani, S., Abbasi, S. 2009. Production of nanoparticles using microorganisms. Crit. Rev. Biotechnol. 29(4): 279–306.

Lahr, R.H., Vikesland, P.J. 2014. Surface-enhanced Raman spectroscopy (SERS) cellular imaging of intracellulary biosynthesized gold nanoparticles. ACS Sustain Chem. Eng. 2(7): 1599–1608.

Lee, W-M., An, Y-J. 2013. Effects of zinc oxide and titanium dioxide nanoparticles on green algae under visible, UVA, and UVB irradiations: No evidence of enhanced algal toxicity under UV pre-irradiation. Chemosphere 91(4): 536–544.

Lei, C., Zhang, L., Yang, K., Zhu, L., Lin, D. 2016. Toxicity of iron-based nanoparticles to green algae. Effect of particle size, crystal phase, oxidation state and environmental aging. Environ. Pollut. 218: 505–512.

Leonardo, T., Farhi, E., Pouget, S., Motellier, S., Boisson, A.M., Banerjee, D., Rebeille, F., den Auwer, C., Rivasseau, C. 2016. Silver accumulation in the green microalga *Coccomyxa actinabiotis*: Toxicity, in situ speciation, and localization investigated using synchrotron XAS, XRD, and TEM. Environ. Sci. Technol. 50(1): 359–367.

Li, F., Liang, Z., Zheng, X., Zhao, W., Wu, M., Wang, Z. 2015. Toxicity of nano-TiO_2 on algae and the site of reactive oxygen species production. Aquat. Toxicol. 158: 1–13.

Li, L., Zhang, Z. 2016. Biosynthesis of gold nanoparticles using green alga *Pithophora oedogonia* with their electrochemical performance for determining carbendazim in soil. Int. J. Electrochem. Sci. 11: 4550–4559.

Nam, S-H., Kwak, J.I., An, Y-J. 2018. Quantification of silver nanoparticle toxicity to algae in soil via photosynthetic and flow-cytometric analyses. Sci. Rep-UK. 8: 292–302. (www.nature.com/scientificreports.)

Navarro, E., Piccapietra, F., Wagner, B., Marconi, F., Kaegi, R., Odzak, N., Sigg, L., Behra, R. 2008. Toxicity of silver nanoparticles to *Chlamydomonas reinhardtii*. Environ. Sci Technol. 42(23): 8959–8964.

Oscar, L.F., Vismaya, S., Arunkumar, M., Thajuddin, N., Dhanasekaran, D., Nithya, C. 2016. Algal nanoparticles: Synthesis and biotechnological potentials. INTECH. 157–182. DOI: 10.5772/62909.

Oukarroum, A., Bras, S., Perreault, F., Popovic, R. 2012. Inhibitory effect of silver nanoparticles in two green algae *Chlorella vulgaris* and *Dunaliella tertiolecta*. Ecotoxicol. Environ. Safety. 78: 80–85.

Parial, D., Pal, R. 2015. Biosynthesis of monodisperse gold nanoparticles by green alga *Rhizoclonium* and associated biochemical changes. J. Appl. Phycol. 27(2): 975–984.

Parial, D., Patra, H.K., Dasgupta, A.K.R., Pal, R. 2012. Screening of differential algae for green synthesis of gold nanoparticles. Eur. J. Phycol. 47(1): 22–29.

Quigg, A., Chin, W-C., Chen, C-S., Zhang, S., Jiang, Y., Miao, A-J., Schwer, K.A., Xu, C., Santschi, P.H. 2013. Direct and indirect toxic effect of engineered nanoparticles on algae: Role of natural organic matter. ACS Sustain Chem. Eng. 1(7): 686–702.

Roy, R., Parashar, A., Bhuvaneshwari, M., Chandrasekaran, N., Mukherjee, A. 2016. Differential effect of P25 TiO$_2$ nanoparticles on freshwater green microalgae: *Chlorella* and S*cenedesmus* species. Aquat Toxicol. 176: 161–171.

Sadowski, Z., Maliszewska, I.H., Grochowalska, B., Polowczyk, I., Kozlecki, T. 2008. Synthesis of silver nanoparticles using microorganisms. Materials Science-Poland. 26(2): 419–424.

Saison, C., Perreault, F., Daigle, J.C., Fortin, C., Claverie, J., Morin, M., Popovic, R. 2010. Effect of core-shell copper oxide nanoparticles on cell culture morphology and photosynthesis (photosystem II energy distribution) in the green alga, *Chlamydomonas reinhardtii*. Aquat. Toxicol. 96(2): 109–114.

Salari, Z., Danafar, F., Dabaghi, S., Ataei, S.A. 2016. Sustainable synthesis of silver nanoparticles using macroalgae *Spirogyra varians* and analysis of their antibacterial activity. J. Saudi. Chem. Soc. 20(4): 459–464.

Sekine, R., Moore, K.L., Matzke, M., Vallotton, P., Jiang, H., Hughes, G.M., Kirby, J.K., Donner, E., Grovenor, C.R.M., Svendsen, C., Lombi, E. 2017. Complementary imagine of silver nanoparticles interactions with green algae: Dark-Field microscopy, electron microscopy, and nanoscale secondary ion mass spectrometry. ACS Nano. 11(11): 10894–10902.

Sharma, A., Sharma, S., Sharma, K., Chetri, S.P.K., Vashishtha, A., Singh, P., Kumar, R., Rathi, B., Agrawal, V. 2016. Algae as crucial organisms in advancing nanotechnology: a systematic review. J. Appl. Phycol. 28(3): 1759–1774.

Sharma, B., Purkayastha, D.D., Hazra, S., Gogoi, L., Bhattacharjee, R.C., Ghosh, N.N., Rout, J. 2014. Biosynthesis of gold nanoparticles using a freshwater green alga *Prasiolacrispa*. Mater Lett. 116: 94–97.

Siddiqi, S.K., Husen, A. 2016. Fabrication of metal and metal oxide nanoparticles by algae and their toxic effects. Nanoscale Res. Lett. 11: 363–374.

Sinha, S.N., Paul, D., Halder, N., Sengupta, D., Patra, S.K. 2015. Green synthesis of silver nanoparticles using fresh water green alga *Pithophora oedogonia* (Mont.) Wittrock and evaluation of their antibacterial activity. Appl. Nanosci. 5: 703–709.

Slade, R., Bauen, A. 2013. Micro-algae cultivation for biofuels: Cost, energy balance, environmental impacts, and future prospects. Biomass Bioenerg. 53: 29–38.

Van Hoecke, K., De Schamphelaere, K.A., Van der Meeren, P., Lucas, S., Janssen, C. 2009. Ecotoxicity of silica nanoparticles to the green alga *Pseudokirchneriella subcapitata*: importance of surface area. Environ. Toxicol. Chem. 27(9): 1167–1173.

Xie, J., Lee, Y.J., Wang, C.I.D., Ting, P.Y. 2007. Identification of active biomolecules in the high yield synthesis of single-crystalline gold nanoplates in algal solution. Small. 3(4): 672–682.

Yousefzadi, M., Rahimi, Z., Ghafori, V. 2014. The green synthesis, characterization and antimicrobial activities of silver nanoparticles synthesized from green alga *Enteromorpha flexuosa* (wulfen). J. Agardh. Mater. Lett. 137: 1–4.

Zheng, S., Zhou, Q., Chen, C., Young, F., Cai, Z., Li, D., Gong, Q., Feng, Y., Wang, H. 2019. Role of extracellular polymeric substances on the behavior and toxicity of silver nanoparticles and ions to green algae *Chlorella vulgaris*. Sci. Total Environ. 660: 1182–1190.

Chapter 6

Mechanism of Microbial Synthesis of Nanoparticles

Magdalena Wypij and *Patrycja Golińska**

Introduction

Nanotechnology is a field of research dealing with synthesis, strategy and manipulation of structures from approximately 1 to 100 nm in size. Metal nanoparticles (NPs), especially silver nanoparticles, have received attention for their extensive applications in the biomedical and physicochemical fields (Singh et al. 2016). Metal nanoparticles are synthesized by physical, chemical and biological methods (Zhang et al. 2016). In physical methods, nanoparticles are prepared by evaporation-condensation using a tube furnace at atmospheric pressure, while in reduction and other types of chemical synthesis, three main components such as metal precursors, reducing agents and stabilizing/capping agents are used. Moreover, the reduction of silver ions includes nucleation and subsequent growth (Deepak et al. 2011). Both physical and chemical techniques involve the use of a high amount of energy or hazardous reagents, which are disadvantages of these approaches (Ahmed et al. 2016). Green synthesis explores the biological pathway and resources such as plant extracts, algae, yeasts, fungi and bacteria for bioproduction of nanoparticles. Thus, biogenic synthesis of nanoparticles offers an interesting alternative to chemical synthesis as it is regarded as safe, cost-effective, sustainable and environment-friendly (Chokriwal et al. 2014, Singh et al. 2016). The biological agents secrete a huge number of enzymes, which are able to hydrolyze metals and thus bring about enzymatic reduction of metals ions (Chokriwal et al. 2014, Gahlawat and Choudhury 2019). Researchers suggest that nanoparticles are synthesized when the microorganisms grab target ions from their environment and turn the metal ions into the element metal through enzymes (Li et al. 2011). Nanoparticles synthesized by biological approach are coated and stabilized by natural molecules such as amino acids, proteins or secondary metabolites, which are released from cells of organisms performing this process (Narasimha et al. 2013, Gurunathan

Department of Microbiology, Nicolaus Copernicus University, 87100 Torun, Poland.
* Corresponding author: golinska@umk.pl

et al. 2014, Gahlawat and Choudhury 2019). Capping proteins attach to the surface of nanoparticles through free amino groups or cysteine residues, which consequently prevents the aggregation of nanoparticles and preserves their properties (Sanjenbam et al. 2014).

Biogenic synthesis of nanoparticles can be performed extra- or intracellularly. Extracellular synthesis takes place outside the organisms through the use of cells, culture supernatants, and cell-free extracts and is more efficient than intracellular synthesis because it is easier to recover nanoparticles from the solution (Husseiny et al. 2007, Abo-State and Partila 2015, Singh et al. 2015). The mechanism of synthesis of metal nanoparticles by microbes has not been completely explored. The biochemical and molecular mechanisms of biological synthesis also remain unknown. Generally, biogenic synthesis of metal nanoparticles from microorganisms was reported by many researchers (Fu et al. 2000, Sharma et al. 2012, Ghorbani et al. 2013, Das et al. 2014, Sathiyanarayanan et al. 2014, Wypij et al. 2018a,b, Avilala and Golla 2019). The reduction of metal ion salts using biological systems leads to the formation of size-controlled, stable, dispersed nanoforms (Klaus et al. 2001, Abo-State and Partila 2015). The control of experimental parameters such as pH, temperature, precursor concentrations, amounts of biomass extract, and reaction time directly affects the shape and size of the nanoparticles obtained (Dahoumane et al. 2017, Wypij et al. 2019). Microorganisms possess different mechanisms to form nanoparticles (Fig. 6.1). However, nanoparticles are usually formed by trapping of metal ions on the surface of or inside the microbial cells, which leads to the formation of the silver nuclei, that grow by further reduction and accumulation of silver ions (Chokriwal et al. 2014, Golinska et al. 2014). The mechanism is probably related to electrostatic interactions between the silver ions and negatively charged carboxylate groups in enzymes present in the cell wall (Abdeen et al. 2014). Silver reduction machinery includes electron shuttle enzymatic silver reduction process (Singh et al. 2015, Gahlawat and Choudhury 2019). It is claimed that NADH-dependent enzymes play an important role in the synthesis of metal nanoparticles by microorganisms.

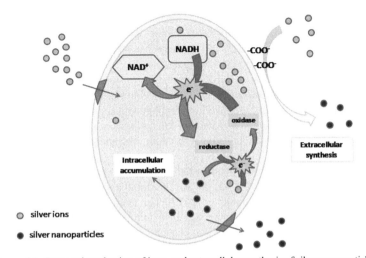

Figure 6.1. Proposed mechanism of intra- and extracellular synthesis of silver nanoparticles.

NADH-dependent nitrate reductase is considered to be involved in synthesis of silver nanoparticles by bacteria and fungi (Kalimuthu et al. 2008, Duran et al. 2010, Karthik et al. 2014). It is also claimed that oxidoreductases can be involved in biosynthesis of metal/metal oxide nanoparticles. They are pH sensitive and work in alternative manner, namely reductases are activated at higher pH while oxidases are activated at lower pH (Jha and Prasad 2010). Moreover, Hosseini-Abari et al. (2013) noticed that biosynthesis of silver nanoparticles can be initiated also by other enzymes such as glucose oxidase, alkaline phosphatase and catalase. The capacity of environmental bacteria and fungi for synthesis of gold nanoparticles suggests that microorganisms reduce Au^{3+} to protein-metal nanoconjugates as a response to toxic stress. It is also evidence that the enzymatic machinery required for reduction process of gold ions is easily available in environmental microorganisms (Kitching et al. 2015). Other heavy metal ions such as Hg^{2+}, Cd^{2+}, Ag^+, Co^{2+}, CrO_4^{2+}, Cu^{2+}, Ni^{2+}, Pb^{2+} or Zn^{2+} also have toxic effects on the survival of microorganisms. It is argued that the formation of heavy metal nanoparticles may be attributed to the metallophilic microorganisms, which developed genetic and proteomic responses to toxic environments. Thus, they are able to strictly regulate metal homeostasis. It is well known that microorganisms such as bacteria and fungi possess metal resistance gene clusters encoding for cell detoxification. Various mechanisms such as complexation, efflux or reductive precipitation are involved in this process (Mergeay et al. 2003, Li et al. 2011).

This chapter emphasizes mechanisms of metal nanoparticle synthesis using bacteria, fungi, algae and viruses. The possible mechanisms involved in fabrication of metal nanoparticles are discussed, as well as deficiencies in this process.

Mechanism of bacteria-mediated synthesis of nanoparticles

Nanoparticles, namely silver nanoparticles, were produced for the first time using *Pseudomonas stutzeri* strain AG259 isolated from a silver mine (Klaus et al. 1999, Gudikandula and Maringanti 2016). To date, a large number of bacterial species have been found to be capable of producing metal nanoparticles, both Gram-positive species such as *Bacillus cereus* (Babu and Gunasekaran 2009), *Halococcus salifodinae* (Srivastava et al. 2013), *Bacillus licheniformis* (Sriram et al. 2012), *Geobacillus stearothermophilus* (Fayaz et al. 2011), *Streptomyces xinghaniensis* (Wypij et al. 2018b), and *Nocardiopsis alba* (Avilala and Golla 2019), and Gram-negative species such as *Shewanella oneidensis* (Debabov et al. 2013), *Proteus mirabilis* (Samadi et al. 2009), *Escherichia coli* (Deplanche et al. 2010), *Klebsiella pneumoniae* (Duraisamy and Yang 2013), and *Cronobacter universalis* (Agrawal and Kulkarni 2017). Bacteria have been successfully used as a biofactory for synthesis of a variety of nanoparticles such as silver, gold, platinum, palladium, titanium, zinc and silver or cadmium sulfide and also uranium oxide (IV) (Table 6.1).

Metal nanoparticles can be synthesized from bacteria extra- and/or intracellularly (Bai et al. 2006, Debabov et al. 2013, Duraisamy and Yang 2013, Srivastava et al. 2013, Gahlawat and Choudhury 2019). The extracellular type of synthesis in bacterial systems is more advantageous than intracellular synthesis, because the production process is simpler and more economical. Intracellular synthesis requires an application of additional suitable detergents or ultrasonic treatment to release nanoparticles from bacterial cells (Deljou and Goudarzi 2016). Metal ions are toxic

Table 6.1. Metal nanoparticles synthesized by microorganisms.

Sources	Nanoparticles	Size (nm)	Shape	Type of synthesis	References
Bacteria					
Bacillus flexus, *Bacillus pseudomycoides,* *Cronobacter universalis,* *Kocuria rosea*	Ag	30–70	spherical and pseudo-spherical	EC	Agrawal and Kulkarni 2017
Nocardiopsis alba	Ag	20–60	spherical	EC	Avilala and Golla 2019
Bacillus cereus	Ag	4–5	spherical	IC	Babu and Gunasekaran 2009
Rhodobacter sphaeroides	ZnS	8	spherical	IC	Bai et al. 2006
Shewanella oneidensis MR-1	Ag_2S	2–16	spherical	EC	Debabov et al. 2013
Escherichia coli	Pd, Pt	-	-	EC	Deplanche et al. 2010
Klebsiella pneumoniae	Ag	15–40	spherical	EC	Duraisamy and Yang 2013
Geobacillus stearothermophilus	Ag and Au	5–35	spherical	EC	Fayaz et al. 2011
Shewanella putrefaciens CN32	UO_2	-	-	EC	Fredrickson et al. 2002
Proteus mirabilis	Ag	10–20	spherical	EC and IC	Samadi et al. 2009
Klebsiella aerogenes	CdS	20–200 28.2–122	crystalline	EC	Holmes et al. 1995
Lactobacillus strains	Ti	40–60	spherical	IC	Prasad et al. 2007
Bacillus licheniformis	Ag and Au	50	spherical	IC	Sriram et al. 2012
Bacillus amyloliquefaciens	CdS	3–4	cubic/hexagonal	EC	Singh 2011
Halococcus salifodinae	Ag	12 and 22	-	IC	Srivastava et al. 2013
Pilimelia columellifera subsp. *pallida*	Ag	4–36	spherical	EC	Wypij et al. 2017
Streptomyces calidiresistens IF11 and IF17	Ag	5–50 and 5–20	spherical	EC	Wypij et al. 2018a

Table 6.1 contd. ...

... Table 6.1 contd.

Sources	Nanoparticles	Size (nm)	Shape	Type of synthesis	References
Bacteria					
Streptomyces xinghaiensis	Ag	5–20	spherical	EC	Wypij et al. 2018b
Fungi					
Fusarium oxysporum	CdS	5–20	-	EC	Ahmad et al. 2002
Fusarium oxysporum	Ag	5–15	-	EC	Ahmad et al. 2003a
Aspergillus fumigatus	ZnO	60–80	spherical	EC	Rajan et al. 2016
Trichoderma harzianum	CdS	3–8	spherical	EC	Bhadwal et al. 2014
Aspergillus fumigatus	Ag	5–25	-	EC	Bhainsa and D'Souza 2006
Rhizopus oryzae	Au	9–10	rod, triangular, hexagonal	IC	Das et al. 2012a
Fusarium oxysporum	Au	8–40	spherical and triangular	IC	Mukherjee et al. 2002
Phoma capsulatum, Phoma putamimum, Phoma citri	Ag	10–80 5–80 5–90	spherical	EC	Rai et al. 2015
Rhodosporidium diobovatum	PbS	2–5	spherical	EC	Seshadri et al. 2011
Saccharomyces cerevisiae	Pd	32	hexagonal	EC	Sriramulu and Sumathi 2018
Colleotrichum sp.	Al$_2$O$_3$	30–50	spherical	EC	Suryavanshi et al. 2017
Algae					
Chlorella vulgaris	Au	2–10	spherical	EC	Annamalai and Nallamuthu 2015
Chlorella vulgaris	Pd	5–20	spherical	EC	Arsiya et al. 2017
Chlorella vulgaris	Ag	9.8	spherical	EC	Da Silva Ferreira et al. 2017

Chlamydomonas reinhardtii	Ag	10–20	spherical	EC	Dahoumane et al. 2014
Scenedesmus sp.	Ag	15–20	spherical	EC	Jena et al. 2014
Shewanella algae	Au	10	triangular, hexagonal, nanoplates	EC	Ogi et al. 2010
Tetraselmis kochinensis	Au	5–35	spherical and triangular	IC	Senapati et al. 2012
Spirulina platensis	Ag	30–50		EC	Sharma et al. 2015
Chlorella vulgaris	Au	800–2000	triangular, truncated triangular, hexagonal	EC	Xie et al. 2007
Viruses					
M13 bacteriophage	TiO$_2$	20–40	nanowires	IC and EC	Chen et al. 2015
Tobacco mosaic virus (TMV)	Au	5	spherical	IC and EC	Kobayashi et al. 2012
M13 bacteriophage	ZnS and CdS	-	-	IC and EC	Mao et al. 2003
Tobacco mosaic virus (TMV)	CdS, PbS	-	-	IC and EC	Shenton et al. 1999

IC, Intracellular; EC, extracellular.

to bacterial cells, and the reduction of the ions and formation of water-insoluble complexes is a defense mechanism developed by bacteria to overcome such toxicity (Iravani 2014). It is claimed that in the presence of silver salts, biomolecules released from bacterial cells into the external medium affect reduction of silver ions to silver nanoparticles or AgNPs, which are formed inside the cells and then secreted outside (Mahdieh et al. 2012). Abundant secretion of proteins that may directly generate synthesis of nanoparticles was proposed by Prabhu and Poulose (2012) as one of the synthesis mechanisms. Some studies of non-enzymatic reduction mechanism suggested that some organic functional groups of microbial cell walls could be responsible for the bioreduction process (Crookes-Goodson et al. 2008). Nucleation of clusters of silver ions during the initial phase of silver nanoparticle synthesis causes an electrostatic interaction between the ions and negatively charged corboxylate groups of the cell wall (Wang et al. 2012). However, the S layer of some bacteria physically masks the negatively charged peptidoglycan sheet of the cell wall and hence can be involved in bacteria–metal surface interaction (Fu et al. 2000, Prakash et al. 2010). Study of *Lactobacillus* sp. strain A09 also showed its capability of nanoparticle synthesis by interaction of silver ions with groups present in the bacterial cell wall (Fu et al. 2000). Another mechanism across the cell wall was proposed in bacteria with electrokinetic potential that create transmembrane proton gradient, which can indirectly drive active symport of sodium along with silver ions from the surroundings. ATP binding employs silver-binding proteins attached to membrane lipids on the external surface of bacteria, which attracts the silver ions initiating silver nanoparticle synthesis (Prakash et al. 2011). Researchers speculated about the crucial role of three gene homologues responsible for silver resistance *silE*, *silP* and *silS* in synthesis of silver nanoparticles. The *silE* gene encodes silE protein, which is responsible for silver uptake by presenting histidine sites for silver ion binding. The presence of silver ions generates silE-based silver-binding machinery of bacteria and leads to Ag^+ uptake. The exact role of *silP* and *silS* genes in production of silver nanoparticles is not known. The ion exposure to bacterial silver reduction machinery generates biomolecules that bind to the ions and reduce them to metallic silver nuclei. Subsequently, these particles grow and form silver nanoparticles of different shapes (Singh et al. 2015).

Among the mechanisms responsible for the reduction of metals are enzymes such as reductases that are present in various microbial cells (Duran et al. 2010). This mechanism was observed in *Bacillus licheniformis*, which effectively synthesized silver nanoparticles in the presence of NADPH-dependent nitrate reductase (Vaidyanathan et al. 2010).

Furthermore, the synthesis of gold nanoparticles using *Escherichia coli* and *Desulfovibrio desulfuricans* was performed with H_2 as the electron donor. The reduction of Au^{3+} ions and bioaccumulation of gold nanoparticles was carried out in the periplasmic space. The authors concluded that the periplasmic hydrogenase of the bacteria was involved in this process, although it was not essential. Moreover, the introduction of additional thiol groups of cysteine residues increased the absorption of gold ions by bacteria and subsequently production of nanoparticles (Deplanche and Macaskie 2008).

Copper is not stable at the nanometer scale and rapidly oxidizes to form copper oxide (CuO), so synthesis of CuNPs is very difficult (Ramanathan et al.

2013). Ramanathan and co-authors (2013) obtained copper nanoparticles by using *Morganella morganii.* The authors suggested that synthesis occurred by uptake of the Cu ions and their subsequent binding to either metal ion reductase or similar protein. Moreover, other researchers reported intracellular synthesis of mercury nanoparticles (HgNPs) using *Pyrobaculum islandicum* and noticed that culture conditions such as pH and concentration of metal substrate affected the efficiency of this process. They found that alkaline pH (8.0) and lower concentration of mercury promoted synthesis of spherical and monodispersed mercury nanoparticles (Kashefi and Lovley 2000). The effect of synthesis conditions on the biosynthesis process of AgNPs by mesophilic bacteria was also reported by Srikar et al. (2016) and Wypij et al. (2019). Srikar et al. (2016) found that the rate of AgNP synthesis increased together with temperature up to 40°C. Similar findings were reported by Wypij et al. (2019). The authors observed increased nanoparticle synthesis with increasing temperature up to optimum temperature of 28°C. However, above these values synthesis was restricted, as at higher temperatures the mesophilic microorganisms or their biomolecules such as enzymes may be inactivated (Srikar et al. 2016, Wypij et al. 2019).

Mechanism of fungi-mediated synthesis of nanoparticles

The biosynthesis of metal nanoparticles using fungi is widespread. As in bacterial systems, the formation of metal nanoparticles by fungi can occur in either the intracellular or extracellular space, but extracellular biosynthesis is prevalent (Kitching et al. 2015). The fungi-derived nanoparticles show monodispersity and well-defined dimensions. Generally, simplicity of fungal cultivation and large amount of biomass produced are additional advantages of these microorganisms as producers of nanoparticles (Mohanpuria et al. 2008, Singh et al. 2016, Gahlawat and Choudhury 2019). The biological synthesis of nanoparticles by fungi has been mainly focused on the production of gold, silver and cadmium sulfide nanoparticles (Ahmad et al. 2002, Ahmad et al. 2003a, Bhainsa and D'Souza 2006, Das et al. 2012a, Bhadwal et al. 2014) but other types of metal nanoparticles were also synthesized (Table 6.1). Among metal sulfide nanoparticles, besides cadmium sulfide (CdS) nanoparticles, lead sulfide (PbS), zinc sulfide (ZnS) and molybdenum sulfide (MoS) nanoparticles were also formed by the fungi (e.g., *Fusarium oxysporum, Trichoderma harzianum, Rhodosporidium diobovatum*) when the appropriate salts were added to the growth medium (Ahmad et al. 2002, Seshadri et al. 2011, Pantidos and Horsfall 2014). The filamentous fungi possess advantages over bacteria due to much higher metal tolerance, wall binding and intracellular metal uptake capabilities (Dias et al. 2002, Velusamy et al. 2016) that result in an efficient and cost-effective production of nanoparticles (Gahlawat and Choudhury 2019). Fungi can accumulate metals by physicochemical and biological mechanisms, including extracellular binding by metabolites and polymers, binding to specific polypeptides, and metabolism-dependent accumulation (Gade et al. 2010). During the synthesis of metal nanoparticles by the fungus, mycelium was exposed to the metal salt solution, which induced the fungus to produce enzymes and metabolites for its own survival. In this process, the toxic metal ions were reduced to non-toxic metallic solid nanoparticles through the catalytic effect of the extracellular enzyme and metabolites released from fungal cells (Vahabi et al. 2011). Fungi are characterized by production of huge amounts of nanoforms due to extensive secretion of various components involved in the reduction and capping of nanoparticles (Ahmad et al. 2003, Gudikandula and Maringanti 2016). Generally, for reduction of silver ions to silver nanoparticles cell mass or extracellular components from cells of many fungi were used, e.g., *Fusarium oxysporum*, *Aspergillus flavus*, *Aspergillus clavatus*, and *Penicillium brevicompactum* (Lee et al. 2002, Marshall et al. 2007, Velusamy et al. 2016). Silver nanoparticle synthesis using proteins from marine *Penicillium fellutanum* as a reducing agent was reported by Kathiresan et al. (2009). Similarly, Kitching et al. (2016) used the cell surface proteins of *Rhizopus oryzae* for *in vitro* fabrication of gold nanoparticles. Mukherjee et al. (2008) studied silver nanoparticle synthesis by using *Trichoderma asperellum* and suggested that bioreduction of metal nanoparticles was conducted by protein extract from fungi that contained amino acid with -SH bonds. The authors claim that most likely cysteine undergoes dehydrogenation on reaction with silver nitrate to produce silver nanoparticles, while the free amino acid groups possibly serve as a capping for silver nanoparticles. Kumar and McLendon (1997) noticed that extracellularly synthesized nanoparticles were stabilized by the proteins and reducing agents secreted by the fungus. They found that at least four high molecular weight proteins released from fungal cells were associated with nanoparticles. One of these proteins was identified as specific NADH-dependent reductase. According to Du et al. (2011), who studied about 200 fungal strains, it was observed that especially two strains, represented by *Verticillium* sp. and *F. oxysporum*, produced metal nanoparticles in high quantity. The authors claimed that NADH-dependent nitrate reductase present in *Fusarium oxysporum* was conjugated with an electron donor (quinine), which reduced the metal ions to elemental forms. Similarly, Nikhil et al. (2009) postulated that NADH-dependent nitrate reductase from *Penicillium brevicompactum* was involved in the reduction of silver ions to metallic silver. They suggest that the reduction may involve the electrons from NADH where NADH-dependent reductase can act as a carrier (Nikhil et al. 2009). Duran and co-authors (2005) studied synthesis of silver nanoparticles using *Fusarium oxysporum* and suggested that the mechanism of synthesis is based on reductase action and also naphthoquinones and anthraquinones that possess excellent redox properties and may act as electron shuttle in metal reductions. Moreover, the cell-wall sugars were also thought to play a major role in the reduction of metal ions (Mukherjee et al. 2001).

Although biosynthesis of silver nanoparticles from fungi and their potential applications have been most often investigated in recent decades, the interest in study of biogenic gold nanoparticles has also increased. The small size of gold nanoparticles makes them reactive and increases their applications in many industries (Pantidos and Horsfall 2014). Researchers claimed that fungal-derived AuNPs were formed extracellularly by trapping and reduction of Au^{3+} ions with proteins present in the cell wall, and intracellularly after diffusion through the cell membrane and reduction by cystolic redox mediators (Das et al. 2012b, Kitching et al. 2015). Moreover, Duran et al. (2011) reported that reduced sugars in the cell wall of *Verticillium* sp. were responsible for the reduction of Au^{3+} ions. They also suggested adsorption of $AuCl^{4-}$ ions on the cell-wall enzymes by electrostatic interaction with positively charged groups such as lysine. Apte et al. (2013) observed biosynthetic redox mediators in synthesis of AuNPs in *Yarrowia lipolytica*. This fungus secreted melanin, which could generate reduction of metal salts to their elemental forms such as nanoforms. Melanin was secreted in a phenol form and reduced Au^{3+} to Au^0, while being oxidized to its quinone form (Apte et al. 2013). Du et al. (2011) noticed that bioreduction of Au^{3+} to gold nanoparticles using *Penicillium* sp. 1-208 cell filtrate was achieved in less than 5 min. The authors assumed that the large amount of extracellular enzymes secreted by fungi in a relatively pure state, free from other cellular proteins, was responsible for synthesis of gold nanoparticles. They also suggested that suspension of fungal biomass in non-growth medium for 2 days prior to cell filtration increased the concentration of extracellular redox proteins (Du et al. 2011).

In fungi-mediated synthesis of metal nanoparticles, the effect of various conditions, mainly pH and temperature, on the quantity and quality of synthesized nanoparticles is significant (Gericke and Pinches 2006, Agnihotri et al. 2009, Pimprikar et al. 2009, Govender et al. 2010). Govender et al. (2010) studied the effect of pH and temperature on synthesis of platinum nanoparticles from *F. oxysporum*. They monitored the activity of a dimeric hydrogenase enzyme (44.5 and 39.4 kDa) from fungi and found that this enzyme had optimum activity at pH 7.5 and 38°C, which could passively reduce H_2PtCl_6 to platinum nanoparticles (Govender et al. 2010). Other authors studied the effect of pH, salt concentration, and reaction time on the particle size and yield of *Cladosporium oxysporum*, which was used for reduction of gold ions into nanoparticles. The highest quantity of AuNPs was obtained with biomass-to-water ratio of 1:5, at 1 mM salt concentration and pH 7 (Bhargava et al. 2016).

Mechanism of algae-mediated synthesis of nanoparticles

Experiments have revealed the huge potential of algae, especially representatives of macroalgae, for synthesis of silver, gold, zinc oxide, and iron oxide nanoparticles (Balasooriya et al. 2017). However, microalgae have been found to have the ability to perform this process. Different groups of algae such as members of *Chlorophyceae*, *Phaeophyceae*, *Cyanophyceae*, *Rhodophyceae* and many more (diatoms and euglenoids) were used for the biological synthesis of metal nanoparticles. Marine algae are explored for the biogenic synthesis of nanosized particles as they contain various biologically active compounds and produce secondary metabolites that allow them to act as "nanofactories" (Fawcett et al. 2017, Gahlawat and Choudhury 2019). Moreover, the ability of algae to accumulate and reduce metal ions makes them more suitable for the biosynthesis of nanoparticles (LewisOscar et al. 2016). Algae belong to eukaryotic aquatic oxygenic photoautotrophs. Their photosynthesis machinery has been evolved from cyanobacteria via endosymbiosis (Velusamy et al. 2016); thus, the researchers believe that algae are as prolific as plants in terms of nanoparticle synthesis, and algae-synthesized nanoparticles have potential in medical applications, mainly as antimicrobial and anti-cancer agents (LewisOscar et al. 2016, Gahlawat and Choudhury 2019).

Although algae have been used for the synthesis of various types of nanoparticles, synthesis of gold and silver nanoparticles has been found to be dominant (Xie et al. 2007, Ogi et al. 2010, Dahoumane et al. 2014, Annamalai and Nallamuthu 2015, Sharma et al. 2015, Da Silva Ferreira et al. 2017) (Table 6.1). It is claimed that *Chlorella* sp. may accumulate various heavy metals such as cadmium, uranium, copper, and nickel. The dried algal cells were found to have a strong binding ability towards tetrachloroaurate (III) ions to form algal-bound gold, which was subsequently reduced to form Au^0.

Nearly 88% of algal-bound gold attained metallic state and the crystals of gold were accumulated in the interior and exterior cell surfaces with tetrahedral, decahedral and icosahedral structures (Hosea et al. 1986, Velusamy et al. 2016). The synthesis of nanoparticles by algal cells was found to be dose-dependent and associated with type of algae used (LewisOscar et al. 2016). As with bacterial- or fungal-mediated synthesis of metal nanoparticles, it is suggested that a vast variety of biomolecules including polysaccharides, peptides and pigments are responsible for the reduction of metals in biosynthesis performed by algae. The synthesized metal nanoparticles were stabilized and capped in aqueous solutions by proteins through amino groups, cysteine residues and sulfated polysaccharides from algal cells (Singaravelu et al. 2007, LewisOscar et al. 2016). As mentioned previously, studies of macroalgae for synthesis of metal nanoparticles are much more frequent, and therefore macroalgal synthesis is better understood. Study of biological synthesis of gold nanoparticles by inactivated biomass of red algae *Chondrus crispus* and green algae *Spyrogira insignis* revealed that bioreduction of gold ions was caused by deprotonated groups present on the algal cell wall that functioned as sorption center (Castro et al. 2013). Other researchers also reported participation of hydroxyl groups of algal pigment named fucoxanthin in synthesis of gold nanoparticles using brown algae *Fucus vesiculosus* (Mata et al. 2009).

Mechanism of virus-mediated synthesis of nanoparticles

Many viruses inhabiting bacteria and plants have been used in nanotechnology (Table 6.1). Such systems are characterized by relative structural and chemical stability, ease of production, and lack of toxicity and pathogenicity in animals or humans (Liu et al. 2005, Velusamy et al. 2016). Viruses display the features of an ideal template for the formation of nano-conjugates with metal nanoparticles. Therefore, scientists believe that further use of viruses in biogenic synthesis of metal nanoparticles makes it possible to develop innovative nano-sized components for various applications (e.g., drug delivery, cancer therapy) (Velusamy et al. 2016). Furthermore, viral materials could be involved in the reduction of toxic environmental contaminants such as metal ions. It was reported that iron nanoparticles produced in the filamentous M13 virus reduced soluble uranium (U^{6+}) to insoluble form (U^{4+}) (Ling et al. 2008). In addition, according to Makarov et al. (2014), an important feature of viruses is a high protein content in their capsid that provides a highly reactive surface for interaction with metallic ions.

Conclusions and future perspectives

Metal nanoparticles have been successfully synthesized using fungi, bacteria, algae and viruses. Green synthesis is an alternative to physicochemical techniques used for the production of various nanoparticles. The methods of extraction of intra- or extracellularly synthesized nanoparticles are inexpensive, simple, and effective on a large scale and do not require complicated equipment. The biosynthesis of metal nanoparticles can be controlled by various parameters and reaction conditions such as pH, temperature and the proportion of salts to reducing agent in order to obtain particles with desired size range and morphology. The optimal reaction conditions also affect activity of enzymes responsible for reduction of metal ions to metal nanoparticles. Probably the mechanism of nanoparticle synthesis begins with the trapping of metal ions on the surface of the cell via electrostatic interactions between the ions and non-ionic hydroxyl, carboxyl and other negatively charged groups present on the cell. Subsequently, ions are reduced by enzymes (reductase) and/or carbonyl groups leading to the formation of metal nuclei, which further grow to form the nanoparticles. The mechanisms of biosynthesis of metal nanoparticles are still not fully understood and much more investigation needs to be done to gain a better understanding of the synthesis pathways using microorganisms.

Nowadays, the bionanotechnology processes have important advantages over conventional physical and chemical methods. Hence, biological synthesis of metal nanoparticles requires better understanding of the biochemical and molecular mechanisms of the reduction of metal ions to nanoparticles. Furthermore, better knowledge of the molecular mechanisms involved in the microbial synthesis of nanoparticles is required. The improvement of biological methods of metal nanoparticle synthesis could lead to large-scale production of nanoparticles in the future.

References

Abdeen, S., Geo, S., Sukanya, S., Praseetha, P.K., Dhanya, R.P. 2014. Biosynthesis of silver nanoparticles from Actinomycetes for therapeutic applications. Int. J. Nano Dimens. 5: 155–162.

Abo-State, M.A.M., Partila, A.M. 2015. Microbial production of silver nanoparticles by *Pseudomonas aeruginosa* cell free extract. J. Eco. Health Env. 3(3): 91–98.

Agnihotri, M., Joshi, S., Kumar, A.R., Zinjarde, S., Kulkarni, S. 2009. Biosynthesis of gold nanoparticles by the tropical marine yeast *Yarrowia lipolytica* NCIM 3589. Mater. Lett. 63: 1231–1234.

Agrawal, P.N., Kulkarni, N.S. 2017. Biosynthesis and characterization of silver nanoparticles. Int. J. Curr. Microbiol. App. Sci. 6(4): 938–947.

Ahmad, A., Mukherjee, P., Mandal, D., Senapati, S., Khan, M.I., Kumar, R., Sastry, M. 2002. Enzyme mediated extracellular synthesis of CdS nanoparticles by the fungus *Fusarium oxysporum*. J. Am. Chem. Soc. 124: 12108–12109.

Ahmad, A., Mukherjee, P., Senapat, S., Mandal, D., Khan, M.I. 2003a. Extracellular biosynthesis of silver nanoparticles using the fungus *Fusarium oxysporum*. Colloids Surf. B: Biointerfaces 28: 313–318.

Ahmad, A., Mukherjee, P., Senapati, S. 2003b. Extracellular biosynthesis of silver nanoparticles using the fungus *Aspergillus oxysporum*. Colloids Surf. B: Biointerfaces 28: 313–318.

Ahmed, S., Ahmad, S.M., Swami, B.L., Ikram, S. 2016. Green synthesis of silver nanoparticles using *Azadirachta indica* aqueous leaf extract. J. Radiat. Res. Appl. Sci. 9(1): 1–7.

Annamalai, J., Nallamuthu, T. 2015. Characterization of biosynthesized gold nanoparticles from aqueous extract of *Chlorella vulgaris* and their anti-pathogenic properties. Appl. Nanosci. 5: 603–607.

Apte, M., Girme, G., Bankar, A., Kumar, R., Zinjarde, S. 2013. 3,4-dihydroxy-L-phenylalanine-derived melanin from *Yarrowia lipolytica* mediates the synthesis of silver and gold nanostructures. J. Nanobiotechnol. 11: 2.

Arsiya, F., Sayadi, M.H., Sobhani, S. 2017. Green synthesis of palladium nanoparticles using *Chlorella vulgaris*. Mater. Lett. 186: 113–115.

Avilala, J., Golla, N. 2019. Antibacterial and antiviral properties of silver nanoparticles synthesized by marine actinomycetes. Int. J. Pharm. Res. 10: 1223–1228.

Babu, M.M.G., Gunasekaran, P. 2009. Production and structural characterization of crystalline silver nanoparticles from *Bacillus cereus* isolate. Colloids and Surf. B: Biointerfaces. 74(1): 191–195.

Bai, H., Zhang, Z., Gong, J. 2006. Biological synthesis of semiconductor zinc sulfide nanoparticles by immobilized *Rhodobacter sphaeroides*. Biotechnol. Lett. 28(14): 1135–1139.

Balasooriya, E.R., Jayasinghe, C.D., Jayawardena, U.A., Ruwanthika, R.W.D., Mendis de Silva, R., Udagama, P.V. 2017. Honey mediated green synthesis of nanoparticles: new era of safe nanotechnology. J. Nanomat. 10.

Bhadwal, A.S., Tripathi, R.M., Gupta, R.K., Kumar, N., Singh, R.P., Shrivastaw, A. 2014. Biogenic synthesis and photocatalytic activity of CdS nanoparticles. RSC Adv. 4: 9484–9490.

Bhainsa, K.C., D'Souza, S.F. 2006. Extracellular biosynthesis of silver nanoparticles using the fungus *Aspergillus fumigatus*. Colloids Surf. B: Biointerfaces 47: 160–164.

Bhargava, A., Jain, N., Khan, M.A., Pareek, V., Dilip, R.V., Panwar, J. 2016. Utilizing metal tolerance potential of soil fungus for efficient synthesis of gold nanoparticles with superior catalytic activity for degradation of rhodamine B. J. Environ. Manage. 183: 22–32.

Castro, L., Blázquez, M.L., Muñoz, J.A., González, F., Ballester, A. 2013. Biological synthesis of metallic nanoparticles using algae. IET Nanobiotechnol. 7: 109–116.

Chen, P.-Y., Dang, X., Klug, M.T., Courchesne, N.-M.D., Qi, J., Hyder M.N., Belcher, A.M., Hammond, P.T. 2015. M13 Virus-enabled synthesis of titanium dioxide nanowires for tunable mesoporous semiconducting networks. Chem. Mater. 27: 1531–1540.

Chokriwal, A., Sharma, M.M., Singh, A. 2014. Biological synthesis of nanoparticles using bacteria and their applications. Am. J. PharmTech. Res. 4(6): 38–61.

Crookes-Goodson, W.J., Slocik, J.M., Naik, R.R. 2008. Bio-directed synthesis and assembly of nanomaterials. Chem. Soc. Rev. 37: 2403–2412.

Dahoumane, S.A., Jeffryes, C., Mechouet, M., Agathos, S.N. 2017. Biosynthesis of inorganic nanoparticles: a fresh look at the control of shape, size and composition. Bioengineering 4(14): 1–16.

Dahoumane, S.A., Wijesekera, K., Filipe, C.D., Brennan, J.D. 2014. Stoichiometrically controlled production of bimetallic gold-silver alloy colloids using micro-alga cultures. J. Colloid Interface Sci. 416: 67–72.

Das, S.K., Dickinson, C., Lafir, F., Brougham, D.F., Marsili, E. 2012a. Synthesis, characterization and catalytic activity of gold nanoparticles biosynthesized with *Rhizopus oryzae* protein extract. Green Chem. 14: 1322–1334.

Das, S.K., Liang, J., Schmidt, M., Laffir, F., Marsili, E. 2012b. Biomineralization mechanism of gold by zygomycete fungi *Rhizopous oryzae*. ACS Nano. 6: 6165–6173.

Das, V.L., Thomas, R., Varghese, R.T., Soniya, E.V., Mathew, J., Radhakrishnan, E.K. 2014. Extracellular synthesis of silver nanoparticles by the *Bacillus* strain CS 11 isolated from industrialized area. 3 Biotech. 4: 121–126.

da Silva Ferreira, V., ConzFerreira, M.E., Lima, L.M.T., Frasés, S., de Souza, W., Sant'Anna, C. 2017. Green production of microalgae-based silver chloride nanoparticles with antimicrobial activity against pathogenic bacteria. Enzyme Microb. Technol. 97: 114–121.

Debabov, V.G., Voeikova, T.A., Shebanova, A.S., Shaitan, K.V., Emelyanova, L.K., Novikova, L.M., Kirpichnikov, M.P. 2013. Bacterial synthesis of silver sulfide nanoparticles. Nanotechnol. Russ. 8: 269–276.

Deepak, V., Umamaheshwaran, P.S., Guhan, K., Nanthini, R.A., Krithiga, B., Jaithoon, N.M., Gurunathan, S. 2011. Synthesis of gold and silver nanoparticles using purified URAK. Colloid Surface B: Biointerfaces 86: 353–358.

Deljou, A., Goudarzi, S. 2016. Green extracellular synthesis of the silver nanoparticles using thermophilic *Bacillus* sp. AZ1 and its antimicrobial activity against several human pathogenetic bacteria. Iran J. Biotechnol. 14(2): 25–32.

Deplanche, K., Macaskie, L.E. 2008. Biorecovery of gold by *Escherichia coli* and *Desulfovibrio desulfuricans*. Biotechnol. Bioeng. 99(5): 1055–1064.

Deplanche, K., Caldelari, I., Mikheenko, I.P., Sargent, F., Macaskie, L.E. 2010. Involvement of hydrogenases in the formation of highly catalytic Pd(0) nanoparticles by bioreduction of Pd(II) using *Escherichia coli* mutant strains. Microbiology. 156: 2630–2640.

Dias, M.A., Lacerda, I.C., Pimentel, P.F., de Castro, H.F., Rosa, C.A. 2002. Removal of heavy metals by an *Aspergillus terreus* strain immobilized in a polyurethane matrix. Lett. Appl. Microbiol. 34: 46–50.

Du, L., Xian, L., Feng, J.-X. 2011. Rapid extra-/intracellular biosynthesis of gold nanoparticles by the fungus *Penicillium* sp. J. Nanopart. Res. 13: 921–930.

Duraisamy, K., Yang, S.L. 2013. Synthesis and characterization of bactericidal silver nanoparticles using cultural filtrate of simulated microgravity grown *Klebsiella pneumoniae*. Enzyme Microb. Technol. 52: 151–156.

Duran, N., Marcato, P.D., Alves, O., Souza, G.I.H.D., Esposito, E. 2005. Mechanistic aspect of biosynthesis of silver nanoparticles by several *Fusarium oxysporum* strains. J. Nanobiotechnol. 3: 8.

Duran, N., Marcato, P.D., De Conti, R. 2010. Potential use of silver nanoparticles on pathogenic bacteria, their toxicity and possible mechanisms of action. J. Braz. Chem. Soc. 21: 949–959.

Duran, N., Marcato, P.D., Duran, M. 2011. Mechanistic aspects in the biogenic synthesis of extracellular metal nanoparticles by peptides, bacteria, fungi and plants. App. Microbiol. Biotechnol. 90: 1609–1624.

Fawcett, D., Verduin, J.J., Shah, M., Sharma, S.B., Poinern, G.E.J. 2017. A review of current research into the biogenic synthesis of metal and metal oxide nanoparticles via marine algae and seagrasses. J. Nanosci. 2017: 1–15.

Fayaz, A.M., Girilal, M., Rahman, M., Venkatesan, R., Kalaichelvan, P.T. 2011. Biosynthesis of silver and gold nanoparticles using thermophilic bacterium *Geobacillus stearothermophilus*. Process Biochem. 46: 1958–1962.

Fredrickson, J.K., Zachara, J.M., Kennedy, D.W., Liu, C., Duff, M.C., Hunter, D.B., Dohnalkova, A. 2002. Influence of Mn oxides on the reduction of uranium (VI) by the metal-reducing bacterium *Shewanella putrefaciens*. Geochim. Cosmochim. Acta. 66: 3247–3262.

Fu, J.K., Liu, Y.Y., Gu, P.Y., Tang, D.L., Lin, Z.Y., Yao, B.X., Weng, S.Z. 2000. Spectroscopic characterization on the biosorption and bioreduction of Ag(I) by *Lactobacillus* sp. A09. Acta Phys-Chim. Sin. 16: 779–782.

Gade, A., Ingle, A., Whiteley, C., Rai, M. 2010. Mycogenic metal nanoparticles: progress and applications. Biotechnol. Lett. 32: 593–600.

Gahlawat, G., Choudhury, A.R. 2019. A review on the biosynthesis of metal and metal salt nanoparticles by microbes. RSC Adv. 9: 12944.

Gericke, M., Pinches, A. 2006. Microbial production of gold nanoparticles. Gold Bull. 39: 22–27.

Ghorbani, H.R. 2013. Biosynthesis of silver nanoparticles using *Salmonella typhirium*. J. Nanostruct. Chem. 3: 29.

Golińska, P., Wypij, M., Ingle, A.P., Gupta, I., Dahm, H., Rai, M. 2014. Biogenic synthesis of metal nanoparticles from actinomycetes: biomedical applications and cytotoxicity. Appl. Microbiol. Biotechnol. 98: 8083–8097.

Govender, Y., Riddin, T.L., Gericke, M., Whiteley, C.G. 2010. On the enzymatic formation of platinum nanoparticles. J. Nanopart. Res. 12: 261–271.

Gudikandula, K., Maringanti, S.C. 2016. Synthesis of silver nanoparticles by chemical and biological methods and their antimicrobial properties. J. Exp. Nanosci. 11(9): 714–721.

Gurunathan, S., Han, J.W., Kwon, D.N., Kim, J.H. 2014. Enhanced antibacterial and anti-biofilm activities of silver nanoparticles against Gram-negative and Gram-positive bacteria. Nanoscale Res. Lett. 9(1): 373.

Holmes, J.D., Smith, P.R., Evans-Gowing, R., Richardson, D.J., Russell, D.A., Sodeau, J.R. 1995. Energy-dispersive X-ray analysis of the extracellular cadmium sulfide crystallites of *Klebsiella aerogenes*. Arch. Microbiol. 163(2): 143–147.

Hosea, M., Greene, B., Mcpherson, R., Henzl, M., Alexander, M.D., Darnall, D.W. 1986. Accumulation of elemental gold on the alga *Chlorella vulgaris*. Inorg. Chim. Acta. 123: 161–165.

Hosseini-Abari, A., Emtiazi, G., Ghasemi, S.M. 2013. Development of an eco-friendly approach for biogenesis of silver nanoparticles using spores of *Bacillus athrophaeus*. World J. Microbiol. Biotechnol. 29: 2359–2364.

Husseiny, M.I., El-Aziz, M.A., Badr, Y., Mahmoud, M.A. 2007. Spectrochimca Acta. Part A: Molecular and Biomolecular Spectroscopy 67: 1003.

Iravani, S. 2014. Bacteria in nanoparticle synthesis: current status and future prospects. Int. Sch. Res. Notices. 2014: 1–18.

Jena, J., Pradhan, N., Nayak, R.R., Dash, B.P., Sukla, L.B., Panda, P.K., Mishra, B.K. 2014. Microalga *Scenedesmus* sp.: a potential low-cost green machine for silver nanoparticle synthesis. J. Microbiol. Biotechnol. 24: 522–533.

Jha, A.K., Prasad, K. 2010. Understanding biosynthesis of metallic/oxide nanoparticles: A biochemical perspective. pp. 23–40. *In*: Kumar, S.A., Thiagarajan, S., Wang, S.F. (eds.). Biocompatible Nanomaterials Synthesis, Characterization and Applications. Nova Science Publishers, Inc. Hauppauge, New York, USA.

Kalimuthu, K., Babu, R.S., Venkataraman, D., Bilal, M., Gurunathan, S. 2008. Biosynthesis of silver nanocrystals by *Bacillus licheniformis*. Colloids Surf. B: Biointerfaces 65: 150–153.

Karthik, L., Kumar, G., Vishnu Kirthi, A., Rahuman, A.A., Bhaskara Rao, K.V. 2014. *Streptomyces* sp. LK3 mediated synthesis of silver nanoparticles and its biomedical application. Bioprocess. Biosyst. Eng. 37: 261–267.

Kashefi, K., Lovley, D.R. 2000. Reduction of Fe(III), Mn(IV), and toxic metals at 100°C by *Pyrobaculum islandicum*. Appl. Environment Microbiol. 66(3): 1050–1056.

Kathiresan, K., Manivannan, S., Nabeel, M.A., Dhivya, B. 2009. Studies on silver nanoparticles synthesized by a marine fungus, *Penicillium fellutanum* isolated from coastal mangrove sediment. Colloids Surf. B: Biointerfaces 71: 133–137.

Kitching, M., Choudhary, P., Inguva, S., Guo, Y., Ramani, M., Das, S.K., Marsili, E. 2016. Fungal surface protein mediated one-pot synthesis of stable and hemocompatible gold nanoparticles. Enzyme Microb. Technol. 95: 76–84.

Kitching, M., Ramani, M., Marsili, E. 2015. Fungal biosynthesis of gold nanoparticles: mechanism and scale up. Microb. Biotechnol. 8(6): 904–917.

Klaus, R.J., Olsson, E., Granqvist, C.G. 2001. Bacteria as workers in the living factory: metal-accumulating and potential for materials science. Trends Biotechnol. 19: 15–20.

Klaus, T., Joerger, R., Olsson, E., Granqvist, C.G. 1999. Silver-based crystalline nanoparticles, microbially fabricated. Proc. Natl. Acad. Sci. 96: 13611–13614.

Kobayashi, M., Tomita, S., Sawada, K., Shiba, K., Yanagi, H., Yamashita, I., Uraoka, Y. 2012. Chiral meta-molecules consisting of gold nanoparticles and genetically engineered tobacco mosaic virus. Opt. Express. 20: 24856–24863.

Kumar, C.V., McLendon, G.L. 1997. Nanoencapsulation of cytochrome c and horseradish peroxidase at the galleries of a-zirconium phosphate. Chem. Mater. 9: 863–870.

Lee, S.W., Mao, C., Flynn, C.E., Belcher, A.M. 2002. Ordering of quantum dots, using genetically engineered viruses. Science. 296: 892–895.

LewisOscar, F., Vismaya, S., Arunkumar, M., Thajuddin, N., Dhanasekaran, D., Nithya, C. 2016. Algal nanoparticles: synthesis and biotechnological potentials. Intech. 7: 158–182.

Li, X., Xu, H., Chen, Z.-S., Chen, G. 2011. Biosynthesis of nanoparticles by microorganisms and their applications. J. Nanomat. 16, Article ID 270974.

Ling, T., Huimin, Y., Shen, Z., Wang, H., Zhu, J. 2008. Virus-mediated FCC iron nanoparticle induced synthesis of uranium dioxide nanocrystals. Nanotechnology 19: 115608.

Liu, L., Cañizares, M.C., Monger, W., Perrin, Y., Tsakiris, E., Porta, C., Shariat, N., Nicholson, L., Lomonossoff, G.P. 2005. Cowpea mosaic virus-based systems for the production of antigens and antibodies in plants. Vaccine. 23: 1788–1792.

Mahdieh, M., Zolanvari, A., Azimeea, A.S., Mahdieh, M. 2012. Green biosynthesis of silver nanoparticles by *Spirulina platensis*. Sci. Iran. Transactions F: Nanotechnol. 19: 926–929.

Makarov, V.V., Love, A.J., Sinitsyna, O.V., Makarova, S.S., Yaminsky, I.V., Taliansky, M.E., Kalinina, N.O. 2014. "Green" nanotechnologies: synthesis of metal nanoparticles using plants. Acta Naturae. 6: 35–44.

Mao, C., Flynn, C.E., Hayhurst, A., Sweeney, R., Qi, J., Georgiou, G., Iverson, B., Belcher, A.M. 2003. Viral assembly of oriented quantum dot nanowires. Proc. Natl. Acad. Sci. 100: 6946–6951.

Marshall, M., Beliaev, A., Dohnalkova, A., David, W., Shi, L., Wang, Z. 2007. C-Type cytochrome-dependent formation of U (IV) nanoparticles by *Shewanella oneidensis*. Plos Biol. 4: 1324–1333.

Mata, Y., Torres, E., Blazquez, M., Ballester, A., Gonzalez, F., Munoz, J. 2009. Gold (III) biosorption and bioreduction with the brown alga *Fucus vesiculosus*. J. Hazard. Mater. 166: 612–618.

Mergeay, M., Monchy, S., Vallaeys, T. 2003. *Ralstonia metallidurans*, a bacterium specifically adapted to toxic metals: towards a catalogue of metal-responsive genes, FEMS Microbiol. Rev. 27(3): 385–410.

Mohanpuria, P., Rana, N.K., Yadav, S.K. 2008. Biosynthesis of nanoparticles: technological concepts and future applications. J. Nanopart. Res. 10(3): 507–517.

Mukherjee, P., Ahmad, A., Mandal, D. 2001. Bioreduction of AuCl ions by the fungus *Verticillium* sp. and surface trapping of the gold nanoparticles formed. Angew. Chem. Int. Ed. 40: 3585–3588.

Mukherjee, P., Roy, M., Mandal, B.P. 2008. Green synthesis of highly stabilized nanocrystalline silver particles by a nonpathogenic and agriculturally important fungus *T. asperellum*. Nanotechnology 19: 103–110.

Mukherjee, P., Senapati, S., Mandal, D., Ahmad, A., Khan, M.I., Kumar, R., Sastry, M. 2002. Extracellular synthesis of gold nanoparticles by the fungus *Fusarium oxysporum*. ChemBiochem. 3: 461–463.

Narasimha, G., Janardhan, A., Mohammad, A.A., Mallikarjuna, K. 2013. Extracellular synthesis, characterization and antibacterial activity of silver nanoparticles by actinomycetes isolative. Int. J. Nano Dimens. 4(1): 77–83.

Nikhil, S.S., Bule, M., Bhambure, R., Singhal, S., Singh, S.K., Szakacs, P.A. 2009. Biosynthesis of silver nanoparticles using aqueous extract from the compactin producing fungal strain. Process Biochem. 44: 939–943.

Ogi, T., Saitoh, N., Nomura, T., Konishi, Y. 2010. Room-temperature synthesis of gold nanoparticles and nanoplates using *Shewanella algae* cell extract. J. Nanopart. Res. 12: 2531–2539.

Pantidos, N., Horsfall, L.E. 2014. Biological synthesis of metallic nanoparticles by bacteria, fungi and plants. J. Nanomed. Nanotechnol. 5(5).

Pimprikar, P.S., Joshi, S.S., Kumar, A.R., Zinjarde, S.S., Kulkarni, S.K. 2009. Influence of biomass and gold salt concentration on nanoparticle synthesis by the tropical marine yeast *Yarrowia lipolytica* NCIM 3589. Colloids Surf. B: Biointerfaces. 74: 309–316.

Prabhu, S., Poulose, E.K. 2012. Silver nanoparticles: mechanism of antimicrobial action, synthesis, medical applications, and toxicity effects. Int. Nano Lett. 2: 32.

Prakash, A., Sharma, S., Ahmad, N., Gosh, A., Sinha, P. 2011. Synthesis of AgNPs by *Bacillus cereus* bacteria and their antimicrobial potential. J. Biomaterials Nanobiotechnol. 2: 156–162.

Prasad, K., Jha, A.K., Kulkarni, A.R. 2007. *Lactobacillus* assisted synthesis of titanium nanoparticles. Nanoscale Res. Lett. 2(5): 248–250.

Rai, M., Ingle, A.P., Gade, A.K., Duarte, M.C., Duran, N. 2015. Three *Phoma* spp. synthesised novel silver nanoparticles that possess excellent antimicrobial efficacy. IET Nanobiotechnol. 9: 280–287.

Rajan, A., Cherian, E., Baskar, G. 2016. Biosynthesis of zinc oxide nanoparticles using *Aspergillus fumigatus* JCF and its antibacterial activity. Int. J. Modern Sci. Technol. 1(2): 52–57.

Ramanathan, R., Field, M.R., O'Mullane, A.P., Smooker, P.M., Bhargava, S.K. 2013. Aqueous phase synthesis of copper nanoparticles: a link between heavy metal resistance and nanoparticle synthesis ability in bacterial systems. Nanoscale. 5: 2300–2306.

Samadi, N., Golkaran, D., Eslamifar, A., Jamalifar, H., Fazeli, M.R., Mohseni, F.A. 2009. Intra/extracellular biosynthesis of silver nanoparticles by an autochthonous strain of *Proteus mirabilis* isolated from photographic waste. J. Biomed. Nanotechnol. 5: 247–253.

Sanjenbam, P., Gopal, J.V., Kannabiran, K. 2014. Anticandidal activity of silver nanoparticles synthesized using *Streptomyces* sp. VITPK1. J. Mycol. Med. 24(3): 211–219.

Sathiyanarayanan, G., Vignesh, V., Saibaba, G., Vinothkanna, A., Dineshkumar, K., Viswanathan, M.B., Selvin, J. 2014. Synthesis of carbohydrate polymer encrusted gold nanoparticles using bacterial exopolysaccharide: a novel and greener approach. RSC Adv. 4: 22817–22827.

Sharma, G., Jasuja, N.D., Kumar, M., Ali, M.I. 2015. Biological synthesis of silver nanoparticles by cell-free extract of *Spirulina platensis*. J. Nanotechnol. 2015: 1–6.

Sharma, N., Pinnaka, A.K., Raje, M., Fnu, A., Bhattacharyya, M.S., Choudhury, A.R. 2012. Exploitation of marine bacteria for production of gold nanoparticles. Microb. Cell Fact. 11: 86.

Shenton, W., Douglas, T., Young, M., Stubbs, G., Mann, S. 1999. Inorganic-organic nanotube composites from template mineralization of tobacco mosaic virus. Adv. Mater. 11: 253–256.

Senapati, S., Syed, A., Moeez, S., Kumar, A., Ahmad A. 2012. Intracellular synthesis of gold nanoparticles using alga *Tetraselmis kochinensis*. Mater. Lett. 79: 116–118.

Seshadri, S., Saranya, K., Kowshik, M. 2011. Green synthesis of lead sulfide nanoparticles by the lead resistant marine yeast, *Rhodosporidium diobovatum*. Biotechnol. Prog. 27(5): 1464–1469.

Singaravelu, G., Arockiamary, J.S., Kumar, V.G., Govindaraju, K. 2007. A novel extracellular synthesis of monodisperse gold nanoparticles using marine alga, *Sargassum wightii* Greville. Colloids Surf. B: Biointerfaces. 57: 97–101.

Singh, B.R. 2011. Synthesis of stable cadmium sulfide nanoparticles using surfactin produced by *Bacillus amyloliquifaciens* strain KSU-109. Colloids Surf. B: Biointerfaces. 85: 207–213.

Singh, R., Shedbalkar, U.U., Wadhwani, S.A., Chopade, B.A. 2015. Bacteriagenic silver nanoparticles: synthesis, mechanism and applications. Appl. Microbiol. Biotechnol. 99: 4579–4593.

Singh, P., Kim, Y-J., Zhang, D., Yang, D-C. 2016. Biological synthesis of nanoparticles from plants and microorganisms. Trends Biotechnol. 34(7): 588–599.

Srikar, S.K., Giri, D.D., Pal, D.B., Mishra, P.K., Upadhyay, S.N. 2016. Green synthesis of silver nanoparticles: a review. Green Sustainable Chem. 6: 34–56.

Sriram, M.I., Kalishwaralal, K., Gurunathan, S. 2012. Biosynthesis of silver and gold nanoparticles using *Bacillus licheniformis*. pp. 33–43. *In*: Soloviev, M. (ed.). Nanoparticles in Biology and Medicine: Methods and Protocols. Springer, Dordrecht.

Sriramulu, M., Sumathi, S. 2018. Biosynthesis of palladium nanoparticles using *Saccharomyces cerevisiae* extract and its photocatalytic degradation behaviour. Adv. Nat. Sci.: Nanosci. Nanotechnol. 9: 025018.

Srivastava, P., Bragança, J., Ramanan, S.R., Kowshik, M. 2013. Synthesis of silver nanoparticles using haloarchaeal isolate *Halococcus salifodinae* BK3. Extremophiles. 17: 821–831.

Suryavanshi, P., Pandit, R., Gade, A., Derita, M., Zachino, S., Rai, M. 2017. *Colletotrichum* sp.-mediated synthesis of sulphur and aluminium oxide nanoparticles and its *in vitro* activity against selected food-borne pathogens. LWT–Food Sci. Technol. 81: 188–194.

Vahabi, K., Mansoori, G.A., Karimi, S. 2011. Biosynthesis of silver nanoparticles by the fungus *Trichoderma reesei*. Insciences J. 1: 65–79.

Vaidyanathan, R., Gopalram, S., Kalishwaralal, K., Deepak, V., Pandian, S.R.K., Gurunathan, S. 2010. Enhanced silver nanoparticles synthesis by optimization of nitrate reductase activity. Colloids Surf. B: Biointerfaces. 75: 335–341.

Velusamy, P., Kumar, G.V., Jeyanthi, V., Das, J., Pachaiappan, R. 2016. Bio-Inspired green nanoparticles: synthesis, mechanism, and antibacterial application. Toxicol. Res. 32(2): 95–102.

Wang, H., Chen, H., Wang, Y., Huang, J., Kong, T., Lin, W., Zhou, Y., Lin, L., Sin, D., Li, Q. 2012. Stable silver nanoparticles with narrow size distribution non-enzymatically synthesized by *Aeromonas* sp. SH10 cells in the presence of hydroxyl ions. Curr. Nanosci. 8: 838–846.

Wypij, M., Czarnecka, J., Dahm, H., Rai, M., Golinska, P. 2017. Silver nanoparticles from *Pilimelia columellifera* subsp. *pallida* SL19 strain demonstrated antifungal activity against fungi causing superficial mycoses. J. Basic. Microbiol. 57(9): 793–800.

Wypij, M., Czarnecka, J., Swiecimska, M., Dahm, H., Rai, M., Golinska, P. 2018b. Synthesis, characterization and evaluation of antimicrobial and cytotoxic activities of biogenic silver nanoparticles synthesized from *Streptomyces xinghaiensis* OF1 strain. World J. Microbiol. Biotechnol. 34(2): 23. doi: 10.1007/s11274-017-2406-3.

Wypij, M., Swiecimska, M., Czarnecka, J., Dahm, H., Rai, M., Golinska, P. 2018a. Antimicrobial and cytotoxic activity of silver nanoparticles synthesized from two haloalkaliphilic actinobacterial strains alone and in combination with antibiotics. J. Appl. Microbiol. 124(6): 1411–1424.

Wypij, M., Świecimska, M., Dahm, H., Rai, M., Golińska, P. 2019. Controllable biosynthesis of silver nanoparticles using actinobacterial strains. Green Proc. Synth. 8: 207–214.

Xie, J., Lee, J.Y., Wang, D.I., Ting, Y.P. 2007. Identification of active biomolecules in the high-yield synthesis of single crystalline gold nanoplates in algal solutions. Small. 3: 672–682.

Zhang, X.F., Liu, Z.G., Shen, W., Gurunathan, S. 2016. Silver nanoparticles: synthesis, characterization, properties, applications, and therapeutic approaches. Int. J. Molecular Sci. 17: 1534.

Chapter 7

Microbially Inspired Nanostructures for Management of Food-Borne Pathogens

Kamel A. Abd-Elsalam,[1,2,*] *Khamis Youssef,*[1,3]
Farah K. Ahmed[4] *and Hassan Almoammar*[5,6]

Introduction

Present sterilization practices involve the use of chlorine/hypochlorite, chlorine dioxide, peracetic acid, hydrogen peroxide, quaternary ammonium salts, ozone, and UV radiation. A new generation of techniques for ensuring food security and quality, particularly microbial control medication, is seemingly within easy reach that can empower the food industry, address current insufficiencies and losses, and reach new levels of well-being, sustainability and economic growth (Eleftheriadou et al. 2017). Another innovation demonstrates that microorganisms on fresh produce and production surfaces can be inactivated by electrolyzed water and water nanostructures produced by electrospraying water vapor (Glaser 2015). Acidic electrolyzed water (AEW) is being studied as a promising disinfecting agent in the food, medical and agricultural industries (Wang et al. 2014a). Results suggest that AEW treatment could altogether decrease populations of *Escherichia coli* O157 H7, *Salmonella*

---

[1] Plant Pathology Research Institute, Agricultural Research Center (ARC), 9 Gamaa St., 12619 Giza, Egypt.
[2] Unit of Excellence in Nano-Molecular Plant Pathology Research, Plant Pathology Research Institute, 9 Gamaa St., 12619 Giza, Egypt.
[3] Department of Agronomy, Londrina State University, 86057-970 Londrina, Parana, Brazil.
[4] Biotechnology English Program, Faculty of Agriculture, Ain Shams University, Shoubra El-Kheima, Cairo 11241, Egypt.
[5] ETH Zürich, Department of Biology, Institute of Microbiology, Vladimir-Prelog-Weg 4. 8093 Zürich, Switzerland.
[6] National Centre for Biotechnology, King Abdulaziz City for Science and Technology (KACST), Saudi Arabia.
* Corresponding author: kamelabdelsalam@gmail.com

typhimurium, and *Listeria monocytogenes* pathogens from the surfaces of lettuce and spinach leaves with increasing time of exposure, while alkaline electrolyzed water did not decrease pathogen levels even after a 5 min treatment on lettuce and spinach (Park et al. 2008). Acidic electrolyzed water has been recognized to be an effective disinfectant for inactivating food-borne pathogens including *Escherichia coli*, *Vibrio parahaemolyticus*, and *Listeria monocytogenes* (Wang et al. 2014b).

The hypothesized mode of action is a decrease in rigidity of the cell wall, core and external film, which prompts rapid leakage of intracellular DNA and proteins (Zeng et al. 2011, Ding et al. 2016). Sun et al. (2012) and Vázquez-Sánchez et al. (2014) concluded that the available chlorine in AEW may be one of the primary factors for the inactivation of *Staphylococcus aureus* biofilms. Acidic electrolyzed water additionally killed biofilms framed by both Gram-negative microscopic organisms (*Vibrio parahaemolyticus*) and Gram-positive microorganisms (*Listeria monocytogenes*) and was seen to inactivate the disengaged cells that are a potential source of secondary contamination (Han et al. 2017). Therefore, AEW is a good sanitizing option and can be applied to control biofilms in food processing facilities as well as to protect food from cross-contamination.

The technique in question uses engineered water nanostructures (EWNS) framed in an electrospray with remarkable properties as a measurement of 25 nm, including reactive oxygen species. The ability of this innovation to inactivate food-borne microorganisms such as *Escherichia coli*, *Salmonella enterica*, and *Listeria innocua* on hard steel surfaces and on natural tomatoes was assessed (Pyrgiotakis et al. 2015).

This chapter highlights the principles and applications of electrolyzed water and EWNS for their antimicrobial activity against air- or food-borne pathogens and their possible mechanism of action.

Principles of electrolysis of water

Among the alternative means for water sanitation, electrolyzed water gained particular attention in the food industry as a novel technology for preventing fruit and vegetable contamination in the postharvest environment (Buck et al. 2002). Electrolyzed water is generated by passing a diluted salt solution through an electrolytic cell; when anode and cathode are not separated by a membrane, neutral electrolyzed water containing several active chemical species such as free radicals is produced. The principles of electrolyzed water generation are based on electrolysis of a diluted solution of salts (e.g., NaCl, KCl, $MgCl_2$), which leads to dissociation of salt ions and formation of anions and cations at anode and cathode, negative and positive electrodes, respectively (Al-Haq et al. 2005). The physical properties and chemical composition of electrolyzed water vary with the concentration of salt solution, electrical current, length of electrolysis and water flow rate (Kiura et al. 2002). Most previous studies have assessed the effect of electrolyzing parameters (flow rate, current intensity, and time for electrolysis) on free chlorine, electric conductivity, and pH of the resulting electrolyzed water (Hsu 2003, Abbasi and Lazarovits 2006, Guentzel et al. 2010, Hussien et al. 2017), while very few researchers have studied the effect of electrolyzing parameters on the ability of electrolyzed water to suppress pathogen unit (Fallanaj et al. 2015). Although salt solutions used in electrolysis play a major role in the effectiveness of electrolyzed water, few salts were studied for their effect on electrolyzed water efficiency; among them are sodium chloride (Hsu 2003, Okull and Laborde 2004, Abbasi and Lazarovits 2006, Guentzel et al. 2010, Whangchai et al. 2013, Khayankarn et al. 2014), potassium chloride (Yaseen et al. 2013) and sodium bicarbonate (Fallanaj et al. 2016). Figure 7.1 shows a diagram of the main compartments of electrolysis and the main steps of the electrolyzing process (using the naturally available salts in tap water without additives).

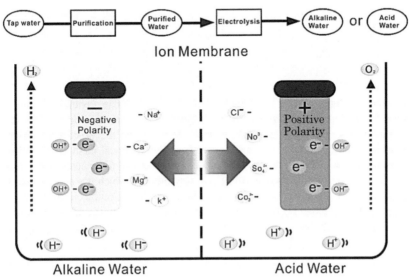

Figure 7.1. Diagram showing the main compartments of electrolysis and the main steps of the electrolyzing process (using the naturally available salts in tap water without additives).

Electrolyzed water applications

The acceptance of electrolyzed water as a sanitizer is evident from its use in a number of applications in agriculture, medical sterilization, food sanitation, livestock management, and other fields that employ antimicrobial techniques (Kim et al. 2000, Huang et al. 2008, Rahman et al. 2016). Electrolyzed water was introduced as decontaminating agent (Shimizu and Hurusawa 1992) and was approved in Japan and USA as a food additive and sanitizing agent (Park et al. 2002, Yoshida et al. 2004). Electrolyzed water shows antimicrobial action against a wide range of microorganisms and most normal kinds of infections, microscopic organisms, fungi, and spores in a moderately short measure of time in food products, food processing surfaces, and non-food surfaces (Hricova et al. 2008, Huang et al. 2008, Ding et al. 2015, Hao et al. 2015, Youssef et al. 2018, Hussien et al. 2018).

Plant diseases

Acidic electrolyzed water helps control foliar diseases of many plants (Fujiwara et al. 2009, Guentzel et al. 2010, Hou et al. 2012, Kusakari et al. 2013, Jia et al. 2015). Acidic electrolyzed water was tested against bacterial spot disease of tomato under greenhouse and field conditions. The viability of propagules of *Xanthomonas campestris pv. vesicatoria*, *Streptomyces scabies* and *Fusarium oxysporum* f. sp. *lycopersici* was reduced significantly. These results indicated a potential use of AEW as a seed surface disinfectant or contact bactericide (Abbasi and Lazarovits 2006). The impacts of foliar treatment of electrolytically ozonated water and acidic electrolyzed oxidizing water was studied on the severity of powdery mildew disease and incidence of visible physiological disorder on cucumber leaves.

Ozonated water is a viable option for controlling powdery mildew disease on cucumber leaves at low initial severity levels, or for prevention of powdery mildew, and acidic electrolyzed oxidizing water can be applied for controlling powdery mildew in treatments against visible physiological disorder (Fujiwara et al. 2009). Acidic electrolyzed oxidizing water is a practical alternative for controlling powdery mildew on gerbera daisies and gives cultivators an extra means to reduce the use of conventional fungicides in greenhouses (Mueller et al. 2003). Oxidizing water was applied for controlling *Botrytis cinerea*, *Sphaerotheca pannosa* and *Peronospora sparsa*, pathogens of two greenhouse rose varieties, and the efficacy of this new technology was verified by comparison with earlier chemical methods. The results showed the feasibility of electrolyzed oxidizing water as a fungicide at the volumes tested, providing significant control of powdery mildew, which has a high incidence rate because of climatic conditions (Fernández et al. 2011). Acidic electrolyzed water plays a positive role in controlling damage by grape downy mildew, *Botrytis cinerea*, anthracnose, grape leaf spot caused by *Alternaria*, and anthracnose of grape, with foliar control effect of 93.5%, 96.2%, 93.20%, 100% and 87.9% respectively (Haihua et al. 2016).

Postharvest decay

Electrolyzed water can be used to protect products from postharvest deterioration caused by some fungi including *Penicillium, Botrytis, Monolinia,* and *Colletotrichum,* and from mycotoxins. *Botrytis cinerea* is an important postharvest pathogen because of the conducive conditions prevailing throughout the postharvest handling chain, including injuries, high humidity, senescing plant tissue, and high sugar content. Major postharvest losses due to *B. cinerea* occur in apple, blackberry, blueberry, currant, grape, kaki, kiwi, pear, pomegranate, quince, raspberry, strawberry, grape and many other fresh fruits (Romanazzi et al. 2016) (Fig. 7.2). In other fruits (e.g., apricot, lemon, orange, peach, plum, sweet cherry), although it is not the main pathogen, it is still capable of causing considerable postharvest losses.

Guentzel et al. (2010) showed that treatment of *B. cinerea* with near-neutral electrolyzed water (10, 25, 50, 75, 100 ppm TRC; pH 6.3–6.5; 10 min contact time) in pure culture resulted in a 6 $\log_{10}$ spores/ml reduction and 100% inactivation. Electrolyzed water is effective in preventing gray rot of peaches due to the germination of *B. cinerea* (Venturini 2013). In a recent study, 13 salt solutions were used to generate alkaline and acidic electrolyzed water and demonstrated a general reduction of gray mold development on 'Crimson Seedless' table grape. Potassium sorbate, sodium carbonate, and sodium metabisulfite were the most effective salts to generate the electrolyzed water under natural infection as they showed higher activity against *Botrytis* mold (Youssef et al. 2018).

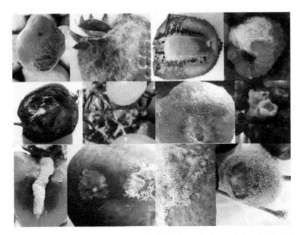

Figure 7.2. Gray mold development on some fruits. From left to right, in the first row: quince, strawberry, kiwi and raspberry. Second row: baby kiwi, table grape, pomegranate and blueberry. Third row: persimmon, peach (infection on the left), orange and sweet cherry (reprinted from Romanazzi et al. 2016).

Electrolyzed water generated by the electrolysis of salt solution through an electrolytic cell showed strong antifungal activity against postharvest disease such as brown rot of peaches caused by *Monolinia fructicola* (Guentzel et al. 2010). The fungicidal effect of electrolyzed oxidizing water on peach was studied. Electrolyzed oxidizing water did not control brown decay in wound-inoculated fruits, but reduced decay incidence and severity in non-wound-inoculated peach (Al-Haq et al. 2001). Spores of *Penicillium expansum*, the essential organism causing patulin in apple juice, were exposed to electrolyzed oxidizing (EO) water in a watery suspension and on injured apples. Full-strength and 50% EO water diminished suitable spore populations by more than 4 and 2 log units, respectively. EO water has potential as an option in contrast to chlorine disinfectants for controlling contamination of apples by *P. expansum* during handling and processing operations (Okull and Laborde 2004). The use of water buoyancy tanks during apple packing increases the danger of contamination of apples by spores of *P. expansum*, which may gather in the recycling water. Routine addition of sanitizers to the water may prevent such contamination. EO water and ClO_2 were effective against *P. expansum* spores. Nonionic surfactants could possibly be used with NaOCl to enhance control of *P. expansum* in buoyancy tanks, but the adequacy of such details ought to be approved under apple packing conditions (Okull et al. 2006). The impacts of electrolyzed salt solutions using thin-film diamond-coated electrodes on *Penicillium* spp. population in organic citrus wash water and on organic product rot were assessed (Fig. 7.3).

Among diverse organic and inorganic salts that were tried, electrolyzed water integrated with sodium bicarbonate proved to be the best treatment in hindering spore germination of *Penicillium* spp. and among the best in reducing *Penicillium* decay, with no deleterious impacts on the fruit (Fallanaj et al. 2013). Similarly, the effect of thin-film diamond-coated electrodes (DiaCell® 101) for the cleansing of water artificially contaminated with *P. digitatum* and *Pseudomonas* spp. was tried. The outcomes demonstrated that *Penicillium* spores and bacterial cells were influenced by flow rate and current thickness. The higher the water flow rate, the more significant the

Figure 7.3. Application of electrolyzed water in packing-house washing tank on citrus fruits.

inactivation of the two microorganisms, which were totally killed at high recirculation flow (Fallanaj et al. 2015).

Hussien et al. (2017) summarized the usefulness of electrolyzed water in *Penicillium* spp. inactivation, but further experiments using amended salt solutions are needed to improve the biocidal effect of electrolysis. In subsequent studies, the effectiveness of several salt solutions integrated with electrolyzed water was tested as sanitizing tools against *Penicillium digitatum*, *P. italicum*, and *P. ulaiense*. Different settings of the electrolyzing device were optimized to enhance sanitation. Results suggested that AEW was more effective than alkaline electrolyzed water in inhibiting growth of fungi and disease development of citrus mold (Hussien et al. 2018). The impacts of ultrasound and EO water on postharvest rot of pineapple cv. Phu Lae were explored using *Fusarium* sp. confined to pineapple fruit. The impact of EO water and ultrasound irradiation against *Fusarium* sp. *in vitro* was investigated. The examination demonstrated that all applications of EO water completely inhibited the spore germination and fungal growth. The potential of EO water in the mix with ultrasound irradiation in pineapple handling systems is high, because of marked synergistic impacts against fungal rot of harvested pineapple fruits (Khayankarn et al. 2013). Likewise, the effect of acidic electrolyzed oxidizing water to control *Fusarium* sp., which causes postharvest decay of pineapple cv. Phu Lae, was assessed. The development of growth inhibition of *Fusarium* sp. *in vitro* was seen after treatment with acidic electrolyzed oxidizing water with different doses (100, 200 and 300 ppm) of free chlorine (Whangchai et al. 2017). Hirayama et al. (2016) applied neutral electrolyzed water through an overhead irrigation system to control strawberry anthracnose caused by *Colletotrichum fructicola*.

Fruit preservation

Disinfecting agents (ethanol, acetic acid, EO water) are used for fruit surface sterilization, mainly when the process of washing is included in postharvest fruit packaging. The efficacy of electrolyzed water in reducing microorganisms including the main food-borne pathogens associated with fresh-cut fruit and vegetables has been reported (Rahman et al. 2011, Gómez-López et al. 2013). The dipping of date fruits in EO water solution using the optimum condition (3% of anolyte for 4 min) significantly reduced the population of surface-associated microorganisms (total count of bacteria, yeast and molds). The treatment has no effect on the sensory descriptors tested: odor, taste, color, and texture (Bessi et al. 2014). The effect of chlorine dioxide electrolyzed water and Berry Very® on blueberry quality during the postharvest storage period was evaluated. Electrolyzed water showed the best results in terms of lower weight loss, postharvest decay and microbial counts, while Berry Very® negatively affected berry color, acidity level and firmness (Chiabrando et al. 2017). Abadías et al. (2008) indicated that diluted neutral electrolyzed water (50 mg/L free chlorine) has a bactericidal power against *E. coli, Salmonella, Listeria innocua* and *Erwinia carotovara* on fresh-cut lettuce, carrot, endive, corn salad, and four-season salad.

Recently, Youssef et al. (2018) concluded that alkaline and acidic electrolyzed water did not influence weight reduction of table grape. It should be noted that the influence of treatments on fruit quality is often ignored since laboratory-scale tests tend to focus on the effectiveness of a treatment to control decay and do not sufficiently take into account the final quality of the product, which is necessary for a potential commercial application. Concerning total soluble solids and titratable acidity, no statistical differences were found among all treatments as compared to the water control with a minor exception. For grape color, in all cases, no significant differences were observed between treatments as compared to the water control except for AEW generated by ammonium molybdate when applied before harvest.

Food-borne bacteria

The viability of EO and acidified chlorinated water was tested in inactivation of *Escherichia coli* O157 H7 and *Listeria monocytogenes* on lettuce. The distinction between the bactericidal action of EO and acidified chlorinated waters was not huge (Park et al. 2001). The impact of electrolyzed water on cell suspensions of pathogenic bacteria was studied. In particular, the effect of strong and weak acidic electrolyzed waters as well as strong and weak alkaline electrolyzed waters on *Vibrio parahaemolyticus, Listeria monocytogenes, Aeromonas hydrophila, Campylobacter jejuni,* and *Escherichia coli* O104 H4 was investigated (Ovissipour et al. 2015). Different fruit washing techniques using tap water, electrolyzed water and rhamnolipid solution produced by *Pseudomonas aeruginosa* LBI were tested so as to repress microbial development on *Eugenia uniflora*. Fruits were washed and then periodically inoculated with microorganisms. It was demonstrated that treatment with rhamnolipids was the most effective, inhibiting the development of fungi and bacteria. The electrolyzed water showed efficient inhibition of bacteria at the initial time, but did not present any inhibitory effect at the final time (Dilarri et al. 2016).

Decontamination of meat microbiota

Meat is very important as a source of vitamins, minerals, and protein that complements a healthy diet. Red and white meats are among the most common types of meat consumed worldwide. Wrong handling of raw meat during slaughtering leads to contamination with *Listeria monocytogenes, Salmonella* sp. or *Escherichia coli,* resulting in food-borne infections (Priyanka et al. 2016). Previous studies have shown that electrolyzed water can be used as a decontaminant agent on eggshells (Fasenko et al. 2009), broiler carcasses (Northcutt et al. 2007), salmon fillets and broilers (Ozer and Demirci 2006, Miks-Krajnik et al. 2017), and raw frozen shrimps (Ye et al. 2014). Arya et al. (2018) summarized that beef, chevon, and pork samples treated with AEW resulted in the highest log reductions of microorganisms: 1.16 (after 4 min), 1.22 (after 12 min), and 1.30 $\log_{10}$ colony forming units (CFU) per ml (after 10 min), respectively. Also, alkaline electrolyzed water treatments resulted in 1.61, 0.96, and 1.52 $\log_{10}$ CFU/mL microbial reductions after 12 min treatment time, respectively. Similarly, Ye et al. (2014) reported that slightly acidic electrolyzed water spray treatment reduced the coliform population on raw frozen shrimps by 0.24–0.72 $\log_{10}$ CFU/g after 3–5 min.

It was reported that *Salmonella* sp., *Listeria* sp., *Campylobacter* sp., *Escherichia coli,* and *Staphylococcus aureus* can cause food poisoning in chicken meat (Kitai et al. 2005, Hong et al. 2008a, Hong et al. 2008b). Combined treatment with alkaline electrolyzed water and mild heat process reduced the *E. coli* O157 H7 and *Salmonella* populations (Koseki and Isobe 2007), while treatment with alkaline electrolyzed water and strongly acidic electrolyzed water significantly reduced *S. enteritidis*, *E. coli*, and *S. aureus*, and reduction of more than 1 log CFU/g was achieved by Shimamura et al. (2016). Low-concentration AEW significantly reduced the natural food-borne microbiota and prolonged the shelf life of fresh meat (Brychcy et al. 2015).

Mode of action of electrolyzed water

The mode of action of electrolyzed water is not fully understood and no comprehensive studies have been carried out yet to better understand its mechanism of action. Electrolyzed water destroys pathogen structures by oxidizing nucleic acids and proteins, without the development of resistance (Al-Haq et al. 2005). As mentioned previously, several studies have assessed the effect of the antimicrobial activity of slightly acidic electrolyzed water against various food-borne pathogens (Wang et al. 2014a,b). Meanwhile, many studies have demonstrated the potential use of slightly acidic electrolyzed water as an alternative sanitizer to reduce microbial contamination on vegetables (Forghani and Oh 2013). However, the underlying inactivation mechanism remains unknown. Slightly acidic electrolyzed water showed an effective bactericidal effect on the *E. coli* which is associated with the physiological and biological changes of apoptosis and necrosis induced by slightly acidic electrolyzed water. Meanwhile, reactive oxygen species (ROS) were formed in the bacterial cells. Finally, *E. coli* inactivation and apoptosis induced by slightly acidic electrolyzed water were observed. These findings illustrate that the bactericidal effect of slightly acidic electrolyzed water against *E. coli* occurred through cellular and biochemical mechanisms of cell necrosis and apoptosis (Ye et al. 2017). Jeong et al. (2006) examined the role of ROS in electrochemical sterilization, and they found that ‾OH was the major destructive species in charge of *E. coli* inactivation in the chloride free electrochemical sanitization process. Electrolysis was performed with the addition of sodium chloride as an electrolyte with a consequent formation of free chlorine and chlorinated organic compounds like chloramines, dichloramines, and trichloromethanes, creating negative effects for handlers and consumers. Free chlorine is also quickly inactivated by the heavy inorganic load present in the wash water of profitable packing houses; therefore, the use in the electrolysis reaction of salts without chlorine might be particularly interesting (Fallanaj et al. 2013, Youssef and Hussien 2020).

Principles of engineered water nanostructures

Pathogenic microorganisms can pollute fruits and vegetables through their broad production chain from field, to transport, to food industry, and home use. Current disinfection strategies include chlorine/hypochlorite, chlorine dioxide, peracetic acid, hydrogen peroxide, quaternary ammonium salts, ozone, and UV light. Another innovation shows inactivation of microorganisms on fresh produce and food production surfaces by water nanostructures produced by electrospraying water vapor (Glaser 2015). Engineered water nanostructure (EWNS) technique depends on production of EWNS using a stage that joins two procedures, namely, specific electrospraying and ionization of water. EWNS blended using electrospray and ionization of water, appear to be a powerful, green, antimicrobial agent for surface and air sterilization, where ROS are created and typified inside the particles during synthesis. We demonstrate that the EWNS produced by electrospraying of water vapor can interact with and inactivate airborne mycobacteria, significantly reducing their concentration.

EWNS synthesis

EWNS are formed by an electrospray device that collects water by condensing atmospheric water vapor on a Peltier-cooled electrode in a procedure called electrospraying. The electrospray device comprises a gold-plated electrode that is chilled by a cooling component to gather water from the climate on its microsize tip, and a concentric cathode floating 5 mm beneath. When initiated with high voltage, the condensed water is electrically charged and drawn toward the ground terminal. EWNS have unique properties: they are 25 nm across, stay airborne in indoor conditions for a considerable length of time, contain ROS, and have solid surface charge (by and large 10e/structure) (Pyrgiotakis et al. 2015).

Physicochemical characterization of EWNS

Surface charge

The yield of the EWNS generation framework was associated directly to a Scanning Mobility Particle Sizer (SMPS, Model 3936, TSI, Shoreview, MN) to gauge the molecule number fixation, and in parallel to a Faraday Aerosol electrometer (TSI, Model 3068B, Shoreview, MN) used to quantify the vaporized current, as depicted in the past distribution (Pyrgiotakis et al. 2012).

Size and lifetime

The atomic force microscope can be used for estimation of EWNS size. Pyrgiotakis et al. (2014b) performed measurements of the size and lifetime of EWNS by using AFM (Asylum MFP-3D and the AC240T probes). The AFM scan rate was 1 Hz and the scanned area 5 μm × 5 μm with 256 scan lines.

Toxicity

Reactive oxygen species can cause oxidative stress in cells and is a significant cause of damage to metal nanoparticles (Tao et al. 2003, Lanone et al. 2009, Sotiriou et al. 2012). When EWNS connect with the alveolar or airway fluid, the ROS are inactivated by natural atoms before they interact with epithelial cells situated beneath the covering fluid. However, EWNS contain ROS that cannot cause any lung damage or inflammation (Pyrgiotakis et al. 2014a). From these results it can be concluded that either EWNS have no injurious impact on the respiratory system or that the inhaled amount of EWNS in the nose and lung was too low to induce a reaction under initial intensive conditions. This is a chemical-free and toxicologically productive strategy, which can be used to diminish the danger of airborne infectious diseases in a wide range of hosts. Further, acute inhalation studies showed that EWNS at high dose showed no lung injury or inflammation (Pyrgiotakis et al. 2014b).

EWNS applications

In the last two decades, nanotechnology has shown that it can enhance our arsenal of methods in the battle against microorganisms that cause disease and spoilage. Indeed, nanotechnology-based approaches, such as antimicrobial food surfaces, nano-enabled sensors, active/intelligent packaging, and novel disinfection platforms, are finding applications within the agri/food/feed sector, bringing great new opportunities to the food industry (Eleftheriadou et al. 2017). The Harvard Center for Nanotechnology and Nanotoxicology built a novel strategy for inactivating microorganisms found in the air and on surfaces or fomites. This technique depends on generating EWNS through electrospraying of water. These EWNS have appeared to adequately inactivate a wide range of food-related microorganisms on food surfaces and food preparation surfaces and in the air (Pyrgiotakis et al. 2012, Pyrgiotakis et al. 2014a, Pyrgiotakis et al. 2014b, Pyrgiotakis et al. 2015, Pyrgiotakis et al. 2016).

The EWNS were successful in inactivation of three distinct classes of microorganisms significant in the food industry: *E. coli, S. enterica* and *L. innocua*. The inactivation potential was increased when the EWNS were applied on surfaces using electrostatic precipitator-based presentation framework. More important, it was evident that use of expanded EWNS resulted in higher inactivation potential. The results make this technique a promising innovation in the battle against food-borne disease (Pyrgiotakis et al. 2015). The EWNS framework was then streamlined to deliver higher concentrations of particles that were focused to be deposited directly on the surface of inoculated organic grape tomatoes (Pyrgiotakis et al. 2016). In this upgraded system, average aerosol doses reached up to 40,000 particles/cm^3. Particles were directed at contaminated tomatoes with an electrostatic precipitator presentation framework. After 45 min exposure, decreases of 2.2, 3.8 and 3.8 log were observed for *S. enterica, L. innocua* and *E. coli* respectively. For comparison, conventional medications, for example, chlorine or peroxyacetic corrosive washes, can accomplish 1 to 3 log reduction of *E. coli* O157 H7, *Listeria* sp. and *Salmonella* spp. within 3 min (Neo et al. 2013). These outcomes from the streamlined EWNS technique are a promising approach for the use of highly charged, nano-size water beads as an elective strategy for food sanitizing.

Recently, EWNS were modified by integrating an additional electrolysis step in their synthesis. Indeed, the resulting eEWNS from the integrated electrolysis electrospray-ionization process, as proven by studies of *E. coli* inactivation using inoculated stainless steel strips, caused 4 log reduction of microorganisms after 45 min of exposure time as compared to 1.9 log reduction for the same length of treatment with EWNS. These results clearly showed that the electrolysis step doubled the inactivation, as evidenced also by the increase in the total ROS content of eEWNS (Vaze et al. 2018). Various nature-inspired antimicrobials, for example, H_2O_2, lysozyme, and citric acid, and their integration, were used to combine an assortment of iEWNS-based nano-sanitizers. Tests for inactivation of food-borne pathogens showed synergistic effects of active ingredient and ROS integrated in the iEWNS structure; a pico- to nanogram-level amount of the active ingredient delivered to the surface using this nano-carrier platform is able to achieve a 5 log decrease of exposure time (Vaze et al. 2019).

EWNS mode of action

The EWNS mechanism of inactivation using a range of methods was explored. The TEM images showed that the control cells (unexposed) appeared normal and had an intact internal structure and cell membrane, whereas the cells exposed to EWNS appeared to have their cell membrane damaged (Pyrgiotakis et al. 2016). Additionally, studies confirmed the EWNS capacity to obliterate the EWNS layer (Pyrgiotakis et al. 2014a). Further, the molecular mechanism of inactivation was assessed using a lipid peroxidation assay (Fig. 7.4).

The bacteria exposed to the EWNS showed a high concentration of lipid peroxide as a result of the lipid membrane oxidation from the ROS. However, when vitamin C, a known antioxidant, was added to the bacteria the lipid peroxidation was not observed (Pyrgiotakis et al. 2014b). This indicates that the ROS presence is one of the primary mechanisms of inactivation. In addition, electron spin resonance showed that the EWNS contain a huge number of ROS, primarily hydroxyl and superoxide radicals (Pyrgiotakis et al. 2014a). It is clear, therefore, that the increased ROS concentration of the optimized EWNS makes a significant contribution to the higher

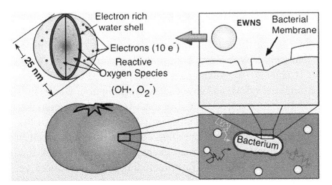

Figure 7.4. Assessment of the efficacy of EWNS produced by electrospraying of water vapor in inactivation of representative food-borne bacteria *Escherichia coli, Salmonella enterica,* and *Listeria innocua* (reprinted from Pyrgiotakis et al. 2015).

observed inactivation rates as compared to the previously reported baseline EWNS. It has been shown that high levels of ROS can cause stress conditions in Gram-negative microorganisms by augmenting irreversible damage to cellular components due to the reduction of cell stability and modification of proteins (Baez and Shiloach 2013). All the control cells seemed ordinary and had an unblemished inner structure and cell film, while the cells exposed to EWNS seemed to have their cell layer damaged. The three pathogens—*E. coli*, *S. enterica*, and *L. innocua*—presented to EWNS showed split and burst films when compared with the control cells, which had intact layers, plainly defined periplasmic space, and intercellular structure. The destruction of the membrane lipid caused by the presence of ROS was evaluated with lipid peroxidation measures, recognizing the ROS as the essential mechanism of inactivation (Pyrgiotakis et al. 2014a, b). One possible mode of bacterial inactivation is ROS-interceded inactivation because of the presence of ROS in the EWNS. EWNS contain gigantic payloads of ROS. ROS are known to attack cell layers, damage DNA, oxidize enzymes and degrade different proteins important for cell or microbial survival. If a droplet happens to bounce into a microbial pathogen, EWNS act like "nanobombs," releasing ROS that can do severe damage and effectively kill the pathogen (Cohen 2016). TEM imaging showed that the inactivation of the pathogen is due to the destruction of the cell membrane (Pyrgiotakis et al. 2016).

Advantages and limitations

Although EWNS reliably kills microbial pathogens, this novel, chemical-free and environment-friendly technique holds incredible potential as a green option in contrast to inactivation strategies at present used in the food industry. Using an optimized electrospray technique, nano-size beads are shaped from water vapor already present in the air. There are no residues to worry about, and in a couple of hours everything dissipates once again into water vapor (Cohen 2016). To address concerns about potential human health impacts from taking in large amounts of the EWNS, Demokritou and team used an animal model to explore lung damage following inhalation. Mice were exposed to EWNS for 4 h at levels 50% higher than the level used for attacking microbial pathogens (60,000 particles/cm^3). The mice showed no respiratory tract toxicity, no expansion in lung inflammation markers, no increment in cell death and no changes in breathing patterns (Pyrgiotakis et al. 2014a). In fact, they proved to be entirely unaffected by the EWNSs they were breathing. The researchers hypothesized that the alveolar lining fluid in the mouse lungs killed the nanobombs before they could come into contact with living cells, although more research may be conducted to affirm this (Cohen 2016). Advantages of the proposed control innovation include convenience, minimal effort, and low energy consumption. Generally speaking, this is a synthetic-free, reasonable, and environment-friendly innovation that can possibly decrease the danger of disease transmission (Pyrgiotakis et al. 2014b). Demokritou and colleagues suggest that electrospraying could be used as a practical method that can be effectively employed at various intensities of intervention. EWNS could be connected to deliver from farm to fork, including postharvest on a vehicle or even as an undetectable fog in the basic supply walkway. Research demonstrates EWNS are suitable for reducing basic food-borne pathogens like *E. coli*, *S. enterica*, and *L. innocua*, as well as yeasts and organisms that cause food to spoil. Such a flexible innovation could help fundamentally decrease the microbial burden on fresh produce and extend its shelf-life, as well as reduce the incidence of food-borne diseases from consumption (Cohen 2016).

Concluding remarks and future challenges

Nanotechnology is a rapidly growing field in the agri/food/feed industries. The use of electrolyzed water for treating plant infections or in aquaculture for cleaning and treating pathogenic microorganisms can provide pesticide-free products for human consumption. Besides, the impact of electrolyzed water on wound healing and disinfection of injuries has been recognized. In addition, the nano-size water may offer a safe, cost-effective approach to killing food-borne pathogens. Engineered water nanostructures can be integrated as a constant sterilization methodology, for example, in commercial centers or stockrooms. Further, since no residues remain, it is a suitable innovation for organic produce. The engineered water nanostructures generation framework will in future be scaled up and larger-scale antimicrobial investigations involving more products and pathogens will be conducted. The innovators suggest that this chemical-free and eco-friendly intervention strategy offers the potential for improvement and application in the food industry as a green option in contrast to existing sterilization strategies. There is still no focus on applying acidic electrolyzed water and engineered water nanostructures for eliminating air- or food-borne pathogen biofilms. The effects of these nanomaterials on human and ecological health should be examined further to guarantee our food security. Finally, application of electrolyzed water combined with a salt solution can be used as an environment-friendly alternative to conventional sanitizing agents and reduce the residues of such sanitizing agents and fungicides used in processing of fresh fruits and vegetables.

Acknowledgments

The present work was supported by the Science and Technology Development Fund (STDF), Egypt (grant no. 5555 Basic & Applied).

References

Abadías, M., Usall, J., Oliveira, M., Alegre, I., Viñas, I. 2008. Efficacy of neutral electrolyzed water (NEW) for reducing microbial contamination on minimally-processed vegetables. Int. J. Food Microbiol. 123(1-2): 151–158.

Abbasi, P.A., Lazarovits, G. 2006. Effect of acidic electrolyzed water on the viability of bacterial and fungal plant pathogens and on bacterial spot disease of tomato. Can. J. Microbiol. 52: 915–923.

Al-Haq, M.I., Seo, Y., Oshita, S., Kawagoe, Y. 2001. Fungicidal effectiveness of electrolyzed oxidizing water on postharvest brown rot of peach. Hortscience 36: 1310–1314.

Al-Haq, M.I., Sugiyama, J., Isobe, S. 2005. Applications of electrolyzed water in agriculture and food industries. Food Sci. Technol. Res. 11: 135–150.

Arya, R., Bryant, M., Degala, H.L., Mahapatra, A.K., Kannan, G. 2018. Effectiveness of a low-cost household electrolyzed water generator in reducing the populations of *Escherichia coli* K12 on inoculated beef, chevon, and pork surfaces. J. Food Process Preserv. e13636. https://doi.org/10.1111/jfpp.13636.

Baez, A., Shiloach, J. 2013. *Escherichia coli* avoids high dissolved oxygen stress by activation of SoxRS and manganese-superoxide dismutase. Microb. Cell Fact. 12: 23.

Bessi, H., Debbabi, H., Grissa, K., Bellagha, S. 2014. Microbial reduction and quality of stored date fruits treated by electrolyzed water. J. Food Qual. 37(1): 42–49.

Brychcy, E., Malik, M., Drozdzewski, P., Ulbin-Figlewicz, N., Jarmoluk, A. 2015. Low-concentrated acidic electrolysed water treatment of pork: inactivation of surface microbiota and changes in product quality. Int. J. Food Sci. Technol. 50: 2340–2350.

Buck, J.W., Van Iersel, M.W., Oetting, R.D. Hung, C.Y. 2002. *In vitro* fungicidal activity of acidic electrolyzed oxidizing water. Plant Dis. 86: 278–281.

Cohen, J.M. 2016. "Nanobombs" protect consumers from foodborne microorganisms. Food Safety Magazine. https://www.foodsafetymagazine.com/magazine-archive1/junejuly2016/e2809cnanobombse2809d-protect-consumers-from-foodborne-microorganisms/.

Chiabrando, V., Peano, C., Giacalone, G. 2017. The efficacy of different postharvest treatments on physico-chemical characteristics, bioactive components and microbiological quality of fresh blueberries during storage period. Food Res. 1(6): 240–248.

Dilarri, G., Luiz Da Silva, V., Pecdra, H.B., Mdntagndlli, R.N., Cdrsd, C.R., Biddia, E.D. 2016. Electrolytic treatment and biosurfactants applied to the conservation of *Eugenia uniflora* fruit. Food Sci. Technol. 36: 456–460.

Ding, T., Ge, Z., Shi, J., Xu, Y-T., Jones, C.L., Liu, D-H. 2015. Impact of slightly acidic electrolyzed water (SAEW) and ultrasound on microbial loads and quality of fresh fruits. LWT-Food Sci. Technol. 60(2): 1195–9.

Ding, T., Xuan, X.T., Li, J., Chen, S.G., Liu, D.H., Ye, X.Q., Shic, J., Xue, S.J. 2016. Disinfection efficacy and mechanism of slightly acidic electrolyzed water on *Staphylococcus aureus* in pure culture. Food Cont. 60: 505–510.

Eleftheriadou, M., Pyrgiotakis, G., Demokritou, P. 2017. Nanotechnology to the rescue: Using nano-enabled approaches in microbiological food safety and quality. Curr. Opin. Biotechnol. 44: 87–93.

Fallanaj, F., Ippolito, A., Ligorio, A., Garganese, F., Zavanella, C., Sanzani, S.M. 2016. Electrolyzed sodium bicarbonate inhibits *Penicillium digitatum* and induces defence responses against green mould in citrus fruit. Postharvest Biol. Technol. 115: 18–29.

Fallanaj, F., Sanzani, S.M., Youssef, K., Zavanella, C., Salerno, M.G., Ippolito, A. 2015. A new perspective in controlling postharvest citrus rots: the use of electrolyzed water. Acta Hort. 1065: 1599–1606.

Fallanaj, F., Sanzani, S.M., Zavanella, C., Ippolito, A. 2013. Salts addition to improve the control of citrus postharvest diseases using electrolysis with conductive diamond electrodes. J. Plant Pathol. 95: 373–383.

Fasenko, G.M., O'dea Christopher, E.E., Mcmullen, L.M. 2009. Spraying hatching eggs with electrolyzed oxidizing water reduces eggshell microbial load without compromising broiler production parameters. Poultry Sci. 88(5): 1121–1127.

Fernández, B.R., Castillo, B., Díaz, E., Camacho, J.H. 2011. Evaluating electrolysed oxidising water as a fungicide using two rose varieties (*Rosa* sp.) in greenhouse conditions. Ingeniería e Investigación 31(2): 91–101.

Forghani, F., Oh, D.H. 2013. Hurdle enhancement of slightly acidic electrolyzed water antimicrobial efficacy on Chinese cabbage, lettuce, sesame leaf and spinach using ultrasonication and water wash. Food Microbiol. 36: 40–45.

Fujiwara, K., Fujii, T., Park, J.-S. 2009. Comparison of foliar spray efficacy of electrolyzed ozonated water and acid electrolyzed oxidizing water for controlling powdery mildew infection on cucumber leaves. Ozone: Sci. Engin. 31: 10–14.

Glaser, J.A. 2015. Nanoscale food protection. Clean Techn. Environ. Policy. 17: 827–832.

Gómez-López, V.M., Marín, A., Medina-Martínez, M.S., Gil, M.I., Allende, A. 2013. Generation of trihalomethanes with chlorine-based sanitizers and impact on microbial: nutritional and sensory quality of baby spinach. Postharvest Biol. Technol. 85: 210–217.

Guentzel, L.L., Lam, K.L., Callan, M.A., Emmons, S.A., Dunham, V.L. 2010. Postharvest management of gray mold and brown rot on surfaces of peaches and grapes using electrolyzed oxidizing water. Int. J. Food Microbiol. 143: 54–60.

Haihua, T.U., Naixiong, T.A.N.G., Nianqian, K.A.N.G., Jian, Z.H.O.U., Aihong, L.I.N. 2016. The application of acidic electrolyzed water to grape cultivation in the southern regions. Asian Agric. Res. 8(7): 1–5.

Han, Q., Song, X., Zhang, Z., Fu, J., Wang, X., Malakar, P.K., Liu, H., Pan, Y., Zhao, Y. 2017. Removal of foodborne pathogen biofilms by acidic electrolyzed water. Front Microbiol. 8: 988.

Hao, J., Li, H., Wan, Y., Liu, H. 2015. Combined effect of acidic electrolyzed water (AcEW) and alkaline electrolyzed water (AlEW) on the microbial reduction of fresh-cut cilantro. Food Cont. 50: 699–704.

Hirayama, Y., Asano, S., Watanabe, K., Sakamoto, Y., Ozaki, M., Ohki, S.T., Tojo, M. 2016. Control of *Colletotrichum fructicola* on strawberry with a foliar spray of neutral electrolyzed water through an overhead irrigation system. J. Gen. Plant Pathol. 82(4): 186–189.

Hong, Y., Ku, K., Kim, M., Won, M., Chung, K., Song, K.B. 2008b. Survival of *Escherichia coli* O157:H7 and *Salmonella typhimurium* inoculated on chicken by aqueous chlorine dioxide treatment. J. Microbiol. Biotechnol. 18: 742–745.

Hong, Y.H., Ku, G.J., Kim, M.K., Song, K.B. 2008a. Effect of aqueous chlorine dioxide treatment on the microbial growth and quality of chicken legs during storage. Int. J. Food Sci. Nutr. 13: 45–50.

Hou, Y.T., Ren, J., Liu, H.J., Li, F.D. 2012. Efficiency of electrolyzed water (EW) on inhibition of *Phytophthora parasitica* var. *nicotianae* growth *in vitro*. Crop Protect. 42: 128–133.

Hricova, D., Stephan, R., Zweifel, C. 2008. Electrolyzed water and its application in the food industry. J. Food Prot. 71: 1934–1947.

Hsu, S.Y. 2003. Effects of water flow rate, salt concentration and water temperature on efficiency of an electrolyzed oxidizing water generator. J. Food Eng. 60: 469–473.

Huang, Y.R., Hung, Y.C., Hsu, Sh.Y., Huang, Y.W., Hwang, D.F. 2008. Application of electrolyzed water in the food industry. Food Cont. 19: 329–345.

Hussien, A., Ahmed, Y., Al-Essawy, A.H., Youssef, K. 2018. Evaluation of different salt-amended electrolysed water to control postharvest moulds of citrus. Tropical Plant Pathol. 43(1): 10–20.

Hussien, A., Al-Essawy, A., Abo Rehab, M., Youssef, K. 2017. Preliminary investigation of alkaline and acidic electrolysed water to control *Penicillium* species of citrus. Citrus Res. Technol. 38(2): 175–183.

Jeong, J., Kim, J.Y., Yoon, J. 2006. The role of reactive oxygen species in the electrochemical inactivation of microorganisms. Environ. Sci. Technol. 40: 6117–6122.

Jia, G.L., Shi, J.Y., Song, Z.H., Li, F.D. 2015. Prevention of enzymatic browning of Chinese yam (*Dioscorea* spp.) using electrolyzed oxidizing water. J. Food Sci. 80: C718–C728.

Khayankarn, S., Jarintorn, S., Srijumpa, N., Uthaibutra, J., Whangchai, K. 2014. Control of *Fusarium* sp. on pineapple by megasonic cleaning with electrolysed oxidising water. MAEJO Int. J. Sci. Technol. 8: 288–296.

Khayankarn, S., Uthaibutra, J., Setha, S., Whangchai, K. 2013. Using electrolyzed oxidizing water combined with an ultrasonic wave on the postharvest diseases control of pineapple fruit cv. 'Phu Lae'. Crop Protect. 54: 43–47.

Kim, C., Hung, Y.C., Brackett, R.E. 2000. Efficacy of electrolyzed oxidizing (EO) and chemically modified water on different types of foodborne pathogens. Int. J. Food Microbiol. 61: 199–207.

Kitai, S., Shimizu, A., Kawano, J., Sato, E., Nakano, C., Kitagawa, H., Fujio, K., Matsumura, K., Yasuda, R., Inamoto, T. 2005. Prevalence and characterization of *Staphylococcus aureus* and enterotoxigenic *Staphylococcus aureus* in retail raw chicken meat throughout Japan. J. Vet. Med. Sci. 67: 269–274.

Kiura, H., Sano, K., Morimatsu, S., Nakano, T., Morita, C., Yamaguchi, M., Maeda, T., Katsuoka, Y. 2002. Bactericidal activity of electrolyzed acid water from solution containing sodium chloride at low concentration, in comparison with that at high concentration. J. Microbiol. Methods 49: 285–293.

Koseki, S., Isobe, S. 2007. Microbial control of fresh produce using electrolyzed water. Jpn. Agric. Res. Q. 41: 273–282.

Kusakari, S., Achiwa, N., Abe, K., Okada, K. 2013. The control effect of acidic electrolyzed water (AEW) on gray mold of strawberry and anthracnose of cucumber in field conditions (in Japanese with English abstract). Ann. Rep. Kansai. Pl Prot. 55: 17–21.

Lanone, S., Rogerieux, F., Geys, J., Dupont, A., Maillot-Marechal, E., Boczkowski, J., Lacroix, G., Hoet, P. 2009. Comparative toxicity of 24 manufactured nanoparticles in human alveolar epithelial and macrophage cell lines. Part Fibre Toxicol. 6: 14.

Miks-Krajnik, M., Feng, L.X.J., Bang, W.S., Yuk, H. 2017. Inactivation of *Listeria monocytogenes* and natural microbiota on raw salmon fillets using acidic electrolyzed water, ultraviolet light or/ and ultrasounds. Food Cont. 74: 54–60.

Mueller, D.S., Hung, Y.-C., Oetting, R.D., van Iersel, M.W., Buck, J.W. 2003. Evaluation of electrolyzed oxidizing water for management of powdery mildew on gerbera daisy. Plant Dis. 87: 965–969.

Neo, S.Y., Lim, P.Y., Phua, L.K., Khoo, G.H., Kim, S.J., Lee, S.C., Yuk, H.G. 2013. Efficacy of chlorine and peroxyacetic acid on reduction of natural microflora, *Escherichia coli* O157: H7, *Listeria monocyotgenes* and *Salmonella* spp. on mung bean sprouts. Food Microbiol. 36(2): 475–480.

Northcutt, J., Smith, D., Ingram, K.D., Hinton, A., Jr., Musgrove, M. 2007. Recovery of bacteria from broiler carcasses after spray washing with acidified electrolyzed water or sodium hypochlorite solutions. Poultry Sci. 86(10): 2239–2244.

Okull, D.O., Laborde, L.F. 2004. Activity of electrolyzed oxidizing water against *Penicilium expansum* in suspension and on wounded apples. J. Food Sci. 69: 23–27.

Okull, D.O., Demirci, A., Rosenberger, D., Laborde, L.F. 2006. Susceptibility of *Penicillium expansum* spores to sodium hypochlorite, electrolyzed oxidizing water, and chlorine dioxide solutions modified with nonionic surfactants. J. Food Protect. 69: 1944–1948.

Ovissipour, M., Al-Qadiri, H.M., Sablani, S.S., Govindan, B.N., Al-Alami, N., Rasco, B. 2015. Efficacy of acidic and alkaline electrolyzed water for inactivating *Escherichia coli* O104:H4, *Listeria monocytogenes*, *Campylobacter jejuni*, *Aeromonas hydrophila*, and *Vibrio parahaemolyticus* in cell suspensions. Food Cont. 53: 117–123.

Ozer, N.P., Demirci, A. 2006. Electrolyzed oxidizing water treatment for decontamination of raw salmon inoculated with *Escherichia coli* O157:H7 and *Listeria monocytogenes* Scott A and response surface modeling. J. Food Engin. 72(3): 234–241.

Park, C.M., Hung, Y.C., Brackett, R.E. 2002. Antimicrobial effect of electrolyzed water for inactivating *Campylobacter jejuni* during poultry washing. Int. J. Food Microbiol. 72: 77–83.

Park, C.M., Hung, Y.C., Doyle, M.P., Ezeike, G.O.I., Kim, C. 2001. Pathogen reduction and quality of lettuce treated with electrolyzed oxidizing and acidified chlorinated water. J. Food Sci. 66: 1368–1372.

Park, E.J., Alexander, E., Taylor, G.A., Costa, R., Kang, D.H. 2008. Effect of electrolyzed water for reduction of foodborne pathogens on lettuce and spinach. J. Food Sci. 73(6): M268–72.

Priyanka, B., Patil, R.K., Dwarakanath, S. 2016. A review on detection methods used for foodborne pathogens. The Indian J. Med. Res. 144(3): 327–338.

Pyrgiotakis, G., McDevitt, J., Bordini, A., Diaz, E., Molina, R., Watson, C., Demokritou, P. 2014a. A chemical free, nanotechnology-based method for airborne bacterial inactivation using engineered water nanostructures. Environ. Sci.: Nano. 1(1): 15–26. https://doi.org/10.1039/C3EN00007A.

Pyrgiotakis, G., McDevitt, J., Gao, Y., Branco, A., Eleftheriadou, M., Lemos, B., Demokritou, P. 2014b. Mycobacteria inactivation using engineered water nanostructures (EWNS). Nanomed. Nanotechnol. Biol. Med. 10(6): 1175e1183. https://doi.org/10.1016/j.nano.2014.02.016.

Pyrgiotakis, G., McDevitt, J., Yamauchi, T., Demokritou, P. 2012. A novel method for bacterial inactivation using electrosprayed water nanostructures. J. Nanopart. Res. 14(8): 1027. https:// doi.org/10.1007/s11051-012-1027-x.

Pyrgiotakis, G., Vasanthakumar, A., Gao, Y., Eleftheriadou, M., Toledo, E., DeAraujo, A., Demokritou, P. 2015. Inactivation of foodborne microorganisms using engineered water nanostructures (EWNS). Environ. Sci. Technol. 49(6): 3737–3745. https://doi.org/10.1021/es505868a.

Pyrgiotakis, G., Vedantam, P., Cirenza, C., McDevitt, J., Eleftheriadou, M., Leonard, S.S. 2016. Optimization of a nanotechnology-based antimicrobial platform for food safety applications using engineered water nanostructures (EWNS). Sci. Rep. 6: 21073. https://doi.org/10.1038/ srep21073.

Rahman, S.M.E., Jin, Y.G., Oh, D.H. 2011. Combination treatment of alkaline electrolyzed water and citric acid with mild heat to ensure microbial safety shelf-life and sensory quality of shredded carrots. Food Microbiol. 28: 484–491.

Rahman, S.M.E., Khan, I., Oh, D. 2016. Electrolyzed water as a novel sanitizer in the food industry: current trends and future perspectives. Comprehensive Rev. Food Sci. Food Safety. 15: 471–490.

Romanazzi, G., Joseph, L.S., Erica, F., Droby, S. 2016. Integrated management of postharvest gray mold on fruit crops. Postharvest Biol. Technol. 113: 69–76.

Shimamura, Y., Shinke, M., Hiraishi, M., Tsuchiya, Y., Masuda, S. 2016. The application of alkaline and acidic electrolyzed water in the sterilization of chicken breasts and beef liver. Food Sci. Nutr. 4(3): 431–440.

Shimizu, Y., Hurusawa, T. 1992. Antiviral, antibacterial, and antifungal actions of electrolyzed oxidizing water through electrolysis. Dental J. 37: 1055–1062.

Sotiriou, G.A., Diaz, E., Long, M.S., Godleski, J., Brain, J., Pratsinis, S.E., Demokritou, P. 2012. A novel platform for pulmonary and cardiovascular toxicological characterization of inhaled engineered nanomaterials. Nanotoxicol. 6(6): 680–690.

Sun, J.L., Zhang, S.K., Chen, J.Y., Han, B.Z. 2012. Efficacy of acidic and basic electrolyzed water in eradicating *Staphylococcus aureus* biofilm. Can J. Microbiol. 58: 448–454.

Tao, F., Gonzalez-Flecha, B., Kobzik, L. 2003. Reactive oxygen species in pulmonary inflammation by ambient particulates. Free Radical Biol. Med. 35(4): 327–340.

Vaze, N., Jiang, Y., Mena, L., Zhang, Y., Bello, D., Leonard, S.S., Morris, A.M., Eleftheriadou, M., Pyrgiotakis, G., Demokritou, P. 2018. An integrated electrolysis–electrospray-ionization antimicrobial platform using engineered water nanostructures (EWNS) for food safety applications. Food Cont. 85: 151–160.

Vaze, N., Pyrgiotakis, G., Mena, L., Baumann, R., Demokritou, A., Ericsson, M., Zhang, Y., Bello, D., Eleftheriadou, M., Demokritou, P. 2019. A nano-carrier platform for the targeted delivery of nature-inspired antimicrobials using engineered water nanostructures for food safety applications. Food Cont. 96: 365–374.

Vázquez-Sánchez, D., Cabo, M.L., Ibusquiza, P.S., Rodríguez-Herrera, J.J. 2014. Biofilm-forming ability and resistance to industrial disinfectants of *Staphylococcus aureus* isolated from fishery products. Food Cont. 39: 8–16.

Venturini, M.C. 2013. Acqua elettrolizzata: tecnologia emergente del settore ortofrutticolo. Alim Bev. 4: 50–55.

Wang, J.J., Sun, W.S., Jin, M.T., Liu, H.Q., Zhang, W., Sun, X.H., Pan, Y.J., Zhao, Y. 2014a. Fate of *Vibrio parahaemolyticus* on shrimp after acidic electrolyzed water treatment. Int. J. Food Microbiol. 179: 50–56.

Wang, J.J., Zhang, Z.H., Li, J.B., Lin, T., Pan, Y.J., Zhao, Y. 2014b. Modeling *Vibrio parahaemolyticus* inactivation by acidic electrolyzed water on cooked shrimp using response surface methodology. Food Cont. 36: 273–279.

Whangchai, K., Khayankarn, S., Uthaibutra, J. 2017. Effect of acidic electrolyzed oxidizing water treatments on the control of postharvest disease and pathogenesis related protein production in pineapple fruit. J. Adv. Agricu. Technol. 4: 240–244.

Whangchai, K., Uthaibutra, J., Phiyanalinmat, S. 2013. Effects of NaCl concentration, electrolysis time, and electric potential on efficiency of electrolyzed oxidizing water on the mortality of *Penicillium digitatum* in suspension. Acta Hortic. 973: 193–198.

Yaseen, T., Ricelli, A., Albanese, P., Carboni, C., Ferri, V., D'Onghia, A. 2013. Use of electrolyzed water to improve fruit quality of some citrus species. Proceeding of Future IPM in Europe, Riva del Garda, Italy, pp. 244–244.

Ye, Z., Qi, F., Pei, L., Shen, Y., Zhu, S., He, D., Wu, X., Ruan, Y., He, J. 2014. Using slightly acidic electrolyzed water for inactivation and preservation of raw frozen shrimp (*Litopenaeus vannamei*) in the field processing. App. Engin. Agric. 30(6): 935–941.

Ye, Z., Wang, S., Chen, T., Gao, W., Zhu, S., He, J., Han, Z. 2017. Inactivation mechanism of *Escherichia coli* induced by slightly acidic electrolyzed water. Sci. Rep. 7(1): 6279.

Yoshida, K., Achiwa, N., Katayose, M. 2004. Application of electrolyzed water for food industry in Japan. Http://ift.confex.com/ift/2004/techprogram/paper_20983.htm.

Youssef, K., Mustafa, Z.M.M., Al-Essawy, A. 2018. Efficacy of alkaline and acidic electrolysed water generated by some salt solutions against gray mold of table grape: pre and postharvest applications. J. Phytopathol. Pest. Manag. 5(1): 1–21.

Youssef, K., Hussien, A. 2020. Electrolysed water and salt solutions can reduce green and blue molds while maintain the quality properties of 'Valencia' late oranges. Postharvest Biol. Technol. 159: 111025.

Zeng, X., Ye, G., Tang, W., Ouyang, T., Tian, L., Ni, Y., Li, P. 2011. Fungicidal efficiency of electrolyzed oxidizing water on *Candida albicans* and its biochemical mechanism. J. Biosci. Bioeng. 112: 86–91.

Chapter 8

Biogenic Synthesis of Nanoparticles and Their Role in the Management of Plant Pathogenic Fungi

Avinash P. Ingle,[1,] Aayushi Biswas,[2] Chhangte Vanlalveni,[3] Ralte Lalfakzuala,[3] Indarchand Gupta,[4,5] Pramod Ingle,[5] Lalthazuala Rokhum[2] and Mahendra Rai[5]*

Introduction

Nanotechnology is a central discipline of contemporary analysis comprising the creation, use and handling of particles ranging from 1 to 100 nm in size. Within this size, all the vital chemical, physical and biological properties of atoms and molecules and their corresponding mass change fundamentally (Kaviya and Viswanathan 2011, Khalil et al. 2013). Nanotechnology is rapidly finding applications in cosmetics, biomedical, health, food and feed, agriculture, energy science, electronics and other industries (Seqqat et al. 2019). The immense growth in these technologies has opened up many newer fields. Because of this wide range of applications, the synthesis of different nanoparticles is a topic of interest to the scientific community. Predominantly, nanoparticles are prepared by a range of chemical and physical methods that are quite expensive and possibly hazardous to the environment, involving chemicals that carry

[1] Department of Biotechnology, Engineering School of Lorena, University of Sao Paulo, Lorena, SP, Brazil.
[2] Department of Chemistry, National Institute of Technology Silchar, Silchar 788010, Assam, India.
[3] Department of Botany, Mizoram University, Aizawl Tanhril 796001, Mizoram, India.
[4] Department of Biotechnology, Institute of Sciences, Nipat Niranjan Nagar, Aurangabad, Maharashtra, India.
[5] Nanobiotechnology Laboratory, Department of Biotechnology, Sant Gadge Baba Amravati University, Amravati, Maharashtra, India.
* Corresponding author: ingleavinash14@gmail.com

innumerable biological risks (Kulkarni and Muddapur 2014). The development of eco-friendly experimental processes for the synthesis of nanoparticles is becoming a vital branch of technology (Kumar et al. 2014, Kumar et al. 2017).

In this context, various attempts have been made to develop green and eco-friendly synthesis of metal and metal oxide nanoparticles using bacteria, fungi, algae, plants, and other biological systems that are discussed in this chapter (Kratošová et al. 2018). The biogenic synthesis of nanoparticles has drawn attention because of its rapid, environment-friendly, non-toxic, cost-effective protocol and because it is a single-step method. The reduction and stabilization of metal ions into respective metal nanoparticles is generally achieved by a combination of biomolecules such as polysaccharides, amino acid, proteins, alkaloids, tannins, phenolics, saponins, terpenoids, and vitamins secreted or present in the extracts of that particular biological system (Vanaja and Annadurai 2012, Dauthal and Mukhopadhyay 2016). As discussed earlier, metal nanoparticles can be effectively used in a variety of applications in many sectors. One of the most important applications of nanoparticles is in the management of various pathogens. Here we have emphasized the use of nanoparticles in the control of plant pathogenic fungi.

Plant pathogenic fungi are responsible for the generation of a number of diseases in a variety of agricultural crops and reduce crop yield all over the world. Therefore, control of fungal diseases in economically important crops affects the economy around the globe. Various fungicides have been traditionally used under the umbrella of agrochemicals to control such plant pathogenic fungi. Moreover, it has been observed that application of such fungicides not only kills pathogens but also affects other beneficial flora and fauna in the field (Akther and Hemalatha 2019). Considering the possible harmful effects of synthetic fungicides, there is an urgent need to develop effective, safe and alternative methods to control plant pathogenic fungi. Hence, researchers are focusing on the development of potent and cost-effective antifungal agents. Various biogenic nanoparticles are found to be effective in the management of plant pathogenic fungi due to their potential antifungal properties (Hassan et al. 2018).

The present chapter focuses on recent advances in the biogenic synthesis of silver, zinc oxide, copper oxide, iron oxide, sulfur and other nanoparticles from a variety of biological systems. Special attention has been given to the application of these biogenic nanoparticles in the management of different plant pathogenic fungi.

Biogenic synthesis of nanoparticles

To date, various biological systems have been successfully exploited for the synthesis of a wide range of nanoparticles. These biological systems mainly include plants, bacteria, fungi, and algae.

Synthesis from plants

A vast range of flora and their respective parts has been used for the synthesis of different metal nanoparticles. Silver nanoparticles (AgNPs) have been preferably synthesized for antimicrobial applications. The green and rapid synthesis of spherical AgNPs with dimensions of 12–50 nm was reported using fruit extract of

Rubus glaucus Benth (Kumar et al. 2017). In another study, it was demonstrated that *Achillea millefolium* leaf extract can be used effectively as reducing and stabilizing agent in the synthesis of AgNPs (Khodadadi et al. 2017). *Marsilea quadrifolia* leaf extract has been exploited for the green synthesis of AgNPs with an average diameter of 22.5 nm (Maji et al. 2017). Further, there is a report of using natural gum obtained from *Acacia nilotica*, commonly known as gum arabic tree. The AgNPs synthesized were found to be in the range of 10–78 nm (Yeasmin et al. 2017). Veisi et al. (2018) demonstrated the biosynthesis of AgNPs using leaf extract of *Thymbra spicata* without additional stabilizers or surfactants. This approach was found to be rapid, single step, eco-friendly and non-toxic. In a similar study, a very rapid approach (only 15 min) was developed for the biogenic synthesis of AgNPs using leaf extract of *Azadirachta indica* at room temperature, without the involvement of any hazardous chemical (Ahmed et al. 2016). Similarly, the synthesis of AgNPs was reported using leaf extracts of *Trapanatans natans* (Saber et al. 2018).

Various other nanoparticles have been synthesized using different plants and their parts, primarily zinc oxide nanoparticles (ZnONPs). A green approach has been proposed where leaf extracts of *Moringa oleifera* and *Pelargonium hortorum* were used as reducing and stabilizing agent for the synthesis of these nanoparticles (Matinise et al. 2017, Rivera-Rangel et al. 2018). In synthesis of ZnONPs, use of plant extract has been found to be the exceptional method. Salam et al. (2014) also used *Ocimum basilicum* L. leaf extract for the synthesis of hexagonal ZnONPs of size 50 nm. Extracts of *Agathosma betulina* and *M. oleifera* leaves were used for the synthesis of ZnONP of various sizes by Thema et al. (2015) and Matinise et al. (2017), respectively.

Another kind of nanoparticles having significant impacts on agricultural applications are copper-based nanoparticles. Green engineering involved in the synthesis of copper oxide nanoparticles (CuONPs) is gaining considerable interest among researchers as an eco-friendly alternative to conventional physical and chemical strategies, as it eliminates the use of cyanogenetic chemicals. CuONPs were reported to be synthesized using leaf extract of *Acalypha indica*. The synthesized nanoparticles were extremely stable and spherical and were within the range of 26–30 nm in size (Sivaraj et al. 2014). Another study involved the use of leaf extract of *Gundelia tournefortii* (Nasrollahzadeh et al. 2015) and *Thymus vulgaris* L. (Nasrollahzadeh et al. 2016) for the biogenic synthesis of CuONPs. Similarly, Groiss et al. (2017) reported an eco-friendly and cost-effective approach for the synthesis of iron oxide nanoparticles using leaf extract of *Cynometra ramiflora*. Masum et al. (2019) demonstrated the synthesis of AgNPs from fruit extract of *Phyllanthus emblica*. The biosynthesis of various nanoparticles using a variety of plants is summarized in Table 8.1.

Bacterial synthesis

Synthesis of metallic nanoparticles by various microorganisms has attracted a great deal of attention. Different strains of bacteria have been successfully exploited for a variety of nanoparticles, e.g., *Bacillus flexus* (Priyadarshini et al. 2013) and *Escherichia coli* (Natarajan et al. 2010) have been reported to be possible candidates for the rapid synthesis of AgNPs. The process of biosorption and bioreduction offers necessary

Table 8.1. Details of nanoparticles synthesized from different plants.

S. N.	Plants	Nanoparticles	Size (nm)	Shape	Reference
1	*Rubus glaucus Benth.*	AgNPs	12–50	Spherical	Kumar et al. 2017
2	*Achillea millefolium*	AgNPs	19–29	Spherical	Khodadadi et al. 2017
3	*Marsilea quadrifolia*	AgNPs	9–42	Spherical	Maji et al. 2017
4	*Acacia nilotica*	AgNPs	10–78	Spherical	Yeasmin et al. 2017
5	*Pelargonium hortorum*	AgNPs	25–150	Various	Rivera-Rangel et al. 2018
6	*Thymbra spicata*	AgNPs	7	Spherical	Veisi et al. 2018
7	*Laminaria japonica*	AgNPs	31	Oval	Kim et al. 2018
8	*Syzygium aromaticum*	AgNPs	5–40	Spherical	Venugopal et al. 2017a
9	*Beta vulgaris*	AgNPs	5–20	Spherical, circular, triangular	Venugopal et al. 2017b
10	*Azadirachta indica*	AgNPs	34	Spherical	Ahmed et al. 2016
11	*Trapa natans*	AgNPs	30–90	Spherical	Saber et al. 2018
12	*Salvia leriifolia*	AgNPs	27	Spherical	Baghayeri et al. 2018
13	*Taraxacum officinale*	AgNPs	5–30	Spherical	Saratale et al. 2018
14	*Solanum viarum*	AgNPs	2–40	Spherical, oval	Biswas et al. 2018a
15	*Sellaginella bryopteris*	AgNPs	4–30	Spherical	Biswas et al. 2018b
16	*Croto caudatus Geisel*	AuNPs	20–50	Spherical	Vijaya-Kumar et al. 2019
17	*Cynometra ramiflora*	Fe_3O_4NPs	5–50	Spherical	Groiss et al. 2017
18	*Gundelia tournefortii*	CuONPs	100	Spherical	Nasrollahzadeh et al. 2015
19	*Thymus vulgaris*	CuONPs	30	Oval	Nasrollahzadeh et al. 2016
20	*Calotropis gigantea*	CuONPs	20–30	Spherical	Sharma et al. 2015b
21	*Aloe barbadensis*	CuONPs	15–30	Spherical	Gunalan et al. 2012a
22	*Acalypha indica*	CuONPs	26–30	Spherical	Sivaraj et al. 2014a
23	*Tabernaemontana divaricata*	CuONPs	48 ± 4	Spherical	Sivaraj et al. 2014b
24	*Gloriosa superba*	CuONPs	5–10	Spherical	Naika et al. 2015
25	*Moringa oleifera*	ZnONPs	2–20	Spherical	Matinise et al. 2017
26	*Moringa oleifera*	ZnONPs	16–31.9	Spherical	Matinise et al. 2017
27	*Ocimum basilicum*	ZnONPs	50	Hexagonal	Salam et al. 2014
28	*Agathosma betulina*	ZnONPs	15.8	Quasi-spherical	Thema et al. 2015

Table 8.1 contd. ...

... Table 8.1 contd.

S. N.	Plants	Nanoparticles	Size (nm)	Shape	Reference
29	*Trifolium pratense*	ZnONPs	≥ 100	Spherical	Dobrucka and Dugaszewska 2016
30	*Camellia sinensis*	ZnONPs	8 ± 0.5	Hexagonal	Nava et al. 2017
31	*Eucalyptus globulus*	ZnONPs	11.6	Spherical	Balaji and Kumar 2017
32	*Tabernaemontana divaricata*	ZnONPs	20–50	Spherical	Raja et al. 2018

insights for understanding the mechanisms underlying the synthesis of nanoparticles. Metal-resistant bacterial strains from electroplating industries have been reported to produce AgNPs. The size and morphology of AgNPs thus produced was typically imaged using high-resolution transmission electron microscopy (HR-TEM); the AgNPs were found to be 4–5 nm and spherical (Ganesh-Babu and Gunasekaran 2009). In another study, synthesis of auriferous bio-nanoparticles of silver through the reduction of aqueous Ag^+ ions using culture supernatants of *Staphylococcus aureus* was reported. This bioreduction of the Ag^+ ions in the solution was monitored in the aqueous component and the spectrum of the solution measured through UV-visible spectrophotometry and characterized by atomic force microscopy (Nanda and Saravanan 2009).

Moreover, a completely unique, rapid, straight-forward and green combination approach for the synthesis of antimonial nanostructures of noble metals such as silver was studied. It involved the use of a combination of culture supernatant of *Bacillus subtilis* and microwave irradiation in water in the absence of a surfactant or soft template. It was found that exposure to culture supernatant of *B. subtilis* and microwave irradiation to silver particles resulted in the formation of AgNPs. The size of the AgNPs produced was within the range of 5–60 nm in diameter (Saifuddin et al. 2009). Mokhtari et al. (2009) demonstrated the synthesis of AgNPs within the range of 1–6 nm using culture supernatant of *Klebsiella pneumoniae*.

Shivaji et al. (2011) used cell-free culture supernatants of five psychrophilic bacteria, namely, *Pseudomonas antarctica, P. proteolytica, P. meridiana, Arthrobacter kerguelensis* and *A. gangotriensis*, and two mesophilic bacteria, *Bacillus indicus* and *B. cecembensis*, for the synthesis of AgNPs. The AgNPs thus synthesized were characterized using different techniques to find their size and stability period. It was reported that the size of these AgNPs was in the range of 6–13 nm and they were stable for 8 months in the dark. This report presented the primary information on the synthesis of AgNPs using culture supernatants of psychrophilic microorganisms and also demonstrated that culture supernatants of species of *Arthrobacter*, a genus that has not been investigated earlier, may also synthesize AgNPs. In addition, biogenic synthesis of silver and gold nanoparticles by *Geobacillus stearothermophilus* has been investigated. The exposure of *G. stearothermophilus* cell-free extract to the respective metal salts resulted in the formation of stable AgNPs and AuNPs (Fayaz et al. 2011). Similarly, other nanoparticles like ZnONPs and CuONPs were synthesized using different kinds of bacteria (Usha et al. 2010, Rajamanickam et al. 2012). Further details about the synthesis of different nanoparticles and their size and shape using various bacteria are summarized in Table 8.2.

Table 8.2. Details of nanoparticles synthesized from different bacteria.

S. N.	Bacteria	Nanoparticles	Size (nm)	Shape	Reference
1	*Bacillus flexus*	AgNPs	20–40	Spherical	Priyadarshini et al. 2013
2	*Escherichia coli*	AgNPs	10	Spherical	Natarajan et al. 2010
3	*Corynebacterium glutamicum*	AgNPs	5–50	Irregular	Sneha et al. 2010
4	*Bacillus cereus*	AgNPs	4–5	Spherical	Ganesh–Babu and Gunasekaran 2009
5	*Staphylococcus aureus*	AgNPs	160–180	**	Nanda and Saravanan 2009
6	*Bacillus subtilis*	AgNPs	5–60	**	Saifuddin et al. 2009
7	*Klebsiella pneumoniae*	AgNPs	5–32	Spherical	Narges et al. 2009
8	*Klebsiella pneumoniae*	AgNPs	50–100	Spherical	Minaeian et al. 2008
9	*Pseudomonas aeruginosa*	AgNPs	10	Spherical	Shivaji et al. 2011
10	*Pseudomonas antarctica*	AgNPs	3.4–33.6	Spherical	
11	*Pseudomonas proteolytica*	AgNPs	2.8–23.1	Spherical	
12	*Arthrobacter gangotriensis*	AgNPs	3.6–22.8	Spherical	
13	*Bacillus indicus*	AgNPs	2.5–13.3	Spherical	
14	*Bacillus cecembensis*	AgNPs	2.8–18.2	Spherical	
15	*Geobacillus stearothermophilus*	AgNPs, AuNPs	5–20	Spherical	Fayaz et al. 2011
16	*Streptomyces hygroscopicus*	AuNPs	30–50	Spherical	Waghmare et al. 2014
17	*Streptomyces* sp.	ZnO/CuO	10–30 12–40	Spherical, rod	Usha et al. 2010
18	*Streptomyces* sp.	ZnO	3.5–17.8	Spherical	Rajamanickam et al. 2012

** Information not available.

Fungal synthesis

Various species of fungi have been also widely studied for the synthesis of nanoparticles. The fungal-mediated synthesis of nanoparticles is much easier than bacterial synthesis because in most cases it is extracellular; in most bacterial species, it is intracellular. Fathima and Balakrishnan (2014) synthesized AgNPs using an endophytic fungus *Fusarium solani* and also designed a mathematical model for optimization of factors influencing the biosynthesis of AgNPs. In another study, synthesis of AgNPs using culture supernatants of *Aspergillus terreus* was reported where reduction of aqueous Ag^+ ions was achieved with culture supernatants of this fungus. TEM characterization revealed that these AgNPs were spherical and range from 1 to 20 nm (Li et al. 2012).

Some reports demonstrated the synthesis of extremely biocompatible and stable AgNPs using different endophytic fungi isolated from a variety of medicinal plants. For instance, Balakumaran et al. (2015) recovered 13 isolates of endophytic fungi from nine different medicinal plants and tested all these strains for the synthesis of AgNPs. The results obtained revealed that among the tested strains *Guignardia mangiferae* showed ability to produce AgNPs extracellularly in the size range of 5–30 nm. In another study, three endophytic fungi, namely, *Aspergillus tamarii* PFL2, *A. niger* PFR6 and *Penicillium ochrochloron* PFR8 were isolated from an ethnomedicinal plant, *Potentilla fulgens* L., and used for the biogenic synthesis of AgNPs. It was found that the AgNPs synthesized using the fungus *A. tamarii* PFL2 had the smallest average particle size (3.5 ± 3 nm), as compared to the AgNPs synthesized using *A. niger* PFR6 and *P. ochrochloron* PFR8 (Devi and Joshi 2015). Moreover, the extracellular formation of spherical AgNPs in the size range of 30–150 nm by *Penicillium chrysogenum* and cubic AgNPs in the size range of 50–200 nm by *P. expansum* has been reported by Mohammadi and Salouti (2014).

Biological synthesis of AgNPs from extracellular extracts of *Rhizopus stolonifer* was reported. The study provided the proof that the molecules of liquid mycelial extract of *R. stolonifer* facilitate the synthesis of AgNPs and highlighted that *R. stolonifer*-mediated synthesis is cost-effective. It was also demonstrated that size of AgNPs can be well controlled by temperature and $AgNO_3$ concentration (Abdel-Rahim et al. 2017). Syed et al. (2013) claimed the synthesis of AgNPs using thermophilic flora of *Humicola* sp. for the first time. Moreover, there are many reports available on the biogenic synthesis of various other nanoparticles using different fungi. Vala (2015) demonstrated the extracellular biosynthesis of AuNPs using marine-derived fungus *Aspergillus sydowii*. Honary et al. (2012) demonstrated the synthesis of CuONPs using *Penicillium aurantiogriseum*, *P. citrinum* and *P. waksmanii*. Generally, it was proposed that various physical parameters and concentration of metal salts used in the synthesis affect the synthesis yield, size, and shape of synthesized nanoparticles. The use of various fungi for the synthesis of nanoparticles is summarized in Table 8.3.

Algal synthesis

Bio-nanotechnology has revolutionized nanomaterial synthesis by providing an artificial platform in the form of biological systems. Among these, microalgae have tremendous potential in the conversion of metal ions to nanoparticles by a detoxification process. Various cyanobacterial species are reported to have the ability to produce AgNPs, for example, *Aphanothece* sp., *Oscillatoria* sp., *Microcoleus* sp., *Aphanocapsa* sp., *Phormidium* sp., *Lyngbya* sp., *Gloeocapsa* sp., *Synechococcus* sp., and *Spirulina* sp. (Sudha et al. 2013). Another study involves the use of Australasian brown marine alga *Cystophora moniliformis* for the biogenic synthesis of AgNPs. The extract of this alga was used as a reducing and stabilizing agent. Temperature-dependent effect on the size of synthesized AgNPs was reported; agglomeration of the nanoparticles was observed at high temperatures. The average size of the AgNPs formed at temperatures < 65°C was 75 nm, whereas it was > 2 μm at higher temperatures (Prasad et al. 2013).

Synthesis of AgNPs using marine alga *Caulerpa racemosa* was demonstrated. Fresh *C. racemosa* collected from the Gulf of Mannar, Southeast coast of the Republic of India, was successfully used for this synthesis (Kathiraven et al. 2015). Extract of *Pithophora oedogonia* has also been used to synthesize AgNPs by reduction of silver nitrate. It was noted that the proposed synthesis method was significantly fast

Table 8.3. Details of nanoparticles synthesized using various fungi.

S. N.	Fungi	Nanoparticles	Size (nm)	Shape	Reference
1	*Fusarium solani*	AgNPs	10	Spherical	Fathima and Balakrishnan 2014
2	*Aspergillus terreus*	AgNPs	1–20	Spherical	Li et al. 2012
3	*Guignardia mangiferae*	AgNPs	5–30	Spherical	Balakumaran et al. 2015
4	*Penicillium ochrochloron*	AgNPs	7.7 ± 4.3	Spherical	Devi and Joshi 2015
5	*Penicillium chrysogenum, Penicillium expansum*	AgNPs	30–150 50–200	Spherical Spherical	Mohammadi and Salouti 2015
6	*Aspergillus oryzae*	AgNPs	19–60	Spherical	Pereira et al. 2014
7	*Penicillium atramentosum*	AgNPs	5–25	Spherical	Sarsar et al. 2015
8	*Raphanus sativus*	AgNPs	4–30	Spherical	Singh et al. 2017
9	*Rhizopus stolonifer*	AgNPs	9.46 ± 2.64	Spherical	Abdel-Rahim et al. 2017
10	*Humicola* spp.	AgNPs	5–25	Spherical	Syed et al. 2013
11	*Inonotus obliquus*	AgNPs	14.7–35.2	Spherical	Nagajyothi et al. 2014
12	*Penicillium citrinum*	AgNPs	80–100	Spherical	Honary et al. 2013
13	*Cryphonectria* spp.	AgNPs	30–70	Spherical	Dar et al. 2013
14	*Aspergillus foetidus*	AgNPs	4–20	Spherical	Roy et al. 2013
15	*Macrophomina phaseolina*	AgNPs	5 40	Spherical	Chowdhury et al. 2014
16	*Penicillium aurantiogriseum, Penicillium citrinum, Penicillium waksmanii*	CuONPs	**	Spherical	Honary et al. 2012
17	*Aspergillus sydowii*	AuNPs	8.7–15.6	Spherical	Vala 2015

** Information not available.

(Sinha et al. 2015). Marine cyanobacterium *Oscillatoria willei* NTDM01 was used for the green synthesis of AgNPs. It has the ability to reduce silver ions and produces highly stable AgNPs. Scanning electron microscopy studies showed that AgNPs thus formed were in the size range of 100–200 nm (Mubarak-Ali et al. 2011). A study explores the intracellular biogenic synthesis of AgNPs using the unicellular green microalga *Scenedesmus* sp. Intracellular nanoparticle biosynthesis was initiated via accumulation of a high amount of Ag^+ ion in the microalgal biomass and subsequent formation of spherical crystalline AgNPs with an average size of 5–10 nm (Jena et al. 2014). Similarly, the biogenic synthesis of AgNPs has been successfully conducted using *Plectonema boryanum* UTEX 485, a threadlike cyanobacterium. The extract of this alga was reacted with aqueous $AgNO_3$ solutions ($\sim$ 560 mg/L Ag ions) at 25–100°C for up to 28 days (Lengke et al. 2006). ZnONPs and other nanoparticles were also synthesized using a variety of algal species, mainly *S. muticum* extract. The ZnONPs thus produced were found to be polygonal, and the average size ranged from 30 to 57 nm (Azizi et al. 2014). Some of the algal species demonstrated to produce nanoparticles are summarized in Table 8.4.

Table 8.4. Biogenic synthesis of nanoparticles using different species of algae.

S. N.	Algae	Nanoparticles	Size (nm)	Shape	Reference
1	*Microcoleus* spp.	AgNPs	40–80	Spherical	Sudha et al. 2013
2	*Cystophora moniliformis*	AgNPs	75	Cubic	Prasad et al. 2013
3	*Pterocladia capillacae*	AgNPs	7	Spherical	El-Rafie et al. 2013
4	*Sargassum longifolium*	AgNPs	20–80	Spherical	Rajeshkumar et al. 2014
5	*Caulerpa racemosa*	AgNPs	5–25	Spherical	Kathiraven et al. 2015
6	*Pithophora oedogonia*	AgNPs	34.03	Spherical	Sinha et al. 2015
7	*Enteromorpha flexuosa*	AgNPs	22	Spherical	Yousefzadi et al. 2014
8	*Limnothrix* sp. 37-2-1	AgNPs	31.86 ± 1	Elongated	Patel et al. 2015
9	*Anabaena* sp. 66-2	AgNPs	24.13 ± 2	Irregular	
10	*Synechocystis* sp. 48-3	AgNPs	14.64 ± 2	Irregular	
11	*Botryococcus braunii*	AgNPs	15.67 ± 1	Spherical	
12	*Coelastrum* sp. 143-1	AgNPs	19.28 ± 1	Spherical	
13	*Limnothrix* sp. 37-2-1	AgNPs	25.65 ± 2	Spherical, elongated	
14	*Spirulina* sp.	AgNPs	13.85 ± 2	Spherical	
15	*Sargassum muticum*	AgNPs	5–15	Spherical	Azizi et al. 2013
16	*Lyngbya majuscula*	AuNPs, AgNPs alloy	5–25	Spherical	Roychoudhury et al. 2015
17	*Spirulina platensis*	AgNPs	5–30	Spherical	Mahdieh et al. 2012
18	*Oscillatoria willei* NTDM01	AgNPs	100–200	Spherical	Mubarak-Ali et al. 2011
19	*Scenedesmus* sp.	AgNPs	5–10	Spherical	Jena et al. 2014
20	*Plectonema boryanum* UTEX 485	AgNPs	50–200	Spherical	Lengke et al. 2006
21	*Nostoc linckia*	AgNPs	5–60	Spherical	Vanlalveni et al. 2018
22	*Microcystis aeruginosa*	CuONPs	551	Spherical, rod, irregular	Sankar et al. 2014
23	*Sargassum muticum*	ZnONPs	30–57	Hexagonal	Azizi et al. 2014
24	*Gracilaria edulis*	AgNPs ZnONPs	55–99 65–95	Spherical, rod	Priyadharshini et al. 2014

Factors influencing biosynthesis of nanoparticles

As discussed earlier, biogenic synthesis of nanoparticles is preferred to physical and chemical methods because it is non-toxic and environment-friendly. However, there are several factors that influence the biosynthesis, size and shape of nanoparticles. These include pH, temperature, concentration of biomass, concentration of metal solution, and incubation/reaction time. Some of these important factors are briefly described here.

pH

pH is considered one of the most important factors that affect the biosynthesis, size and texture of the nanoparticle (Armendariz et al. 2004, Bawaskar et al. 2010). The extracts of different biological systems show different pH; extracts of different plant parts have different pH (Sathishkumar et al. 2009). In addition, it was suggested that it is possible to control size and shape of nanoparticles by altering the pH of the solution media (Suchomel et al. 2018). There are many reports that demonstrated that the pH affects the shape and size of the biosynthesized nanoparticle, especially AgNPs from *Chrysosporium tropicum* and *Fusarium oxysporum* (Soni and Prakash 2011) and *Zingiber officinale* (Aziz and Jassim 2018), and AuNPs from *Avena sativa* (Armendariz et al. 2004). It was also claimed that nanoparticles synthesized at lower pH were larger than those synthesized at higher pH (Dubey et al. 2010).

Temperature

Temperature is another important factor that affects the biosynthesis of nanoparticles (Bawaskar et al. 2010). Conventional physical and chemical methods of nanoparticle synthesis require very high temperatures (350°C or more). Biosynthesis can be carried out at ambient temperature (less than 100°C) (Patra and Baek 2004), but temperature variations still play a key role in determining the size and shape of nanoparticles. Liu et al. (2017) reported that the size of AgNPs synthesized using leaf extract of *Cinnamomum camphora* slightly increases when there is rise in reaction temperature from 70 to 80°C, whereas it decreases with further rise in temperature (from 80 to 90°C); hence, they concluded that high temperature is responsible for the decrease in size. Similarly, Dwivedi and Gopal (2010) demonstrated that the rate of AuNP synthesis is higher at higher temperatures and they further revealed that high temperature is responsible for synthesis of nano-rod and platelet-shaped AuNPs; on the other hand, spherical AuNPs were formed at lower temperatures. There is an optimum temperature for the biosynthesis of each type of nanoparticle. The synthesis of AgNPs using extract of *Trichoderma longibrachiatum* was attempted at 23, 28, and 33°C; it was reported that 28°C is the optimum temperature that showed synthesis of AgNPs, whereas there was no synthesis at the other two temperatures (Elamawi et al. 2018).

Concentration of biomass

Elamawi et al. (2018) inoculated various concentrations of fungal biomass (1, 5, 10, 15, and 20 g) in 100 ml of Milli Q water to get the desired fungal cell extract. It was demonstrated that formation of AgNPs increases with increase in the quantity of fungal biomass. The authors also suggested that agitation is very important for the synthesis of nanoparticles. In another study performed on synthesis of AgNPs using *F. oxysporum* at varying biomass concentration (3 g, 5 g, 7 g, 9 g, 11 g, 13 g, 15 g, 17 g and 19 g/100 ml of water), it was reported that the weight range of biomass from 3 to 9 g and 13 to 19 g decreased AgNP size. However, 11 g was found to be the optimal weight for synthesis of the smallest nanoparticles (Husseiny et al. 2015).

Reaction and storage time

Most studies revealed that various characteristics of nanoparticles such as quality, shape and size are greatly affected by incubation time of the reaction mixture (Elamawi et al. 2018). Optimum incubation time is required to complete the synthesis process and its subsequent stabilization by the capping of biomolecules present in the extract of the biological agent used (Veerasamy et al. 2011). Even after stabilization, in some cases, longer storage of nanoparticles may lead to aggregation and shrinkage (Baer 2011).

Concentration of metal solution

Concentration of metal solution used is also considered an exciting factor in the biosynthesis of nanoparticles. The available reports suggest that variation in concentration of metal solution mostly affected the size of biosynthesized nanoparticles. Varying size was reported for biosynthesized AgNPs produced from *Aspergillus oryzae* at various concentrations of silver nitrate used (1–10 mM) (Phanjom and Ahmed 2017). Similarly, smallest size of AgNPs was produced from the extract of *F. oxysporum* at the 0.01 M concentration of silver nitrate as compared to other two concentrations, i.e., 0.1 M and 0.001 M (Husseiny et al. 2015). These results suggest that optimum concentration of metal solution is necessary to get nanoparticles of desired size, or it might be possible to control the size of nanoparticles using various concentrations of metal solution.

Possible mechanisms involved in biogenic synthesis

As we know, plants, fungi, and bacteria are the most important biological systems used for the synthesis of various nanoparticles. Although no confirmed mechanism has been investigated for the synthesis of nanoparticles, various possible hypothetical mechanisms have been proposed by many researchers for each of the above-mentioned biological systems (Kitching et al. 2015, Sharma et al. 2015a, Velusamy et al. 2016). However, the common mechanism for biosynthesis of nanoparticles involved reduction of aqueous metal ions via the donation of electrons from a particular compound or biomolecule present in the extracellular metabolites of extract of biological system, and certain compounds play an important role in transferring these electrons to aqueous metal ions to make up the deficiency and get reduced to their neutral form, which is commonly referred to as its nano form or nanoparticle (Fig. 8.1). In some cases, involvement of more than one biomolecule or a group of biomolecules was proposed in the biosynthesis and stabilization of nanoparticles. Some of the important and widely accepted mechanisms are discussed here.

Mechanisms of bacterial synthesis

Different mechanisms have been suggested for the synthesis of nanoparticles using bacteria. Bacteria are generally known to synthesize nanoparticles by reduction of metal ions both intra- and extracellularly. In intracellular synthesis, metals ions are trapped on the surface of or inside the bacterial cell followed by their reduction to

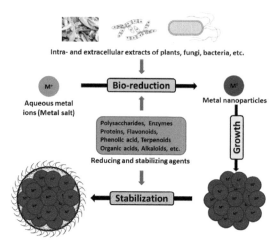

Figure 8.1. Schematic representation of common mechanism involved in the biosynthesis of metal NPs using various biological systems (adapted and modified from Velusamy et al. 2016, an open access article, unrestricted use granted under the terms of the Creative Commons Attribution Non-Commercial License).

zero-valent nanoparticles by different biomolecules (Ghashghaei and Emtiazi 2015). The entrapment and transportation of metal ions in the bacterial cell may be facilitated by electrostatic interaction of the positively charged metal ions and negatively charged cell walls. In intracellular synthesis, nanoparticles can be separated from cell debris by ultrasonication or centrifugation (Akbulut et al. 2012). However, a matrix and protein casing contributes significantly to the stability of nanoparticles, so it is not necessary to add further surfactants to the system (Velusamy et al. 2016), as is often required in chemical synthesis. Stable nanoparticles do not aggregate, which benefits their application in various practices. It is not even necessary to separate nanoparticles from cells and these nanoparticle-doped cells may be directly used in catalytic applications (Kratošová et al. 2016, Konvičková et al. 2018, Holišová et al. 2019). On the other hand, different metabolites such as NADH-dependent reductase and sulfur-containing protein secreted by the bacteria in the surrounding medium are responsible for the extracellular reduction of metal ions into nanoparticles (Luo et al. 2018).

According to one theory, the synthesis of metal nanoparticles by bacteria is the outcome of detoxification pathways, where the number of toxic metal ions is taken up through cationic membrane transport systems that normally transport metabolically important cations. The specialized mechanism developed to counteract this kind of uptake prevents excessive accumulation of toxic metals (Crookes-Goodson et al. 2008). Lin et al. (2005) demonstrated that the hydroxyl group of saccharides and the ionized carboxyl group of amino acid residues present on the membrane of bacterium acts as binding sites for Au(III), and the free aldehyde group of the hemiacetalic hydroxyl group from reducing sugars (hydrolysates of the polysaccharides) serves as the electron donor, which collectively reduces the Au(III) to Au(0) (i.e., AuNPs). Many reports confirmed that heavy metal resistance genes present in some bacteria might play an important role in nanoparticle formation. The zinc resistance gene (czcC gene) present in resistant strain of *Pseudomonas stutzeri* and *Brevundimonas diminuta* is reported to enable the production of zinc nanoparticles (Mirhendi et al. 2013).

Mechanisms of fungal synthesis

As far as fungal synthesis is concerned, it is proposed that the extracellular metabolites including different enzymes secreted by fungi for their own survival when exposed to different environmental stresses are mostly responsible for the reduction of metals ions to metallic solid nanoparticles through the catalytic effect (Yadav et al. 2015, Elamawi et al. 2018). In one of the hypothetical mechanisms, it was proposed that NADH-dependent nitrate reductase enzyme secreted by *Fusarium oxysporum* is responsible for the reduction of aqueous silver ions into Ag nanoparticles (Ahmad et al. 2003). A similar mechanism has been proposed by Ingle et al. (2008) for the synthesis of AgNPs from *Fusarium acuminatum*. These authors also pointed out the involvement of cofactor NADH and nitrate reductase enzyme in the biosynthesis of AgNPs because they reported the presence of nitrate reductase in fungal cell-free extract using specific substrate and discs for nitrate purchased from Hi-Media Pvt. Ltd. Mumbai, India. Duran et al. (2005) and Kumar et al. (2007) proposed similar possible mechanisms for biosynthesis of AgNPs from *F. oxysporum*. The former study reported the role of anthraquinone and NADPH-nitrate reductase in the biosynthesis of AgNPs and it was proposed that the electron required to make up the deficiency of aqueous silver ions (Ag^+) and convert it into Ag neutral (Ag^0, i.e., AgNPs) was donated by both quinone and NADPH. However, in the latter study, it was demonstrated that the reduction of -NADPH to -NADP$^+$ and the hydroxyquinoline possibly acts as an electron shuttle transferring the electron generated during the reduction of nitrate to Ag^+ ions, converting them to Ag^0.

Apart from these, the role of proteins, particularly amino acids having -SH bonds and most likely cysteine, is also confirmed in the biosynthesis of metal nanoparticles. It is proposed that such amino acids undergo dehydrogenation on reaction with metal salts like silver nitrate and produce AgNPs (Mukherjee et al. 2008). Moreover, various amide groups (I, II and III) are also reported to play an important role in the biosynthesis of metal NPs. According to Das et al. (2009), surface protein molecules bound on fungal mycelium participate in the reduction of gold ions to AuNPs, as was confirmed from the FTIR analysis of both fungal extracts before and after treatment with aqueous gold ions ($AuCl_4$). It was observed that fungal extract after treatment with $AuCl_4$ (AuNPs) showed the presence of amide I, II and III groups instead of carbonyl groups, which are initially present in the extract. Similarly, Sanghi and Verma (2009) anticipated the involvement of amide I and amide II groups in synthesis of AgNPs from *Coriolus versicolor*.

Various other studies performed on mycosynthesis proposed the role of other enzymes and proteins. But among all these mechanisms for extracellular mycosynthesis, the hypothetical mechanism involving the role of NADH-dependent nitrate reductase enzyme is mostly proposed and accepted.

In addition to extracellular mechanisms, there are a few mechanisms proposed for intracellular mycosynthesis of metal nanoparticles. In intracellular fungal synthesis, metal nanoparticles commonly formed below the cell surface owing to the reduction of metal ions by enzymes present in the cell membrane. Generally, two-step mechanisms have been proposed for intracellular mycosynthesis of nanoparticles. In the first step, an aqueous metal ion gets attached to the fungal cell surface by electrostatic interaction between lysine residues and metal ions (Riddin et al. 2006). In the second step, actual synthesis of nanoparticles occurs by enzymatic reduction of metal ions, which leads to aggregation and formation of nanoparticles. It was also demonstrated that cell-wall sugars play an important role in the reduction of metal ions to metal nanoparticles (Mukherjee et al. 2001). Both the above-mentioned mechanisms for extracellular and intracellular mycosynthesis are proposed for a few metals only, but it is assumed that the reduction of other metals may occur in a similar pattern.

Mechanisms of phytosynthesis

Several studies have been initiated on the elucidation and investigation of the mechanism involved in phytosynthesis of metallic nanoparticles and various theories have been proposed, but the actual mechanism behind the phytosynthesis is not yet identified (Mittal et al. 2014). However, the common hypothetical mechanism for phytosynthesis mainly required three constituents, i.e., reducing agents, stabilizing agents and solvent medium, and the phytochemicals present in the plant extract play the dual role of reducing and stabilizing agent (Makarov et al. 2014, Zayed and Eisa 2014). Owing to the complex nature and huge number of phytochemicals present in plant extract, it is very difficult to identify a specific compound that acts as reducing and stabilizing agents in synthesis of nanoparticles. The role of polyphenols (flavonoids, phenolic acid and terpenoids), organic acid and proteins has been proposed in phytosynthesis of nanoparticles (Dauthal and Mukhopadhyay 2016). Such phytochemicals perform bioreduction of aqueous metal ions to form respective zerovalent metal, which further leads to the agglomeration of metal atoms to respective metal nanoparticles (Aljabali et al. 2018).

Some of the studies performed suggested similar hypothesized mechanisms for phytosynthesis of AgNPs and AuNPs using various plants such as *Pelargonium graveolens* (geranium) (Shankar et al. 2003a, Shankar et al. 2003b), *Azadirachta indica* (neem) (Shankar et al. 2004a), and *Cymbopogon flexuosus* (lemongrass) (Shankar et al. 2004b). The presence of proteins and other secondary metabolites was reported in the extract of different plant leaf extract used for synthesis of metal nanoparticles. Further, it was proposed that terpenoid is responsible for the reduction of silver ions and oxidized to carbonyl groups to form AgNPs. It was also demonstrated that proteins present in leaf extract of geranium play an important role in stabilization of AuNPs, thereby capping the nanoparticles (Shankar et al. 2003a). Similarly, it was hypothesized that flavonoids of seed extract of fenugreek may a play major role as reducing agents in the reduction of aqueous Au ions, whereas the carboxylate group present in proteins can act as a surfactant to attach onto the surface of AuNPs, and the stabilization occurs via electrostatic attraction (Aromal and Philip 2012).

Another mechanism anticipated that the reducing sugars present in the leaf extract of *A. indica* might be responsible for reduction of metal ions to respective nanoparticles (Shankar et al. 2004a). A similar mechanism has been proposed in case of phytosyntheszied Au nanostructures using extract of lemongrass where aldose group in reducing sugars is reported to be responsible for reduction of Au ions (Shankar et al. 2004b). Moreover, aromatic amine, amide (I) group, secondary alcohols, and phenolic groups are proposed to be responsible for phytosynthesis of AgNPs using *Coleus aromaticus* leaf extract (Vanaja and Annadurai 2012). The possible role of various phenolic acids (e.g., gallic acid, caffeic acid, ellagic acid), terpenoids, proteins, organic acids, and other phytochemicals in synthesis of various metal NPs has been critically reviewed by Dauthal and Mukhopadhyay (2016).

Management of plant pathogenic fungi

Management and control of fungal pathogens in economically important plants and crops is of vital importance. There are various methods available for the control of plant pathogens. However, currently the application of various nanomaterials is considered the most promising approach in the management of plant pathogenic fungi (Rai and Kratošová 2015). In this context, the role of some important nanoparticles in the management of plant pathogenic fungi has been discussed below.

Silver nanoparticles (AgNPs)

AgNPs are the most widely studied nanoparticles and are reported to have noteworthy antifungal activity against plant pathogens. Gajbhiye et al. (2009) reported the comparative antifungal activity of AgNPs synthesized from *Alternaria alternata* in combination with fluconazole, tested against *Phoma glomerata, P. herbarum, Fusarium semitectum,* and *Trichoderma* sp. Enhanced activity of fluconazole was reported in presence of AgNPs. The maximum inhibition was reported against *P. glomerata* followed by *Trichoderma* sp. and *P. herbarum*. However, this combination does not show significant inhibition of *F. semitectum* (Gajbhiye et al. 2009). Krishnaraj et al. (2012) demonstrated the antifungal activity of AgNPs synthesized by green approach from *Acalypha indica* leaf extract as reducing agents. The AgNPs were tested against plant pathogenic fungi *A. alternata, Sclerotinia sclerotiorum, Macrophomina phaseolina, Rhizoctonia solani, Botrytis cinerea* and *Curvularia lunata*.

Lee et al. (2013) reported AgNP synthesis using cow milk and its antifungal activity against phytopathogens. The proteins present in the cow milk might be responsible for the reduction of ionic silver to form AgNPs. Significant inhibition of phytopathogenic fungus by AgNPs (average size 30–90 nm) was observed. The tested fungi were inhibited at varying levels: *Colletotrichum coccodes* (87.1%), *Monilinia* sp. (86.5%) and *Pyricularia* sp. (83.5%). Narayanan and Park (2014) reported the antifungal activity of AgNPs synthesized using turnip leaf extract (*Brassica rapa* L.) against wood-rotting fungal pathogens. Antifungal activity of *Bacillus*-mediated AgNPs against *Fusarium* sp. was demonstrated by Gopinath and Velusamy (2013).

Akther and Hemalatha (2019) demonstrated the synthesis of AgNPs using the extract of endophytic fungus isolated from the medicinal plant *Solanum nigrum*. Further, they evaluated the efficacy of AgNPs thus synthesized against different *A. niger, Rhizoctonia solani, F. graminearum* and *Fusarium udum*. Among these plant pathogenic fungi, *A. niger* and *F. graminearum* are well-known producers of potent mycotoxins, i.e., ochratoxins; *R. solani* is a soil-borne pathogen that causes root rot, collar rot and wire stem; *F. graminearum* is found to cause *Fusarium* head blight or scab of wheat and barley and also produces a group of vomitoxins that contaminates wheat and barley and causes liver diseases; and *F. udum* is responsible for setting wilt disease in pigeon pea plants. The findings obtained revealed that AgNPs effectively inhibited the growth of all these fungal pathogens; however, the level of efficacy may vary depending on the pathogens.

Copper nanoparticles (CuNPs)

Among various nanoparticles, CuNPs are reported to be among the promising antifungal agents against a wide range of plant pathogenic fungi. It has been demonstrated that biologically synthesized CuNPs from an endophytic actinomycete *Streptomyces capillispiralis* showed promising antifungal activity against phytopathogenic fungi. The study involved the effective use of CuNPs against *Alternaria* spp., *A. niger*, *Pythium* spp. and *Fusarium* spp. (Hassan et al. 2018).

Ponmurugan et al. (2016) demonstrated the antifungal activity of CuNPs biosynthesized extracellularly by using *Streptomyces griseus* and tested against *Poria hypolateritia* Berk. ex Cooke root rot causing pathogenic fungi in tea plants. Nano-copper was found to be the second most effective fungicide after carbendazime among the tested phytopathogens (Ponmurugan et al. 2016). Biologically synthesized CuNPs have been shown to have significant antifungal activity against plant pathogens. Shende et al. (2015) have reported the antimicrobial activity of biologically synthesized CuNPs using citron juice (*Citrus medica* Linn.). The *in vitro* antimicrobial activity was evaluated by Kirby-Bauer disk diffusion method against plant pathogenic fungi. *Fusarium culmorum* was found to be the most sensitive fungi to the CuNPs, followed by *F. oxysporum* and *F. graminearum*.

Banik and Pérez-de-Luque (2017) studied the *in vitro* efficacy of CuNPs on plant pathogenic fungi. Commercial CuNPs in combination with copper oxychloride showed 76% inhibition in oomycete *Phytophthora cinnamomi* at 50 mg/ml. Synergistically, CuNPs with copper oxychloride inhibited mycelial growth and sporulation in *A. alternata*. This antifungal activity of CuNPs can be used in the development of nano-copper-based fungicide for agriculture purposes.

Zinc oxide nanoparticles (ZnONPs)

Sclerotinia homoeocarpa is a causal agent of dollar spot on cool-season turf grasses. Li et al. (2017) studied the effect of ZnONPs and AgNPs on *S. homoeocarpa*. Exposure of this causal agent to these nanoparticles significantly induced the stress response gene expression, glutathione S-transferase (*Shgst1*) and superoxide dismutase 2 (*ShSOD2*) expressions. Increased nucleic acid content was observed after treatment with AgNPs. For the first time, involvement of zinc transporter gene (*Shzrt1*) in the accumulation of ZnONPs and AgNPs inside filamentous plant pathogenic fungi was reported. This mechanism could be useful in designing effective nano-based antifungal strategies against plant pathogens (Li et al. 2017). Biologically synthesized ZnONPs using *Parthenium hysterophorus* L. have been shown to have size-dependent antifungal activity against many plant pathogenic fungi. The size of ZnONPs was controlled by varying the plant extract concentration. As compared to 84 ± 2 nm ZnONPs, there was greater inhibition of *Aspergillus flavus* by 27 ± 5 nm ZnONPs (24.66 ± 0.57 mm at a concentration of 50 µg/ml, which is more than positive control, i.e., 19.66 ± 0.57 mm). Maximum inhibition was reported for *Aspergillus niger* and *A. flavus*, and comparatively less inhibition was recorded in the case of *F. culmorum*.

It was confirmed that smaller ZnONPs have greater antifungal activity against fungal pathogens (Gunalan et al. 2012b, Rajiv et al. 2013).

Jamdagni et al. (2018) reported the antifungal activity of aqueous flower extract of *Nyctanthes arbor-tristis*-mediated ZnONPs against selected phytopathogens *Alternaria* sp., *Fusarium* sp., *Botrytis* sp., *Penicillium* and *Aspergillus* sp. Zabrieski et al. (2015) demonstrated the influence of metal ion chelators such as Fe-chelator on the antifungal effect of metal oxide nanoparticles. *Pythium ultimum* was found to be more sensitive to CuONPs than *P. aphanidermatum*. The presence of ZnONPs in media inhibited the growth of *Pythium* sp., but the growth resumed after transfer to media lacking ZnONPs.

Sulfur nanoparticles (SNPs)

Choudhury and Goswami (2013) reviewed the antifungal activity of SNPs. Nano-sized fabrication of elemental sulfur and its surface modification can offer a retrieve use of sulfur in various fields. Most of the bacteria and fungi are resistant to the conventionally available sulfur products. Thus, there is a need for some advanced mode for controlling bacterial and fungal pathogens. Compounds containing nano-sulfur were tested against *Erysiphe cichoracearum, which causes* powdery mildew *of okra. They included* nano-sulfur synthesized at the Indian Agricultural Research Institute (IARI), commercial sulfur (Merck), commercial nano-sulfur (M K Impex, Canada) and Sulfur 80 WP (Corel Insecticide). All the tested sulfur fungicides significantly reduced the germination of conidia of *E. cichoracearum*. The order of conidial germination was recorded from lower to higher as IARI nano-sulfur (4.56%) followed by Canadian nanosulfur (14.17%), Merck sulfur (15.53%), Sulfur 80 WP (15.97%), and control (23.09%). Cleistothecial appendages were also disrupted in contact with nano-sulfur and the cleistothecia became sterile. This indicated that the sulfur synthesized at IARI could be used for powdery mildew in okra. Rao and Paria (2013) showed the efficiency of SNPs against two phytopathogens, *Fusarium solani* (isolated from an infected tomato leaf, responsible for early blight and Fusarium wilt diseases) and *Venturia inaequalis* (responsible for the apple scab disease). Small SNPs (~ 35 nm) were shown to have higher antifungal activity.

Iron oxide nanoparticles

Devi et al. (2019) reported the green synthesis of iron oxide nanoparticles from *Platanus orientalis* leaf extract for the first time. These iron oxide nanoparticles were highly effective against *A. niger* and *Mucor piriformis*. The application of iron oxide nanoparticles in the field of antimicrobial activity is still in its infancy. Green approach for the synthesis of iron oxide nanoparticles using polyphenol derivative of plants, i.e., tannic acid, as a reducing and capping agent was reported by Parveen et al. (2018). The antifungal activity of these nanoparticles was evaluated against *Trichothecium roseum, Cladosporium herbarum, P. chrysogenum, A. alternata,* and *A. niger*. The activity was shown in the form of inhibition of spore germination and determining zones of inhibition of fungal pathogens. The significant antimycotic activity was shown by the iron oxide nanoparticles. The highest inhibition of spore germination was observed in *T. roseus* (87.74%) followed by *C. herbarum* (84.89%). The highest activity index was observed for *P. chrysogenum* (0.81). The zones of inhibition were recorded as *P. chrysogenum* (28.67 mm) followed by *A. niger* (26.33 mm), *T. roseum* (22.67 mm), and *A. alternata* (21.33 mm). Table 8.5 summarizes the details about biological nanoparticles used against selective phytopathogenic fungi.

Table 8.5. Biological nanoparticles used against selective phytopathogenic fungi.

Biogenic nanoparticles	Source of synthesis	Tested phytopathogenic fungi	References
Copper nanoparticles	*Streptomyces capillispiralis*	*Alternaria* spp., *Aspergillus niger*, *Pythium* spp. and *Fusarium* spp.	Hassan et al. 2018
	Streptomyces griseus	*Poria hypolateritia* Berk. ex Cooke	Ponmurugan et al. 2016
	Citrus medica Linn.	*Fusarium culmorum, F. oxysporum* and *F. graminearum*	Shende et al. 2015
	Commercial, in combination with copper oxychloride	*Phytophthora cinnamomi, Alternaria alternata*	Banik and Pérez-de-Luque 2017
Silver nanoparticles	*Alternaria alternata* in combination with fluconazole	*Phoma glomerata,* *P. herbarum,* *F. semitectum* and *Trichoderma* sp.	Gajbhiye et al. 2009
	Brassica rapa L.	*Gloeophyllum abietinum,* *G. trabeum, Chaetomium globosum* and *Phanerochaete sordida*	Narayanan and Park 2014
	Bacillus sp. strain GP-23	*Fusarium oxysporum*	Gopinath and Velusamy 2013
	Acalypha indica	*Alternaria alternata, Sclerotinia sclerotiorum, Macrophomina phaseolina, Rhizoctonia solani, Botrytis cinerea* and *Curvularia lunata*	Krishnaraj et al. 2012
Zinc oxide nanoparticles	*Parthenium hysterophorus* L.	*Aspergillus flavus, A. niger* and *Fusarium culmorum*	Gunalan et al. 2012b, Rajiv et al. 2013
		Sclerotinia homoeocarpa	Li et al. 2017
	Nyctanthes arbor-tristis	*Aspergillus niger* (ITCC 7122), *Botrytis cinerea* (ITCC 6192), *Fusarium oxysporum* (ITCC 55) and *Penicillium expansum* (ITCC 6755), *Alternaria alternata* (ITCC 6531)	Jamadagni et al. 2018
Sulfur nanoparticles	CTAB- and SDBS-mediated	*Fusarium solani* and *Venturia inaequalis*	Rao and Paria 2013
Iron oxide nanoparticles	*Platanus orientalis*	*Aspergillus niger* and *Mucor piriformis*	Devi et al. 2019
	Plant-derived tannic acid-mediated	*Trichothecium roseum, Cladosporium herbarum, Penicillium chrysogenum, Alternaria alternata* and *Aspergillus niger*	Parveen et al. 2018

Conclusions

Nanobiotechnology is one of the emerging branches of science having potential applications in agriculture and other fields. Biogenic approaches proposed for the synthesis of various kinds of nanoparticles are economically viable, eco-friendly and safe, as these are free from the use of toxic chemicals or organic solvents. Moreover, biogenic nanoparticles possess strong antifungal properties and hence play an important role in the control of a variety of plant pathogenic fungi. Physical parameters such as temperature, pH, and concentration of salt used also have significant effects on the size, shape, and yield of resulting nanoparticles.

It is proposed that many biomolecules secreted by biological systems, such as proteins/enzymes, amino acids, polysaccharides, alkaloids, alcoholic compounds, and vitamins, could be involved in bioreduction, formation, and stabilization of metal nanoparticles. But the exact mechanisms of biogenic nanoparticle synthesis using different biological systems have not yet been clearly and thoroughly understood. Therefore, further studies are required to understand the exact molecular mechanism involved in the biogenic synthesis of nanoparticles to identify the biomolecules responsible for the reduction and stabilization of nanoparticles. Similarly, research studies are needed to optimize the various reaction conditions in order to have better control over nanoparticle size, shape, and polydispersity, which play a key role in their bioactivities. In addition, studies should focus on the ecotoxicity of such nanoparticles to assess its impact on environment and human health. This information will be imperative for future use of such nanoparticles as alternative agents for the management of plant pathogenic fungi and crop protection.

Acknowledgment

MR is thankful to the University Grants Commission, New Delhi, for the award of BSR Faculty Fellowship.

References

Abdel-Rahim, K., Mahmoud, S.Y., Mohamed-Ali, A., Almaary, K.S., Mustafa, M.A., Moussa, S. 2017. Extracellular biosynthesis of silver nanoparticles using *Rhizopus stolonifer* Saudi. J. Biol. Sci. 24: 208–216.

Ahmad, A., Mukherjee, P., Senapati, S., Mandal, D., Khan, M.I., Kumar, R., Sastry, M. 2003. Extracellular biosynthesis of silver nanoparticles using the fungus *Fusarium oxysporum*. Colloids Surf. B Biointerfaces 28: 313–318.

Ahmed, S., Saifullah, Ahmad, M., Swami, B.L., Ikram, S. 2016. Green synthesis of silver nanoparticles using *Azadirachta indica* aqueous leaf extract. J. Radiat. Res. Appl. Sci. 9: 1–7.

Akbulut, O., Mace, C.R., Martinez, R.V., Ashok, A., Nie, Z., Patton, M.R., Whitesides, G.M. 2012. Separation of nanoparticles in aqueous multiphase systems through centrifugation. Nano Lett. 12: 4060–4064.

Akther, T., Hemalatha, S. 2019. Mycosilver nanoparticles: synthesis, characterization and its efficacy against plant pathogenic fungi. BioNanoScience 9: 296–301.

Aljabali, A.A.A., Akkam, Y., Al-Zoubi, M.S., Al-Batayneh, K.M., Al-Trad, B., Abo-Alrob, O., Alkilany, A.M., Benamara, M., Evans, D.J. 2018. Synthesis of gold nanoparticles using leaf extract of *Ziziphus zizyphus* and their antimicrobial activity. Nanomaterials 8: 174. doi: 10.3390/nano8030174.

Armendariz, V., Herrera, I., Peralta-Videa, J.R., Jose-yacaman, M., Troiani, H., Santiago, P., Gardea-Torresdey, J.L. 2004. Size controlled gold nanoparticle formation by *Avena sativa* biomass: use of plants in nanobiotechnology. J. Nanopart. Res. 6(4): 377–382.

Aromal, S.A., Philip, D. 2012. Green synthesis of gold nanoparticles using *Trigonella foenum-graecum* and its size-dependent catalytic activity. Spectrochim. Acta A: Mol. Biomol. Spectrosc. 97: 1–5.

Aziz, W.J., Jassim, H.A. 2018. A novel study of pH influence on Ag nanoparticles size with antibacterial and antifungal activity using green synthesis. World Sci. News 97: 139–152.

Azizi, S., Ahmad, M.B., Namvar, F., Mohamad, R. 2014. Green bio-synthesis and characterization of zinc oxide nanoparticles using brown marine macro alga *Sargassum muticum* aqueous extract. Mater. Lett. 116: 275–277.

Azizi, S., Namvar, F., Mahdavi, M., Ahmad, M.B., Mohamad, R. 2013. Biosynthesis of silver nanoparticles using brown marine macroalga, *Sargassum muticum* aqueous extract. Materials 6: 5942–5950.

Baer, D.R. 2011. Surface characterization of nanoparticles: critical needs and significant challenges. J. Surf. Anal. 17(3): 163–169.

Baghayeri, M., Mahdavi, B., Abadi, Z.H.M., Farhadi, S. 2018. Green synthesis of silver nanoparticles using water extract of *Salvia leriifolia*: Antibacterial studies and applications as catalysts in the electrochemical detection of nitrite. Appl. Organometal. Chem. 32(2): e4057. https://doi.org/10.1002/aoc.4057.

Balaji, S., Kumar, M.B. 2017. Facile green synthesis of zinc oxide nanoparticles by *Eucalyptus globulus* and their photocatalytic and antioxidant activity. Adv. Powder Technol. 28(3): 785–797.

Balakumaran, M.D., Ramachandran, R., Kalaichelvan, P.T. 2015. Exploitation of endophytic fungus, *Guignardia mangiferae* for extracellular synthesis of silver nanoparticles and their *in vitro* biological activities. Microbiol. Res. 178: 9–17.

Banik, S., Pérez-de-Luque, A. 2017. *In vitro* effects of copper nanoparticles on plant pathogens, beneficial microbes and crop plants. Spanish J. Agri. Res. 15: e1005. https://doi.org/10.5424/sjar/2017152-10305.

Bawaskar, M., Gaikwad, S., Ingle, A., Rathod, D., Gade, A., Duran, N., Marcato, P.D., Rai, M. 2010. A new report on mycosynthesis of silver nanoparticles by *Fusarium culmorum*. Curr. Nanosci. 6: 376–380.

Biswas, A., Chawngthu, L., Vanlalveni, C., Hnamte, R., Lalfakzuala, R., Rokhum, L. 2018a. Biosynthesis of silver nanoparticles using *Selaginella bryopteris* plant extracts and studies of their antimicrobial and photocatalytic activities. J. Bionanosci. 12: 227–232.

Biswas, A., Vanlalveni, C., Adhikari, P.P., Lalfakzuala, R., Rokhum, L. 2018b. Green biosynthesis, characterisation and antimicrobial activities of silver nanoparticles using fruit extract of *Solanum viarum*. IET Nanobiotechnol. 12(7): 933–938.

Choudhury, R.S., Goswami, A. 2013. Supramolecular reactive sulphur nanoparticles: a novel and efficient antimicrobial agent. J. Appl. Microbiol. 114(1): 1–10.

Chowdhury, S., Basu, A., Kundu, S. 2014. Green synthesis of protein capped silver nanoparticles from phytopathogenic fungus *Macrophomina phaseolina* (Tassi) Goid with antimicrobial properties against multidrug-resistant bacteria. Nanoscale Res. Lett. 9: 1–11.

Crookes-Goodson, W.J., Slocik, J.M., Naik, R.R. 2008. Bio-directed synthesis and assembly of nanomaterials. Chem. Soc. Rev. 37: 2403–2412.

Dar, M.A., Ingle, A., Rai, M. 2013. Enhanced antimicrobial activity of silver nanoparticles synthesized by *Cryphonectria* sp. evaluated singly and in combination with antibiotics. Nanomedicine NMB 9: 105–110.

Das, S.K., Das, A.R., Guha, A.K. 2009. Gold nanoparticles: microbial synthesis and application in water hygiene management. Langmuir 25: 8192–8199.

Dauthal, P., Mukhopadhyay, M. 2016. Noble metal nanoparticles: plant-mediated synthesis, mechanistic aspects of synthesis, and applications. Ind. Eng. Chem. Res. 55(36): 9557–9577.

Devi, H.S., Boda, M.A., Shah, M.A., Parveen, S., Wani, A.H. 2019. Green synthesis of iron oxide nanoparticles using *Platanus orientalis* leaf extract for antifungal activity. Green Process Synth. 8: 38–45.

Devi, L.S., Joshi, S.R. 2015. Ultrastructures of silver nanoparticles biosynthesized using endophytic fungi. J. Microsc. Ultrastruct. 3(1): 29–37.

Dobrucka, R., Dugaszewska, J. 2016. Biosynthesis and antibacterial activity of ZnO nanoparticles using *Trifolium pratense* flower extract. Saudi J. Biol. Sci. 23: 517–523.

Dubey, S.P., Lahtinen, M., Sillanpaa, M. 2010. Tansy fruit mediated greener synthesis of silver and gold nanoparticles. Process Biochem. 45(7): 1065–1071.

Duran, N., Marcato, P.D., Alves, O.L., De-Souza, G.I.H., Esposito, E. 2005. Mechanistic aspects of biosynthesis of silver nanoparticles by several *Fusarium oxysporum* strains. J. Nanobiotechnol. 3: 1–8.

Dwivedi, A.D., Gopal, K. 2010. Biosynthesis of silver and gold nanoparticles using *Chenopodium album* leaf extract. Colloids Surf. A 369: 27–33.

Elamawi, R.M., Al-Harbi, R.E., Hendi, A.A. 2018. Biosynthesis and characterization of silver nanoparticles using *Trichoderma longibrachiatum* and their effect on phytopathogenic fungi. Egypt J. Biol. Pest Control 28: 28; https://doi.org/10.1186/s41938-018-0028-1.

El-Rafie, H.M., El-Rafie, M.H., Zahran, M.K. 2013. Green synthesis of silver nanoparticles using polysaccharides extracted from marine macro algae. Carbohydr. Polym. 96: 403–410.

Fathima, B.S., Balakrishnan, R.M. 2014. Biosynthesis and optimization of silver nanoparticles by endophytic fungus *Fusarium solani*. Mater. Lett. 132: 428–431.

Fayaz, A.M., Girilal, M., Rahman, M., Venkatesan, R., Kalaichelvan, P.T. 2011. Biosynthesis of silver and gold nanoparticles using thermophilic bacterium *Geobacillus stearothermophilus* Process Biochem. 46: 1958–1962.

Gajbhiye, M., Kesharwani, J., Ingle, A., Gade, A., Rai, M. 2009. Fungus-mediated synthesis of silver nanoparticles and their activity against pathogenic fungi in combination with fluconazole. Nanomedicine: NBM 5: 382–386.

Ganesh-Babu, M.M., Gunasekaran, P. 2009. Production and structural characterization of crystalline silver nanoparticles from *Bacillus cereus* isolate. Colloids Surf. B 74: 191–195.

Ghashghaei, S., Emtiazi, G. 2015. The methods of nanoparticle synthesis using bacteria as biological nanofactories, their mechanisms and major applications. Curr. Bionanotechnol. 1: 3–17.

Gopinath, V., Velusamy, P. 2013. Extracellular biosynthesis of silver nanoparticles using *Bacillus* sp. GP-23 and evaluation of their antifungal activity towards *Fusarium oxysporum*. Spectrochimica Acta Part A: Mol. Biomol. Spect. 106: 170–174.

Groiss, S., Selvaraj, R., Varadavenkatesan, T., Vinayagam, R. 2017. Structural characterization, antibacterial and catalytic effect of iron oxide nanoparticles synthesized using the leaf extract of *Cynometra ramiflora*. J. Mol. Struct. 1128: 572–578.

Gunalan, S., Sivaraj, R., Venckatesh, R. 2012a. Aloe barbadensis Miller mediated green synthesis of mono-disperse copper oxide nanoparticles: Optical properties. Spectrochimica Acta Part A: Mol. and Biomol. Spect. 97: 1140–1144.

Gunalan, S., Sivaraj, R., Rajendran, V. 2012b. Green synthesized ZnO nanoparticles against bacterial and fungal pathogens. Prog. Nat. Sci.: Mater. Int. 22(6): 693–700.

Hassan, S.E.D., Salem, S.S., Fouda, A., Awad, M.A., El-Gamal, M.S., Abdo, A.M. 2018. New approach for antimicrobial activity and bio-control of various pathogens by biosynthesized copper nanoparticles using endophytic actinomycetes. J. Radiation Res. Appl. Sci. 11: 262–270.

Holišová, V., Urban, M., Kolenčík, M., Němcová, Y., Schröfel, A., Peikertová, P., Slabotinský, J., Kratošová, G. 2019. Biosilica-nanogold composite: Easy-to-prepare catalyst for soman degradation. Arabian J. Chem. 12(2): 262–271.

Honary, S., Barabadi, H., Gharaei-Fathabad, E., Naghibi, F. 2013. Green synthesis of silver nanoparticles induced by the fungus *Penicillium citrinum*. Tropical J. Pharm. Res. 12: 7–11.

Honary, S., Barabadi, H., Gharaei-Fathabad, E., Naghibi, F. 2012. Green synthesis of copper oxide nanoparticles using *Penicillium aurantiogriseum*, *Penicillium citrinum* and *Penicillium waksmanii*. Digest J. Nanomater. Biostruct. 7: 999–1005.

Husseiny, S.M., Salah, T.A., Anter, H.A. 2015. Biosynthesis of size controlled silver nanoparticles by *Fusarium oxysporum*, their antibacterial and antitumor activities. Beni-Surf. Uni. J. Basic Appl. Sci. 4(3): 225–231.

Ingle, A., Gade, A., Pierrat, S., Sonnichsen, C., Rai, M. 2008. Mycosynthesis of silver nanoparticles using the fungus *Fusarium acuminatum* and its activity against some human pathogenic bacteria. Curr. Nanosci. 4: 141–144.

Jamdagni, P., Khatri, P., Rana, J.S. 2018. Green synthesis of zinc oxide nanoparticles using flower extract of *Nyctanthes arbortristis* and their antifungal activity. J. King Saud Uni. Sci. 30(2): 168–175.

Jena, J., Pradhan, N., Nayak, R.R., Dash, B.P., Sukla, L.B., Panda, P.K., Mishra, B.K. 2014. Microalga *Scenedesmus* sp.: A potential low-cost green machine for silver nanoparticle synthesis. J. Microbiol. Biotechnol. 24: 522–533.

Kathiraven, T., Sundaramanickam, A., Shanmugam, N., Balasubramanian, T. 2015. Green synthesis of silver nanoparticles using marine algae *Caulerpa racemosa* and their antibacterial activity against some human pathogens. Appl. Nanosci. 5(4): 499–504.

Kaviya, S.S.J., Viswanathan, B. 2011. Green synthesis of silver nanoparticles using *Polyalthia longifolia* leaf extract along with D-sorbitol. J. Nanotechnol. 2011: 152970. http://dx.doi.org/10.1155/2011/152970.

Khalil, K.A., Fouad, H., Elsarnagawy, T., Almajhdi, F.N. 2013. Preparation and characterization of electrospun PLGA/silver composite nanofibers for biomedical applications. Int. J. Electrochem. Sci. 8: 3483–3493.

Khodadadi, B., Bordbar, M., Nasrollahzadeh, M. 2017. *Achillea millefolium* L. extract mediated green synthesis of waste peach kernel shell supported silver nanoparticles: Application of the nanoparticles for catalytic reduction of a variety of dyes in water. J. Colloid Interface Sci. 493: 85–93.

Kim, D.Y., Saratale, R.G., Shinde, S., Syed, A., Ameen, F., Ghodake, G. 2018. Green synthesis of silver nanoparticles using *Laminaria japonica* extract: characterization and seedling growth assessment. J. Clean Prod. 172: 2910–2918.

Kitching, M., Ramani, M., Marsili, E. 2015. Fungal biosynthesis of gold nanoparticles: Mechanism and scale up. Microb. Biotechnol. 8: 904–917.

Konvičková, Z., Holišová, V., Kolenčík, M., Niide, T., Kratošová, G. 2018. Phytosynthesis of colloidal Ag-AgCl nanoparticles mediated by *Tilia* sp. leachate, evaluation of their behaviour in liquid phase and catalytic properties. Colloid Polym. Sci. 296: 677–687.

Kratošová, G., Holišová, V., Konvičková, Z., Ingle, A.P., Gaikwad, S., Škrlová, K., Prokop, A., Rai, M., Plachá, D. 2018. From biotechnology principles to functional and low-cost metallic bionanocatalysts. Biotechnol. Adv. 37: 154–176.

Kratošová, G., Natšinová, M., Holišová, V., Obalová, L., Chromčáková, Ž., Vávra, I. 2016. Transmission electron microscopy observation of bionanogold used for preliminary N$_2$O decomposition testing. Adv. Sci. Lett. 22: 631–636.

Krishnaraj, C., Ramachandran, R., Mohan, K., Kalaichelvan, P.T. 2012. Optimization for rapid synthesis of silver nanoparticles and its effect on phytopathogenic fungi. Spectrochimica Acta Part A: Mol. Biomol. Spect. 93: 95–99.

Kulkarni, N., Muddapur, U. 2014. Biosynthesis of metal nanoparticles: A review. J. Nanotechnol. 2014: 510246. http://dx.doi.org/10.1155/2014/510246.

Kumar, D.A., Palanichamy, V., Roopan, S.M. 2014. Green synthesis of silver nanoparticles using *Alternanthera dentata* leaf extract at room temperature and their antimicrobial activity. Spectrochim Acta Part A: Mol. Biomol. Spectrosc. 127: 168–171.

Kumar, S.A., Abyaneh, M.K., Gosavi, S.W., Kulkarni, S.K., Pasricha, R., Ahmad, A., Khan, M.I. 2007. Nitrate reductase-mediated synthesis of silver nanoparticles from AgNO$_3$. Biotechnol. Lett. 29: 439–445.

Kumar, B., Smita, K., Cumbal, L. 2017. Green synthesis of silver nanoparticles using Andean blackberry fruit extract. Saudi J. Biol. Sci. 24: 45–50.

Lee, K.J., Park, S.H., Govarthanan, M., Hwang, P.H., Seo, Y.S., Cho, M., Lee, W.H., Lee, J.Y., Kamala-Kannan, S., Oh, B.T. 2013. Synthesis of silver nanoparticles using cow milk and their antifungal activity against phytopathogens. Mater. Lett. 105: 128–131.

Lengke, M., Fleet, M., Southam, G. 2006. Biosynthesis of silver nanoparticles by filamentous cyanobacteria from a silver(I) nitrate complex. Langmuir 10: 1021–1030.

Li, G., He, D., Qian, Y., Guan, B., Gao, S., Cui, Y., Yokoyama, K., Wang, L. 2012. Fungus-mediated green synthesis of silver nanoparticles using *Aspergillus terreus*. Int. J. Mol. Sci. 13: 466–476.

Li, J., Sang, H., Guo, H., Popko, J.T., He, L., White, J.C., Dhankher, O.P, Jung, G., Xing, B. 2017. Antifungal mechanisms of ZnO and Ag nanoparticles to *Sclerotinia homoeocarpa*. Nanotechnology 28(15): 155101. doi:10.1088/1361-6528/aa61f3.

Lin, Z., Wu, J., Xue, R., Yang, Y. 2005. Spectroscopic characterization of Au^{3+} biosorption by waste biomass of *Saccharomyces cerevisiae*. Spectrochim Acta A: Mol. Biomol. Spect. 61: 761–765.

Liu, H., Zhang, H., Wang, J., Wei, J. 2017. Effect of temperature on the size of biosynthesized silver nanoparticle: Deep insight into microscopic kinetics analysis. Arabian J. Chem. doi: 10.1016/j. arabjc.2017.09.004.

Luo, K., Jung, S., Park, K.H., Kim, Y.R. 2018. Microbial biosynthesis of silver nanoparticles in different culture media. J. Agric. Food Chem. 66(4): 957–962.

Mahdieh, M., Zolanvari, A., Azimee, A.S., Mahdieh, M. 2012. Green biosynthesis of silver nanoparticles by *Spirulina platensis*. Scientia Iranica 19: 926–929.

Maji, A., Beg, M., Mandal, A.K., Das, S., Jha, P.K., Kumar, A., Sarwar, S., Hossain, M., Chakrabrti, P. 2017. Spectroscopic interaction study of human serum albumin and human hemoglobin with *Mersilea quadrifolia* leaves extract mediated silver nanoparticles having antibacterial and anticancer activity. J. Mol. Struct. 1141: 584–592.

Makarov, V.V., Love, A.J., Sinitsyna, O.V, Makarova, S.S., Yaminsky, I.V, Taliansky, M.E., Kalinina, N.O. 2014. Green nanotechnologies: Synthesis of metal nanoparticles using plants. Acta Naturae 6(1): 35–44.

Masum, M.M.I., Siddiqa, M., Ali, K.A., Zhang, Y., Abdallah, Y., Ibrahim, E., Qiu, W., Yan, C., Li, B. 2019. Biogenic synthesis of silver nanoparticles using *Phyllanthus emblica* fruit extract and its inhibitory action against the pathogen *Acidovorax oryzae* strain RS-2 of rice bacterial brown stripe. Front. Microbiol. 10: 820. doi: 10.3389/fmicb.2019.00820.

Matinise, N., Fuku, X.G., Kaviyarasu, K., Mayedwa, N., Maaza, M. 2017. ZnO nanoparticles via *Moringa oleifera* green synthesis: physical properties & mechanism of formation Appl. Surf. Sci. 406: 339–347.

Minaeian, S., Shahverdi, A.R., Nohi, A.S., Shahverdi, H.R. 2008. Extracellular biosynthesis of silver nanoparticles by some bacteria. J. Sci. I.A.U. 17: 1–4.

Mirhendi, M., Emtiazi, G., Roghanian, R. 2013. Production of nano zinc, zinc sulphide and nanocomplex of magnetite zinc oxide by *Brevundimonas diminuta* and *Pseudomonas stutzeri*. IET Nanobiotechnol. 7(4): 135–139.

Mittal, J., Jain, R., Sharma, M.M. 2014. Phytofabrication of nanoparticles through plant as nanofactories. Adv. Nat. Sci: Nanosci. Nanotechnol. 5: 043002. doi:10.1088/2043-6262/5/4/043002.

Mohammadi, B., Salouti, M. 2015. Extracellular biosynthesis of silver nanoparticles by *Penicillium chrysogenum* and *Penicillium expansum*. Synth. React. Inorg. Met. Org. Chem. 245(6): 844–847.

Mokhtari, N., Shahram, D., Seyedali, S., Reza, A., Khosro, A., Saeed, S., Sara, M., Hamid, R.S., Ahmad, R.S. 2009. Biological synthesis of very small silver nanoparticles by culture supernatant of *Klebsiella pneumonia*: the effects of visible-light irradiation and the liquid mixing process. Mater. Res. Bull. 44: 1415–1421.

Mubarak-Ali, D., Sasikala, M., Gunasekaran, M., Thajuddin, N. 2011. Biosynthesis and characterization of silver nanoparticles using marine cyanobacterium, *Oscillatoria willei* NTDM01. Digest J. Nanomater. Biostruct. 6: 385–390.

Mukherjee, P., Ahmad, A., Mandal, D., Senapati, S., Sainkar, S.R., Khan, M.I., Ramani, R., Parischa, R., Ajayakumar, P.V., Alam, M., Sastry, M., Kumar, R. 2001. Bioreduction of $AuCl_4^-$ ions by the fungus *Verticillium* sp. and surface trapping of the gold nanoparticles formed. Angew. Chem. Int. Ed. 40: 3585–3588.

Mukherjee, P., Roy, M., Mandal, B.P., Dey, G.K., Mukherjee, P.K., Ghatak, J., Tyagi, A.K., Kale, S.P. 2008. Green synthesis of highly stabilized nanocrystalline silver particles by a nonpathogenic and agriculturally important fungus *T. asperellum*. Nanotechnology 19: 103–110.

Mukherjee, P., Senapati, S., Mandal, D., Ahmad, A., Khan, M.I., Kumar, R., Sastry, M. 2002. Extracellular synthesis of gold nanoparticles by the fungus *Fusarium oxysporum*. ChemBioChem. 3: 461–463.

Nagajyothi, P.C., Sreekanth, T.V.M., Lee, J., Lee, K.D. 2014. Mycosynthesis: antibacterial antioxidant and antiploriferative activities of silver nanoparticles synthesized from *Inonotus obliquus* (Chaga mushroom) extract. J. Photochem. Photobiool. B 130: 299–304.

Naika, H.R., Lingarajua, K., Manjunath, K., Kumar, D., Nagaraju, G., Suresh, D., Nagabhushan, H. 2015. Green synthesis of CuO nanoparticles using *Gloriosa superba* L. extract and their antibacterial activity. J. Taibah Uni. Sci. 9: 7–12.

Nanda, A., Saravanan, M. 2009. Biosynthesis of silver nanoparticles from *Staphylococcus aureus* and its antimicrobial activity against MRSA and MRSE. Nanomedicine 5(4): 452–456.

Narayanan, K.B., Park, H.H. 2014. Antifungal activity of silver nanoparticles synthesized using turnip leaf extract (*Brassica rapa* L.) against wood rotting pathogens. European J. Plant Pathol. 140(2): 185–192.

Nasrollahzadeh, M., Maham, M., Sajadi, S.M. 2015. Green synthesis of CuO nanoparticles by aqueous extract of *Gundelia tournefortii* and evaluation of their catalytic activity for the synthesis of N-monosubstituted ureas and reduction of 4-nitrophenol. J. Colloid Interface Sci. 455: 245–253.

Nasrollahzadeh, M., Sajadi, S.M., Rostami-Vartooni, A., Hussin, S.M. 2016. Green synthesis of CuO nanoparticles using aqueous extract of *Thymus vulgaris* L. leaves and their catalytic performance for N-arylation of indoles and amines. J. Colloid Interface Sci. 466: 113–119.

Natarajan, S., Selvaraj, Ramachandra, M.V. 2010. Microbial production of silver nanoparticles. Digest J. Nanomater. Biostruct. 5: 135–140

Nava, O.J., Luque, P.A., Gomez-Gutierrez, C.M, Vilchis-Nestor, A.R., Castro-Beltran, A., Mota-Gonzalez, M.L., Olivas, A. 2017. Influence of *Camellia sinensis* extract on zinc oxide nanoparticle green synthesis. J. Mol. Struct. 1134: 121–125.

Parveen, S., Wani, A.H., Shah, M.A., Devi, H.S., Bhat, M.Y., Koka, J.A. 2018. Preparation, characterization and antifungal activity of iron oxide nanoparticles. Microbial Pathogen. 115: 287–292.

Patel, V., Berthold, D., Puranik, P., Gantar, M. 2015. Screening of cyanobacteria and microalgae for their ability to synthesize silver nanoparticles with antibacterial activity Biotechnol. Reports 5: 112–119.

Patra, J.K., Baek, K.H. 2004. Green nanobiotechnology: Factors affecting synthesis and characterization techniques. J. Nanomater. 2014: 417305. doi: http://dx.doi.org/10.1155/2014/417305.

Pereira, L., Dias, N., Carvalho, J., Fernandes, S., Santos, C., Lima, N. 2014. Synthesis, characterization and antifungal activity of chemically and fungal-produced silver nanoparticles against *Trichophyton rubrum*. J. Appl. Microbiol. 117: 1601–1613.

Phanjom, P., Ahmed, G. 2017. Effect of different physicochemical conditions on the synthesis of silver nanoparticles using fungal cell filtrate of *Aspergillus oryzae* (MTCC No. 1846) and their antibacterial effect. Adv. Nat. Sci: Nanosci. Nanotechnol. 8: 045016. https://doi.org/10.1088/2043-6254/aa92bc.

Ponmurugan, P., Manjukarunambika, K., Elango, V., Gnanamangai, B.M. 2016. Antifungal activity of biosynthesized copper nanoparticles evaluated against red root-rot disease in tea plants. J. Exp. Nanosci. 11(13): 1019–1031.

Prasad, T.N.V.K.V., Kambala, V.S.R., Naidu, R. 2013. Phyconanotechnology: synthesis of silver nanoparticles using brown marine algae *Cystophora moniliformis* and their characterization. J. Appl. Phycol. 25(1): 177–182.

Priyadarshini, S., Gopinath, V., Priyadharsshini, N.M., Ali, M.D., Velusamy, P. 2013. Synthesis of anisotropic silver nanoparticles using novel strain. Colloids Surf. B: Biointerfaces 102: 232–237.

Priyadharshini, R.I., Prasannaraj, G., Geetha, N., Venkatachalam, P. 2014. Microwave-mediated extracellular synthesis of metallic silver and zinc oxide nanoparticles using macro-algae (*Gracilaria edulis*) extracts and its anticancer activity against human PC3 cell lines. Appl. Biochem. Biotechnol. 174: 2777–2790.

Rai, M., Kratošová, G. 2015. Management of phytopathogens by application of green nanobiotechnology: Emerging trends and challenges. J. Agri. Sci. Debrecen 66: 15–22.

Raja, A., Ashokkumar, S., Pavithra-Marthandam, R., Jayachandiran, J., Khatiwada, C.P., Kaviyarasu, K., Ganapathi-Raman, R., Swamin, M. 2018. Eco-friendly preparation of zinc oxide nanoparticles using *Tabernaemontana divaricata* and its photocatalytic and antimicrobial activity. J. Photochem. Photobiol. B: Biol. 181: 53–58.

Rajamanickam, U., Mylsamy, P., Viswanathan, S., Muthusamy, P. 2012. Biosynthesis of zinc nanoparticles using actinomycetes for antibacterial food packaging. Int. Conf. Nutri. Food Sci. IPCBEE, vol. 39; IACSIT.

Rajeshkumar, S., Malarkodi, C., Paulkumar, K., Vanaja, M., Gnanajobitha, G., Annadurai, G. 2014. Algae mediated green fabrication of silver nanoparticles and examination of its antifungal activity against clinical pathogens. Int. J. Metals 2014: 1–8.

Rajiv, P., Rajeshwari, S., Venckatesh, R. 2013. Bio-Fabrication of zinc oxide nanoparticles using leaf extract of *Parthenium hysterophorus* L. and its size-dependent antifungal activity against plant fungal pathogens. Spectrochimica Acta Part A: Mol. Biomol. Spect. 112: 384–387.

Rao, J.K., Paria, S. 2013. Use of sulfur nanoparticles as a green pesticide on *Fusarium solani* and *Venturia inaequalis* phytopathogens. RSC Adv. 3: 10471–10478.

Riddin, T.L., Gericke, M., Whiteley, C.G. 2006. Analysis of the inter- and extracellular formation of platinum nanoparticles by *Fusarium oxysporum* f. sp. *lycopersici* using response surface methodology. Nanotechnology 17: 3482–3489.

Rivera-Rangel, R.D., González-Muñoz, M.P., Avila-Rodriguez, M., RazoLazcano, T.A., Solans, C. 2018. Green synthesis of silver nanoparticles in oil-in-water microemulsion and nano-emulsion using geranium leaf aqueous extract as a reducing agent. Colloids Surf. A Physicochem. Eng. Asp. 536: 60–67.

Roy, S., Mukherjee, T., Chakraborty, S., Kumar, T. 2013. Biosynthesis, characterization & antifungal activity of silver nanoparticles synthesized by the fungus *Aspergillus foetidus* MTCC-8876. Digest J. Nanomater. Biostruct. 8: 197–205.

Roychoudhury, P., Ghosh, S., Pal, R. 2015. Cyanobacteria mediated green synthesis of gold-silver nanoalloy. J. Plant Biochem. Biotechnol. 25: 73–78.

Saber, M.M., Mirtajani, S.B., Karimzadeh, K. 2018. Green synthesis of silver nanoparticles using *Trapa natans* extract and their anticancer activity against A431 human skin cancer cells. J. Drug Deliv. Sci. Technol. 47: 375–379.

Saifuddin, N., Wong, C.W., Yasumira, A.A.N. 2009. Rapid biosynthesis of silver nanoparticles using culture supernatant of bacteria with microwave irradiation. EJ Chem. 6: 61–70.

Salam, H.A., Sivaraj, R., Venckatesh, R. 2014. Green synthesis and characterization of zinc oxide nanoparticles from *Ocimum basilicum* L. var. purpurascens Benth.-Lamiaceae leaf extract. Mater. Lett. 131: 16–18.

Sanghi, R., Verma, P. 2009. Biomimetic synthesis and characterization of protein capped silver nanoparticles. Bioresour. Technol. 100: 502–504.

Sankar, R., Prasath, B.B., Nandakumar, R., Santhanam, P., Shivashangari, K.S., Ravikumar, V. 2014. Growth inhibition of bloom forming cyanobacterium *Microcystis aeruginosa* by green route fabricated copper oxide nanoparticles. Environ. Sci. Pollu. Res. 21: 14232–14240.

Saratale, R.G., Benelli, G., Kumar, G., Kim, D.S., Saratale, G.D. 2018. Bio-fabrication of silver nanoparticles using the leaf extract of an ancient herbal medicine, dandelion (*Taraxacum officinale*), and antimicrobial activity against phytopathogens. Environ. Sci. Pollut. Res. 25: 10392–10406.

Sarsar, V., Selwal, M.K., Selwal, K.K. 2015. Biofabrication, characterization and antibacterial efficacy of extracellular silver nanoparticles using novel fungal strain of *Penicillium atramentosum* KM. J. Saudi Chem. Soc. 19: 682–688.

Sathishkumar, M., Sneha, K., Yun, Y.S. 2009. Palladium nanocrystal synthesis using *Curcuma longa* tuber extract. Int. J. Mater. Sci. 4: 11–17.

Seqqat, R., Blaney, L., Quesada, D., Kumar, B., Cumbal, L. 2019. Nanoparticles for environment, engineering, and nanomedicine. J. Nanotechnol. 2019: 2850723. https://doi.org/10.1155/2019/2850723.

Shankar, S.S., Ahmad, A., Sastry, M. 2003a. Geranium leaf assisted biosynthesis of silver nanoparticles. Biotechnol. Prog. 19: 1627–1631.

Shankar, S.S., Ahmad, A., Pasricha, R., Sastry, M. 2003b. Bioreduction of chloroaurate ions by geranium leaves and its endophytic fungus yields gold nanoparticles of different shapes. J. Mater. Chem. A 13: 1822–1826.

Shankar, S.S., Rai, A., Ahmad, A., Sastry. M. 2004a. Rapid synthesis of Au, Ag, and bimetallic Au core-Ag shell nanoparticles using Neem (*Azadirachta indica*) leaf broth. J. Colloid Interface Sci. 275: 496–502.

Shankar, S.S., Rai, A., Ankamwar, B., Singh, A., Ahmad, A., Sastry, M. 2004b. Biological synthesis of triangular gold nanoprisms. Nat. Mater. 3: 482–488.

Sharma, D., Kanchi, S., Bisetty, K. 2015a. Biogenic synthesis of nanoparticles: A review. Arabian J. Chem. doi: http://doi.org/10.1016/j.arabjc.2015.11.002.

Sharma, J.K., Akhtar, M.S., Ameen, S., Srivastava, P., Singh, G. 2015b. Green synthesis of CuO nanoparticles with leaf extract of *Calotropis gigantea* and its dye-sensitized solar cells applications. J. Alloys Comp. 632: 321–325.

Shende, S., Ingle, A.P., Gade, A., Rai, M. 2015. Green synthesis of copper nanoparticles by *Citrus medica* Linn. (Idilimbu) juice and its antimicrobial activity. World J. Microbiol. Biotechnol. 31(6): 865–873.

Shivaji, S., Madhu, S., Singh, S. 2011. Extracellular synthesis of antibacterial silver nanoparticles using psychrophilic bacteria. Process Biochem. 46: 1800–1807.

Singh, T., Jyoti, K., Patnaik, A., Singh, A., Chauhan, R., Chandel, S.S. 2017. Biosynthesis, characterization and antibacterial activity of silver nanoparticles using an endophytic fungal supernatant of *Raphanus sativus*. J. Genet. Eng. Biotechnol. 15(1): 31–39.

Sinha, S.N., Paul, D., Halder, N., Sengupta, D., Patra, S.K. 2015. Green synthesis of silver nanoparticles using fresh water green alga *Pithophora oedogonia* (Mont.) Wittrock and evaluation of their antibacterial activity. Appl. Nanosci. 5: 703–709.

Sivaraj, R., Rahman, P.K., Rajiv, P., Narendhran, S., Venckatesh, R. 2014a. Biosynthesis and characterization of *Acalypha indica* mediated copper oxide nanoparticles and evaluation of its antimicrobial and anticancer activity. Spectrochimica Acta Part A: Mol. and Biomol. Spect. 129: 255–258.

Sivaraj, R., Rahman, P.K., Rajiv, P., Salam, H.A., Venckatesh, R. 2014b. Biogenic copper oxide nanoparticles synthesis using *Tabernaemontana divaricate* leaf extract and its antibacterial activity against urinary tract pathogen. Spectrochimica Acta Part A: Mol. and Biomol. Spect. 133: 178–181.

Sneha, K., Sathishkumar, M., Mao, J., Kwak, I.S., Yun, Y.S. 2010. *Corynebacterium glutamicum* mediated crystallization of silver ions through sorption and reduction processes. Chem. Eng. J. 162: 989–996.

Soni, N., Prakash, S. 2011. Factors affecting the geometry of silver nanoparticles synthesis in *Chrysosporium tropicum* and *Fusarium oxysporum*. American J. Nanotechnol. 2(1): 112–121.

Suchomel, P., Kvitek, L., Prucek, R., Panacek, A., Halder, A., Vajda, S., Zboril, R. 2018. Simple size-controlled synthesis of Au nanoparticles and their size-dependent catalytic activity. Sci. Reports 8: 4589. doi: 10.1038/s41598-018-22976-5.

Sudha, S.S., Rajamanickam, K., Rengaramanujam, J. 2013. Microalgae mediated synthesis of silver nanoparticles and their antimicrobial activity against pathogenic bacteria. Ind. J. Exp. Biol. 52: 393–399.

Syed, A., Saraswati, S., Kundu, G.C., Ahmad, A. 2013. Biological synthesis of silver nanoparticles using the fungus *Humicola* sp. and evaluation of their cytotoxicity using normal and cancer cell lines. Spectrochim. Acta A Mol. Biomol. Spectrosc. 114: 144–147.

Thema, F.T., Manikandan, E., Dhlamini, M.S., Maaza, M. 2015. Green synthesis of ZnO nanoparticles via *Agathosma betulina* natural extract. Mater. Lett. 161: 124–127.

Usha, R., Prabu, E., Palaniswamy, M., Venil, C.K., Rajendran, R. 2010. Synthesis of metal oxide nanoparticles by *Streptomyces* sp. for development of antimicrobial textiles. Global J. Biotechnol. Biochem. 5: 153–160.

Vala, A.K. 2015. Exploration on green synthesis of gold nanoparticles by a marine-derived fungus *Aspergillus sydowii*. 34(1): 194–197.

Vanaja, M., Annadurai, G. 2012. *Coleus aromaticus* leaf extract mediated synthesis of silver nanoparticles and its bactericidal activity. Appl. Nanosci. 3: 217–223.

Vanlalveni, C., Rajkumari, K., Biswas, A., Adhikari, P.P., Lalfakzuala, R., Rokhum, L. 2018. Green synthesis of silver nanoparticles using *Nostoc linckia* and its antimicrobial activity: a novel biological approach. BioNanoScience 8: 624–631.

Veerasamy, R., Xin, T.Z., Gunasagaran, S., Xiang, T.F.W., Yang, E.F.C. 2011. Biosynthesis silver nanoparticles using mangosteen leaf extract and evaluation of their antimicrobial activities. J. Saudi Chem. Soc. 15: 113–120.

Veisi, H., Azizi, S., Mohammadi, P. 2018. Green synthesis of the silver nanoparticles mediated by *Thymbraspicata* extract and its application as a heterogeneous and recyclable nanocatalyst for catalytic reduction of a variety of dyes in water. J. Clean. Prod. 170: 1536–1543.

Velusamy, P., Kumar, G.V., Jeyanthi, V., Das, J., Pachaiappan, R. 2016. Bio-inspired green nanoparticles: Synthesis, mechanism, and antibacterial application. Toxicol. Res. 32(2): 95–102.

Venugopal, K., Ahmad, H., Manikandan, E., Thanigai-Arul, K., Kavitha, K., Rajagopal, K., Moodley, M.K., Balabhaskar, R., Bhaskar, M. 2017b. The impact of anticancer activity upon *Beta vulgaris* extract mediated biosynthesized silver nanoparticles (AgNPs) against human breast (MCF-7), lung (A549) and pharynx (Hep-2) cancer cell lines. J. Photochem. Photobiol. B Biol. 173: 99–107.

Venugopal, K., Rather, H.A., Rajagopal, K., Shanthi, M.P., Sheriff, K., Illiyas, M., Rather, R.A., Manikandan, E., Uvarajan, S., Bhaskar, M., Maaza, M. 2017a. Synthesis of silver nanoparticles (Ag NPs) for anticancer activities (MCF 7 breast and A549 lung cell lines) of the crude extract of *Syzygium aromaticum*. J. Photochem. Photobiol. B. 167: 282–289.

Vijaya-Kumar, P., Mary, S., Kala, J., Prakash, K.S. 2019. Green synthesis of gold nanoparticles using *Croton Caudatus Geisel* leaf extract and their biological studies. Mater. Lett. 236: 19–22.

Waghmare, S.S., Deshmukh, A.M., Sadowski, Z. 2014. Biosynthesis, optimization, purification and characterization of gold nanoparticles. African J. Microbiol. Res. 8: 138–146.

Yadav, A., Kon, K., Kratosova, G., Duran, N., Ingle, A.P., Rai, M. 2015. Fungi as an efficient mycosystem for the synthesis of metal nanoparticles: progress and key aspects of research. Biotechnol. Lett. 37: 2099–2120.

Yeasmin, S., Datta, H.K., Chaudhuri, S., Malik, D., Bandyopadhyay, A. 2017. *In-vitro* anti-cancer activity of shape controlled silver nanoparticles (AgNPs) in various organ specific cell lines. J. Mol. Liq. 242: 757–766.

Yousefzadi, M., Rahimi, Z., Ghafori, V. 2014. The green synthesis, characterization and antimicrobial activities of silver nanoparticles synthesized from green alga *Enteromorpha flexuosa* (wulfen). J. Agardh. Mater. Lett. 137: 1–4.

Zabrieski, Z., Morrell, E., Hortin, J., Dimkpa, C., McLean, J., Britt, D., Anderson, A. 2015. Pesticidal activity of metal oxide nanoparticles on plant pathogenic isolates of *Pythium*. Ecotoxicology 24(6): 1305–1314.

Zayed, M.F., Eisa, W.H. 2014. *Phoenix dactylifera* L. leaf extract phytosynthesized gold nanoparticles; controlled synthesis and catalytic activity. Spectrochim. Acta Part A Mol. Biomol. Spectrosc. 121: 238–244.

Chapter 9

Microbe-mediated Synthesis of Zinc Oxide Nanoparticles and Its Biomedical Applications

Happy Agarwal, Amatullah Nakara, Soumya Menon and
Venkat Kumar Shanmugam*

Introduction

Nanotechnology is the manipulation of material such that it gains certain unique properties not exhibited by its bulk counterparts. Nanomaterials are materials whose components are in the size range of 1 to 100 nm. Nanoparticles possess unique physical and chemical properties that are dependent on their size, shape, structure, and stability. Owing to characteristics not exhibited by the bulk material, nanoparticles have emerged as a novel drug molecule. They are suitable for catalysis, bio-sensing, medical imaging, remediation, agriculture, drug delivery, healthcare, and many other multi-disciplinary applications. They have been extensively used in these fields because of their high stability, solubility, biocompatibility and therapeutic properties (Khan et al. 2017, Jeevanandam et al. 2018). Different metal and metal oxide nanoparticles are being employed in cosmetic formulations, textile industries, and medical and pharmaceutical applications.

Owing to the rising demand for nanoparticles on the grounds of their extensive applications, a wide range of physical and chemical techniques have been developed and optimized for the production of nanoparticles with different morphologies. Traditionally, nanoparticles have been synthesized using various physicochemical techniques such as sol-gel synthesis, micro-emulsion hydrothermal synthesis, solvothermal synthesis, microwave, thermal decomposition, precipitation, laser ablation, and surfactant-assisted hydrothermal growth, which are also employed

School of Bio-Sciences and Technology, Vellore Institute of Technology, Vellore 632014, Tamil Nadu, India.
* Corresponding author: venkatkumars@vit.ac.in; drvenkatshanmugam@gmail.com

for synthesis of zinc oxide nanoparticles (ZnONPs) (Aminuzzaman et al. 2018). These physicochemical routes often employ materials with a potential threat of carcinogenicity and environmental toxicity. A green approach for nanoparticle synthesis is preferred to physical and chemical methods since it requires ambient temperatures and an approximately neutral pH. Researchers are trying to replace the physical and chemical methods of nanoparticle synthesis with biological methods employing plants, algae, and microbes such as bacteria and fungi. Biosynthesis of nanoparticles using such organic sources eliminates the need to use toxic and expensive chemicals as reducing and capping agents. Therefore, biosynthesis of nanoparticles has particularly proved to be useful for medical and therapeutic applications since it provides an eco-friendly, rapid, non-toxic and cost-effective technique of synthesis of nanoparticles with better catalytic and therapeutic activities on a large scale (Mittal et al. 2013, Shah et al. 2015).

The natural environment contains high levels of metals and metal oxides and, for their survival, microbes have evolved mechanisms to cope with the metal stress. Because of this reducing property of microbes, they can be used as reducing agents for nanoparticle synthesis from precursors (Yusof et al. 2019). Extracellular or intracellular microbial extracts are typically used depending on the type, morphology, and application of the nanoparticles to be synthesized (Iravani 2014).

The present chapter focuses on microbe-mediated nanoparticle synthesis employing various bacteria and fungi. It gives a brief literature review highlighting the use of various microbial species for biosynthesis of ZnONPs having different morphological features and applications. Moreover, it provides a concise overview of the mechanisms of antibacterial, anti-inflammatory, anti-cancerous and wound healing applications of nanoparticles synthesized using the biological route.

ZnONPs and their properties

Nanoparticles exhibit different properties from their corresponding bulk counterparts as well as between different nanoforms of the same metal/metal oxide nanoparticles because of their smaller ratio of surface area to volume. Changes in these physical characteristics can cause a significant change in their stability as well as reactivity with biological macromolecules (Schwirn et al. 2014). ZnONPs have been synthesized using plants, microbes (bacteria and fungi), and algae, as well as other biocompatible chemicals.

ZnONPs synthesized using such green methods also demonstrate significant interesting physicochemical properties that govern their interactions with biological molecules. Some notable biological and chemical properties include their antibacterial (Ramesh et al. 2015, Ogunyemi et al. 2019), antifungal (Rajiv et al. 2013, Shobha et al. 2019), antioxidant (Suresh et al. 2015, Iqbal et al. 2019), anti-inflammatory (Nagajyothi et al. 2015, Mohammad et al. 2019), anti-diabetic (Bala et al. 2015, Bayrami et al. 2019), anti-cancerous (Baskar et al. 2015, Karthikeyan et al. 2019), pesticidal (Kirthi et al. 2011) and photocatalytic properties (Hong et al. 2009, Khalafi et al. 2019). They also demonstrate significant physical and optical properties such as magnetic properties (Garcia et al. 2007), piezoelectric properties (He et al. 2007), oxygen-sensing and thermal conducting properties (Singh 2010), and UV-blocking properties (Cross et al. 2007).

These unique properties of ZnONPs are due to their ability to produce reactive oxygen species (ROS) inside cancerous cells, ability to activate tumor suppressor genes p53, p21 and JNK, inhibition of cytokine activity, regulation of some enzymes involved in vital metabolic pathways of cells, disruption of cell membrane integrity in certain pathogenic microbial species, and upregulation of fibroblast activity in damaged cells. All these properties attributed to ZnONPs are due to their smaller size allowing for better penetration and facilitation of action inside cells, and ZnONPs can be used in a dose- and time-dependent manner to furnish novel properties in the system in which they are incorporated.

Microbe-mediated synthesis of ZnONPs

Various species of bacteria and fungi have been exploited for the eco-friendly synthesis of ZnONPs. Both prokaryotic and eukaryotic microbes can be used. These NPs have been extensively synthesized using extracts of various bacterial and fungal cultures. They are generally synthesized using precipitation method employing zinc-containing precursors, usually zinc oxide or zinc acetate that is reduced to ZnONPs form by freshly grown cultures of the microbes. Culture solutions inoculated with the precursor usually are observed to have white coalescent clusters after transformation (Jayaseelan et al. 2012, Rajan et al. 2016). Two types of synthesis methods are commonly used: intracellular and extracellular. In intracellular synthesis, the microbial biomass is incubated for a certain time period with the precursor solution; in extracellular synthesis, the filtrates are treated with precursors. Pathways of NP synthesis by microbes involve common cellular biochemical processes, metal ion flux in and out of cells, nucleation of metal oxides, and microbial resistance mechanisms towards toxic metals and their oxides (Naveed Ul Haq et al. 2017).

A probiotic bacterium *Lactobacillus plantarum* VITES07 is reported to have been used for the synthesis of spherical, polydispersed ZnONPs having size 7–19 nm using zinc sulfate as precursor (Selvarajan and Mohanasrinivasan 2013). *Acinetobacter schindleri* S1Z7 (a rod-shaped bacterium) was used for the extracellular biosynthesis of ZnONPs. The ZnONPs thus formed were found to be 20–100 nm, spherical and polydispersed. The ZnONPs synthesized by *A. schindleri* were found to have antibacterial activity against *Escherichia coli, Salmonella enterica, Staphylococcus aureus* and *Vibrio parahaemolyticus* (Busi et al. 2016). ZnONPs synthesized by *Aeromonas hydrophilia* were found to have an average particle size of 57.72 nm, spherical and oval shapes, and a crystalline nature. They were reported to have antibacterial activity against *Pseudomonas aeruginosa, S. aureus, Streptococcus pyogenes, Enterococcus faecalis* and *E. coli* and antifungal activity against *Aspergillus flavus, Aspergillus niger* and *Candida albicans* (Jayaseelan et al. 2012).

Another study reported the extracellular mycosynthesis of ZnONPs by the fungus *Alternaria alternata* using zinc sulfate as a reducing agent. The ZnONPs synthesized by this method were reported to have spherical, triangular and hexagonal structures and a size range of 45–150 nm with an average particle size of 75 ± 5 nm. They were reported to have a decrease in mitochondrial dehydrogenase activity and an increase in DNA fragmentation, indicating the cytotoxic effects of the biosynthesized NPs (Sarkar et al. 2014). ZnONPs were also reported to be synthesized from the extracellular proteins from a cell-free filtrate of the fungal isolate *Aspergillus aeneus* NJP12 using zinc acetate as a precursor. The synthesized ZnONPs were found to be spherical and had a size range of 100–140 nm (Jain et al. 2013). The extracellular cell-free filtrate of *Aspergillus fumigatus* TFR-8 has also been reported to be used in the synthesis of ZnONPs with zinc nitrate as a precursor. The ZnONPs synthesized were reported to have spherical and hexagonal structures and a size range of 1.2–6.8 nm with an average size of 3.8 nm. The NPs had a crystalline nature and were monodispersed. These NPs were reported to cause a significant improvement in chlorophyll content and plant growth as well as overall plant biomass in clusterbean crop (Raliya and Tarafdar 2013). Similarly, *Aspergillus niger* was reported to be used for the extracellular synthesis of ZnONPs using zinc nitrate as a precursor. The synthesized NPs were spherical and compactly arranged with a size range of 84–91 nm. They showed antibacterial activity against *S. aureus* and *E. coli* and showed catalytic activity towards degradation of Bismarck Brown dye (Kalpana et al. 2018).

Another study reported the synthesis of ZnONPs using *Bacillus cereus* cells as a biotemplate with raspberry and plate-like NP structures with a size range of 20–30 nm and with zinc acetate as a precursor (Hussein et al. 2019). *Bacillus licheniformis* has been used for the biosynthesis of ZnO nano-flowers using zinc acetate dihydrate as a precursor. The size of these nano-flowers ranged from 250 nm to 1 μm with a mean size of 620 nm and they had good crystallinity. The synthesized ZnONPs had good photostability and possessed good photocatalytic activity by photodegrading a pollutant dye (Tripathi et al. 2014). ZnONPs of size range 20–30 nm and quasi-spherical shapes have been synthesized using *Bacillus subtilis* suspensions as eco-friendly reducing and capping agents. The synthesized NPs showed good recyclability and better efficiency as catalysts in thiophene synthesis (Shamsuzzaman et al. 2014). ZnONPs having an average size of 19.4 nm and a hexagonal wurtzite structure were synthesized in presence of zinc acetate as a precursor using another *Bacillus* species, *B. thuringiensis*. They demonstrated a reduction in fecundity and a prolonged total development period of an insect pest *Callosobruchus maculatus* and its mortality at higher concentrations, thereby proving its biopesticidal effects (Malaikozhundan et al. 2017). ZnONPs have also been synthesized using *Candida albicans* as a reducing and capping agent. The size of the NPs ranged from 15 to 25 nm and they were quasi-spherical. They were reported to have good catalytic properties for the fast and effective synthesis of steroidal pyrazolines (Shamsuzzaman et al. 2017). ZnONPs have been synthesized from zinc sulfate solution using *Enterococcus faecalis* culture. The NPs were spherical and monodispersed and had an average size of 16–96 nm. They showed antibacterial activity against *E. coli*, *Klebsiella pneumoniae*, and *S. aureus* (Ashajyothi et al. 2014). *Lactobacillus sporogens* has been used for the biosynthesis of ZnONPs having a hexagonal structure and a size range of 5–15 nm (Prasad and Jha 2009). *Pseudomonas aeruginosa* rhamnolipids have been used in the biosynthesis of anti-oxidant ZnONPs using zinc nitrate as a precursor. The NPs had a size range of 35–80 nm and were spherical (Singh et al. 2014). Green synthesis of ZnONPs (16.35 nm) has been reported using the Gram-negative bacterium *Pseudomonas stutzeri*. The NPs showed antibacterial activity against *E. coli* and *S. aureus* and also showed magnetic properties (Mirhendi et al. 2013). Biogenic ammonia from ureolytic bacterium *Serratia ureilytica* has been reported to be used in the synthesis of ZnONPs (300–600 nm) of spherical to flower-like shapes using zinc acetate as a precursor. The NPs exhibited an increase in tensile strength in

NP-coated cotton fabrics and also exhibited antibacterial activity against *E. coli* and *S. aureus* (Dhandapani et al. 2014). Another study reported the extracellular biosynthesis of ZnONPs from the actinobacterium *Rhodococcus pyridinivorans* NT2 using zinc sulfate as a precursor. The NPs were roughly spherical and had an average size of 100–120 nm. Textiles coated with the NPs synthesized using this method showed good UV-blocking properties and improved self-cleaning properties and also exhibited antibacterial activity against *Staphylococcus epidermidis*. The anthraquinone released from the biosynthesized NPs exhibited an ability to kill HT-29 cancer cells (Kundu et al. 2014).

Biomedical applications

Antibacterial property

The high surface-to-volume ratio of nano ZnO as well as enhanced physicochemical properties enables better nanoparticle–cell interaction. ZnONPs have demonstrated a concentration- and time-dependent antibacterial activity against a wide range of Gram-positive as well as Gram-negative bacteria. NPs enter the cell through pores in the cell membrane or through transmembrane ion channels called porins that facilitate passive diffusion of NPs into the cell. The mode of ingestion of the NP depends upon its size and concentration. NPs can also be internalized by endocytosis (Simkó et al. 2011, Sirelkhatim et al. 2015). One of the major antibacterial mechanisms of ZnONPs is disruption of cell membrane integrity due to the electrostatic interaction between the NPs and cell surface, leading to the formation of membrane blebs and irregular cell surfaces (Agarwal et al. 2018). Teichoic and lipoteichoic acids on the peptidoglycan layer of Gram-positive and Gram-negative bacteria respectively confer a negative charge to bacterial cell walls and provide attachment sites for ZnONPs. They also act as a chelating agent by chelating Zn^{2+} from ZnO, which is then transported into the cell. NP attachment and incorporation alter the resting membrane potential of the cell membrane and kindle the depolarization of the cell membrane (Warren and Payne 2015). This disrupts the external morphology of the bacterial cells, which leads to leakage of intracellular components such as reducing sugars, proteins, and DNA, eventually resulting in a decline in cell viability (He and Chen 2010).

Studies have reported the formation of ROS after ZnONP internalization into the bacterial cells, primarily O_2^-, OH⁻ and H_2O_2, which are strong oxidizing agents. This was inferred by a significant increase in the transcription levels of oxidative stress genes and stress-response genes (Xie et al. 2011). The ROS are anticipated to be produced as a result of the redox reaction of the electrons trapped at the oxygen valence sites of the NPs. The ROS further disrupts or alters protein synthesis cycle or DNA replication, causing cell death. It also causes membrane damage by creating holes on the membrane surface, which increases its overall surface area. This increased surface area assists in more absorption of ROS on the surface, causing an autocrine stimulation of ROS (Mittal et al. 2014). Bacterial cellular components such as proteins and nucleic acids promote the dissolution of ZnO by forming ionic salts with Zn^{2+} ions. Though Zn^{2+} ions are essential for microbial metabolism, an excess of Zn^{2+} due to internalization of the NPs disrupts the Zn^{2+} homeostasis of the microbe, leading to intracellular cytotoxicity (Joe et al. 2017). There is also a consequential increase in the Zn^{2+} levels in the bacterial cytoplasm that causes a loss of proton motive force and subsequently membrane leakage (Sirelkhatim et al. 2015). Another mechanism reported to induce bacterial cell death is the inhibition of vital enzymes involved in essential metabolic pathways. Zn^{2+} may distort the active site of cardinal enzymes or bind to them via various macromolecular interactions, thereby bringing about a conformational change in their structure. This inhibits or inactivates polymerases, dehydrogenases, and phosphatases, which are involved in essential metabolic pathways of the organism (Maret 2013). ZnONP concentration above threshold levels in bacterial cells has been shown to inhibit vital enzymes such as glutathione reductase and thiolperoxidase. Glutathione reductase is a crucial enzyme involved in the glutathione redox cycle that helps in maintaining glutathione balance in the cell. Reduced glutathione helps resist oxidative stress by reacting with ROS. Thiolperoxidase is responsible for catalyzing the reduction of ROS such as H_2O_2. By inhibiting vital enzymes, ZnONPs consequently inhibit the destruction of ROS, favoring cell death (Kumar et al. 2011, Ji et al. 2015). It is also shown that ZnONPs selectively inhibit the adenylyl cyclase pathway (cAMP-dependent pathway) in *Vibrio cholerae*, thereby downregulating cAMP levels. cAMP (cyclic AMP) is a secondary messenger and an ATP derivative that is involved in various intracellular signal transduction cascades like DNA transcription, regulation of gene expression and enzyme activation (Salem et al. 2015). The various mechanisms of the antibacterial activity of ZnONPs are summarized in Fig. 9.1.

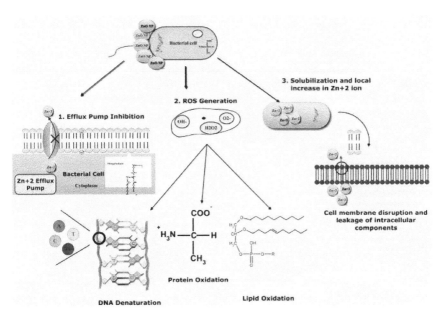

Figure 9.1. Anti-bacterial activity of ZnONPs. Printed with permission from Elsevier self-publication, Chemico-Biological Interactions 286(2018): 60–70, license no. 4655741037987.

Anti-inflammatory property

Inflammation is caused by an external injury or pathogenic infections. It is a localized physical condition in which swelling (oedema), redness, heat sensation, and pain advances in the inflamed area. The redness and heat sensation are caused by the increased number of erythrocytes migrating to the site of inflammation caused by an excess of blood flow through the dilated blood vessels. These effects are mediated by secretion of excessive coagulation factors, cytokines and other cytotoxic inflammatory mediators that intensify the advancement of inflammation through vasodilation and extravasation of complement factors and neutrophils to the site. Macrophages and neutrophils play a key role in auto-regulation of inflammatory processes (Ahmed 2011). On ingestion, NPs enter the circulatory system and interact with the proteins present in the blood plasma. The interaction leads to the formation of a protein corona around the NPs composed mainly of serum proteins, since they have a high affinity to bind to the NPs (Walkey and Chan 2012, Vinluan and Zheng 2015). NP uptake by cells involves adhesive interactions due to electrostatic or Van der Waals forces. Small NPs are readily endocytosed by cellular vesicles at higher concentrations. Neutrophils and macrophages also carry out phagocytosis and pinocytosis to engulf the NPs (Kuhn et al. 2014). ZnONPs have been shown to suppress lipopolysaccharide (LPS)-induced cyclooxygenase 2 (COX2) gene activation in macrophages in a dose-dependent manner. LPS found in the cell wall of Gram-negative bacteria induces expression of COX2 gene, which in turn causes the release of prostaglandin-E2 (PG-E2), an inflammation promoter, and other inflammatory lipids. LPS also induces macrophages to secrete leukotrienes and pro-inflammatory cytokines such as tumor necrosis factor α (TNF-α) and interleukin 1β (IL-1β). Therefore, ZnONPs inhibit release of the inflammatory cytokines as well as PG-E2 (Britt et al. 2012). They are also proven to suppress LPS-induced nuclear factor κB (NF-κB) expression and nuclear translocation in RAW 264.7 macrophages. NF-κB is responsible for switching on the expression of cytokines and other genes that affect cell proliferation. This subsequently blocks IL-1β and TNF-α production (pro-inflammatory cytokines) that are involved in the promotion of differentiation and proliferation of mast cells (Kim and Jeong 2015). ZnONPs are also shown to block the caspase-1 enzyme in activated mast cells that release various inflammatory mediators like cytokines, chemokines, and leukotrienes in immune disorders. Caspase-1 is responsible for the conversion of pro-IL 18 and pro-IL 1β to their active forms IL-18 and IL-1β cytokines respectively (Liang et al. 2010). LPS along with IFN-γ, a cytokine, upregulate mRNA and protein levels of inducible nitric oxide synthase (iNOS); that in turn leads to a significant increase in nitric oxide (NO) production in the cell. iNOS expression is involved in chronic inflammatory diseases *in vivo*. NO is highly toxic to cells as it causes local tissue destruction as well as instigates toxic chemical reaction in tissues, becoming a direct cause of inflammation. ZnONPs have also been shown to bring about a significant decline in NO production in LPS plus IFN-γ stimulated macrophages (Nagajyothi et al. 2015). Epithelial cells secrete thymic stromal lymphopoietin (TSLP) in response to inflammatory stimuli such as external or internal injuries, invasion by pathogenic microbes and already existing cytokines. TSLP induces production of IL-1, IL-13, and TNF-α in mast cells. ZnONPs are known to significantly reduce TSLP production in inflamed epithelial cells (Allakhverdi et al. 2007). Another mechanism of inflammation attenuation by ZnONPs is the regulation of p53 expression. Persistence of p53 activation (a tumor suppressor gene) is known to induce activation of high mobility group protein 1 (HMG-1). HMG-1 plays a role in selective cytokine activation as well as chemotaxis of neutrophils and macrophages to the site of inflammation, thereby amplifying the inflammatory response (Yan et al. 2013).

Anti-cancer property

ZnONPs are a novel and efficient tool for anti-cancer therapy. ZnONPs have demonstrated a strong preferential ability to kill cancerous T cells compared to normal healthy cells. Normal resting T cells show a better resistance to ZnONP cytotoxicity. T cells activated through TCR and CD28 co-stimulation pathways result in excessive cell division and proliferation. Cellular processes accompanying T cell activation such as increased membrane protein (CD40L) expression and cell cycle progression result in certain physicochemical changes in the cell that facilitate strong NP interaction with the cell and consequently better intracellular uptake. This experiment proves the selective targeting of proliferating cancerous cells by ZnONPs (Hanley et al. 2008, Foell et al. 2007). Because of rapid and uncontrolled proliferation of tumor cells, the blood and lymphatic vessels developed as a result of angiogenesis undergo improper development and consequently have pores ranging up to a few micrometers. Therefore, NPs can easily diffuse through the blood vessels towards a cancer cell. Owing to the poorly developed vasculature, NPs have enhanced retention time inside the cancer cell (Rasmussen et al. 2010). Efficiency of targeting of ZnONPs to vulnerable areas in the cancer system can be enhanced by doping (Thurber et al. 2012) or synthesis of nanocomposites (Bisht et al. 2016). Strong NP–cell interaction paves a way for the uptake of ZnONPs through phagocytosis or pinocytosis or passive transport through the cell membrane. NP entry into the cell increases the Zn^{2+} ion concentration and disrupts the Zn^{2+} equilibrium in the cell. Zn^{2+} is a co-factor for a large number of mammalian enzymes. This leads to deregulation in the activity of the enzymes for which Zn^{2+} is a co-factor and affects crucial cellular processes, including DNA replication, damage repair, protein functions, and metabolic cycles, manifesting cellular dysfunction and ultimately apoptosis. It is also observed that a decrease in pH facilitates dissolution of Zn^{2+} ions as the endosome progresses to the lysosome stage. This enhances the dissolution of Zn^{2+} in the lysosome, causing its destabilization (Shen et al. 2013, Bisht and Rayamajhi 2016). The ROS-generating mechanism is attributed to the semiconducting nature of ZnO. Being a semiconductor, the valence and conduction band of ZnO are separated by an energy gap. In the case of nano ZnO, the energy gap decreases significantly, and the electrons can jump from valence band to conduction band. Electrons and holes in case of nano ZnO move to the NP surface. Here, the particle size determines the number of reactive sites on its surface. Holes being powerful oxidants break water molecules in the cell to H^+ and OH^-, while the electrons are powerful reducing agents and react with the dissolved and absorbed O_2 molecules in the cell, generating O_2^-. These are ROS themselves that further combine to give H_2O_2 (Rasmussen et al. 2010). ROS production that exceeds the antioxidant defence system of the cell results in a state of oxidative stress in the cell, leading to lipid peroxidation and protein denaturation, impairment of essential metabolic pathways, depolarization of mitochondrial membrane due to membrane phospholipid damage, followed by damage of DNA due to DNA fragmentation and damage of other vital cellular components, ultimately leading to cell apoptosis. These ROS can further react with cellular biomolecules to produce reactive nitrogen species (RNS) such as highly toxic and reactive nitric oxide and peroxynitrite ($ONOO^-$) (Manke et al. 2013, Song et al. 2010). ROS also react with DNA, causing DNA-protein cross-links or DNA single-stranded breaks due to the formation of DNA adducts. This leads to loss of mitochondrial membrane potential, which activates the mitochondrial apoptotic pathway releasing apoptotic proteins in the cytosol and, eventually, cell death by apoptosis (Shi et al. 2004, Sharma et al. 2012). Experimental studies in MCF-7 breast cancer cell line also report an increase in p53, p21, and JNK (all tumor suppressor genes) after treatment with ZnONPs. p53/p21 is responsible for controlling cell cycle checkpoints at various phases in a cell cycle. If DNA damage is detected, it activates downstream signaling to correct the defect or initiates apoptosis if the defect cannot be corrected. p53 initiates apoptosis in cancer cells that have sustained DNA damage due to the various mechanisms mentioned above by ZnONPs. NP-induced apoptosis was shown to occur through both extrinsic and intrinsic apoptotic pathways by upregulation of pro-apoptotic genes p53, p21 and JNK and downregulation of anti-apoptotic genes AKT1 and Bcl-2 (Moghaddam et al. 2017). ZnONPs are also shown to arrest cell cycle at the G2/M checkpoint, bringing about epigenetic changes and thus activating the mitochondrial pathway of apoptosis in human epidermal keratinocytes (Gao et al. 2016). Studies have also shown that ZnONPs are able to cause impairment of autophagic flux leading to an increased count of autophagosomes and autolysosomes and increase in a number of autophagic markers in A549 cells, favoring autophagy in cancer cells (Zhang et al. 2017). The mechanisms of anti-cancer activity of ZnONPs are summarized in Fig. 9.2.

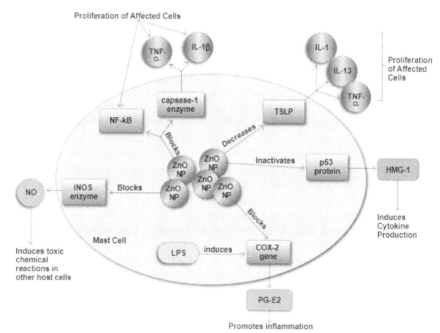

Figure 9.2. Anti-inflammatory activity of ZnONPs. Reproduced from Elsevier open access article. https://doi.org/10.1016/j.biopha.2018.11.116.

Wound healing property

ZnO nanocomposites have extensively been used in various pharmaceutical drugs and wound dressings owing to their competent wound healing properties. The antibacterial and anti-inflammatory activities of ZnONPs primarily contribute to its wound healing property. The major mechanism behind the wound healing activity is the selective production of low concentration of ROS (Kaushik et al. 2019). ZnONPs are known to selectively induce ROS production in damaged cells. At lower concentrations, ROS induces epithelial cell and smooth muscle cell proliferation and migration and also induces vascular endothelial growth factor (VEGF) in human keratinocytes that influence wound closure and epidermal repair (Sen et al. 2002). Polycaprolactone membranes containing ZnONPs have been demonstrated to enhance fibroblast proliferation due to the production of ROS at low concentration and a subsequent increase of p38 mitogen-activated protein kinase (MAPK), which acts as a growth kinase during the wound healing process. This fibroblast proliferation and differentiation leads to the formation of myofibroblasts that secrete and organize the extracellular matrix (ECM) and participate in contracting the wound (Augustine et al. 2014). ZnO nanocomposite hydrogels are shown to enhance fibroblast activity and assist wound healing by their antibacterial activity against various Gram-negative bacteria (Raguvaran et al. 2017). The astringent property of ZnONPs facilitates the coagulation of soluble proteins, aiding the wound healing activity. Zn being a co-factor for the matrix metalloproteinase (MMP) enzyme helps in upregulation of MMPs during tissue remodeling. MMPs in addition to degrading and remodeling the ECM also regulate cell–cell and cell–matrix signaling through the release of growth factors. MMPs are also involved in releasing biologically active fragments of some degraded proteins that play a role in keratinocyte migration to the wounded site (Rath et al. 2016). Nanofibres containing ZnONPs showed a noteworthy fibrous connective tissue proliferation (epithelialization) and a sequential regeneration of the keratin layer in Wistar rat models (Rath et al. 2016). ZnONP-based bandages have been shown to induce faster re-epithelialization, blood clotting and enhanced collagen deposition, which is a major component of the connective tissue in the skin (Kumar et al. 2012). ZnONP-containing gels have been shown to have a significant antibacterial effect against some opportunistic pathogens and to release ZnONPs in a controlled manner with high overall stability; they are thus useful in wound dressings (Vasile et al. 2014).

Conclusion and future perspectives

The microbe-mediated synthesis of zinc oxide nanoparticles and their antibacterial, anti-inflammatory, anti-cancer and wound healing activities have been discussed in this chapter. It accentuates the significance of microbe-mediated ZnONP synthesis and its range of applications from existing literature. It also emphasizes the various mechanisms of action under each property and their interrelationships. The antibacterial property of ZnONPs is predominantly due to the disorganization of cell membrane integrity following interactions with the cell membrane leading to blebbing and consecutive cell death. Inactivation of key enzymes involved in essential metabolic pathways of the microbe and excessive toxic ROS production are other causes of microbial cell death mediated by ZnONPs. The prime cause of the anti-inflammatory property of ZnONPs is their inhibition of various inflammatory cytokines and chemokines and other inflammation promoters. They also decrease nitric oxide over-production and regulate expression of some tumor suppressor genes. The anti-cancer property of ZnONPs is facilitated by upregulation of tumor suppressor genes and cancer cell apoptosis by ROS generation and disturbance in Zn^{2+} homeostasis. The wound healing property of the nanoparticles is due to their stimulation of enhanced fibroblast activity, upregulation of growth kinases and promotion of re-epithelialization by certain enzymes.

The antibacterial, anti-inflammatory, anti-cancerous and wound healing properties of ZnONPs can be exploited for use in drug delivery systems, drug formulations, remediation, cosmetics, and other medical and industrial applications. ZnONPs have an excellent therapeutic application as components of some therapeutic drugs or drug carriers because they increase the bioavailability of drug molecules and have a therapeutic potential themselves. This property of ZnONPs could be employed to increase therapeutic efficacy of drug molecules. ZnONPs also are potential candidates for application as antibacterial and anti-cancerous agents with limited potential in a dose-dependent manner. The dose could be optimized to increase efficiency as well as prevent cytotoxic side effects. ZnONPs must be further explored as they hold the promise of remarkable applications in biomedical, diagnostic and therapeutic fields.

Acknowledgment

We would like to thank Dr. Thirunavukkarasu Chinnasamy for granting permission to reuse Fig. 9.2.

References

Agarwal, H., Menon, S., Venkat Kumar, S., Rajeshkumar, S. 2018. Mechanistic study on antibacterial action of zinc oxide nanoparticles synthesized using green route. Chem. Biol. Interact. 286(march): 60–70. Https://doi.org/10.1016/j.cbi.2018.03.008.

Ahmed, A.U. 2011. An overview of inflammation: mechanism and consequences. Front. Biol. 6(4): 274. Https://doi.org/10.1007/s11515-011-1123-9.

Allakhverdi, Z., Corneau, M.R., Jessup, H.K., Yoon, B.R., Brewer, A., Chartier, S., Paquette, N., Ziegler, S.F., Sarfati, M., Delespesse, G. 2007. Thymic stromal lymphopoietin is released by human epithelial cells in response to microbes, trauma, or inflammation and potently activates mast cells. J. Exp. Med. 204(2): 253–58. Https://doi.org/10.1084/jem.20062211.

Aminuzzaman, M., Ying, L.P., Goh, W-S., Watanabe, A. 2018. Green synthesis of zinc oxide nanoparticles using aqueous extract of *Garcinia mangostana* fruit pericarp and their photocatalytic activity. Bull. Mater. Sci. 41(2): 50. Https://doi.org/10.1007/s12034-018-1568-4.

Ashajyothi, C., Manjunath, Rahul Narasanna, Chandrakanth R. Kelmani. 2014. Antibacterial activity of biogenic zinc oxide nanoparticals synthesised from *Enterococcus faecalis*. Int. J. Chemtech. Res. 6(5): 3131–36.

Augustine, R., Dominic, E.A., Reju, I., Kaimal, B., Kalarikkal, N., Thomas, S. 2014. Electrospun polycaprolactone membranes incorporated with ZnO nanoparticles as skin substitutes with enhanced fibroblast proliferation and wound healing. Rsc. Adv. 4(47): 24777–85. Https://doi.org/ 10.1039/c4ra02450h.

Bala, N., Saha, S., Chakraborty, M., Maiti, M., Das, S., Basu, R., Nandy, P. 2015. Green synthesis of zinc oxide nanoparticles using *Hibiscus subdariffa* leaf extract: effect of temperature on synthesis, anti-bacterial activity and anti-diabetic activity. Rsc. Adv. 5(7): 4993–5003. Http://dx.doi.org/10.1039/c4ra12784f.

Baskar, G., Chandhuru, J., Sheraz Fahad, K., Praveen, A.S., Chamundeeswari, M., Muthukumar, T. 2015. Anticancer activity of fungal L-asparaginase conjugated with zinc oxide nanoparticles. J. Mater. Sci. Mater. Med. 26(1): 43. Https://doi.org/10.1007/s10856-015-5380-z.

Bayrami, A., Alioghli, S., Rahim Pouran, S., Habibi-Yangjeh, A., Khataee, A., Ramesh, S. 2019. A facile ultrasonic-aided biosynthesis of ZnO nanoparticles using *Vaccinium arctostaphylos*

L. Leaf extract and its antidiabetic, antibacterial, and oxidative activity evaluation. Ultrason. Sonochem. 55: 57–66. Http://www.sciencedirect.com/science/article/pii/s1350417719301488.

Bisht, G., Rayamajhi, S. 2016. ZnO nanoparticles: a promising anticancer agent. Nanobiomedicine 3: 9. Https://www.ncbi.nlm.nih.gov/pubmed/29942384.

Bisht, G., Rayamajhi, S., Paudel Sn, K.C.B., Karna, D., Shrestha, B.G. 2016. Synthesis, characterization, and study of *in vitro* cytotoxicity of Zno-Fe$_3$O$_4$ magnetic composite nanoparticles in human breast cancer cell line (MDA-MB-231) and mouse fibroblast (NIH 3t3). Nanoscale Res. Lett. 11(1): 537. Https://www.ncbi.nlm.nih.gov/pubmed/27914092.

Boroumand Moghaddam, A., Moniri, M., Azizi, S., Abdul Rahim, R., Bin Ariff, A., Navaderi, M., Mohamad, R. 2017. Eco-friendly formulated zinc oxide nanoparticles: induction of cell cycle arrest and apoptosis in the MCF-7 cancer cell line. Genes 8(10): 281. Https://www.ncbi.nlm.nih.gov/pubmed/29053567.

Britt Jr., R.D., Locy, M.L., Tipple, T.E., Nelin, L.D., Rogers, L.K. 2012. Lipopolysaccharide-induced cyclooxygenase-2 expression in mouse transformed clara cells. Cell. Physiol. Biochem. 29(1-2): 213–22. Http://www.ncbi.nlm.nih.gov/pmc/articles/pmc3487147/.

Busi, S., Rajkumari, J., Pattnaik, S., Vasudevan, V.V., Hnamte, S. 2016. Extracellular synthesis of zinc oxide nanoparticles using *Acinetobacter* SIZ7 and its antimicrobial property against foodborne pathogens. J. Microbiol. Biotechnol. Food Sci. 5.

Cross, S.E., Innes, B., Roberts, M.S., Tsuzuki, T., Robertson, T.A., Mccormick, P. 2007. Human skin penetration of sunscreen nanoparticles: *in-vitro* assessment of a novel micronized zinc oxide formulation. Skin Pharmacol. Physiol. 20(3): 148–54. Https://www.karger.com/doi/10.1159/000098701.

Dhandapani, P., Siddarth, A.S., Kamalasekaran, S., Maruthamuthu, S., Rajagopal, G. 2014. Bio-approach: ureolytic bacteria mediated synthesis of zno nanocrystals on cotton fabric and evaluation of their antibacterial properties. Carbohydr. Polym. 103: 448–55. Http://dx.doi.org/10.1016/j.carbpol.2013.12.074.

Foell, J., Hewes, B., Mittler, R.S. 2007. T cell costimulatory and inhibitory receptors as therapeutic targets for inducing anti-tumor immunity. Curr. Cancer Drug Targets 7(1): 55–70. Http://www.eurekaselect.com/node/58740/article.

Gao, F., Ma, N., Zhou, H., Wang, Q., Zhang, H., Wang, P., Hou, H., Wen, H., Li, L. 2016. Zinc oxide nanoparticles-induced epigenetic change and G2/M arrest are associated with apoptosis in human epidermal keratinocytes. Int. J. Nanomedicine 11: 3859–74. Https://www.ncbi.nlm.nih.gov/pubmed/27570453.

Garcia, M.A., Merino, J.M., Fernandez Pinel, E., Quesada, A., De La Venta, J., Ruiz Gonzalez, M.L., Castro, G.R., Crespo, P., Llopis, J., Gonzalez-Calbet, J.M., Hernando, A. 2007. Magnetic properties of ZnO nanoparticles. Nano Lett. 7(6): 1489–94. Https://doi.org/10.1021/nl070198m.

Hanley, C., Layne, J., Punnoose, A., Reddy, K.M., Coombs, I., Coombs, A., Feris, K., Wingett, D. 2008. Preferential killing of cancer cells and activated human T cells using ZnO nanoparticles. Nanotechnology 19(29): 295103. Https://www.ncbi.nlm.nih.gov/pubmed/18836572.

He, J.H., Hsin, C.L., Liu, J., Chen, L.J., Wang, Z.L. 2007. Piezoelectric gated diode of a single ZnO nanowire. Adv. Mater. 19(6): 781–84. Https://doi.org/10.1002/adma.200601908.

He, Y., Chen, C-Y. 2010. Quantitative analysis of viable, stressed and dead cells of *Campylobacter jejuni* strain 81-176. Food Microbiol. 27(4): 439–46. Http://www.sciencedirect.com/science/article/pii/s0740002009002755.

Hong, R.Y., Li, J.H., Chen, L.L., Liu, D.Q., Li, H.Z., Zheng, Y., Ding, J. 2009. Synthesis, surface modification and photocatalytic property of ZnO nanoparticles. Powder Technol. 189(3): 426–32. Http://www.sciencedirect.com/science/article/pii/s0032591008003677.

Hussein, M.Z., Azmin, W.H.W.N., Mustafa, M., Yahaya, A.H. 2009. *Bacillus cereus* as a biotemplating agent for the synthesis of zinc oxide with raspberry- and plate-like structures. J. Inorg. Biochem. 103(8): 1145–50. Http://dx.doi.org/10.1016/j.jinorgbio.2009.05.016.

Iqbal, J., Abbasi, B.A., Mahmood, T., Kanwal, S., Ahmad, R., Ashraf, M. 2019. Plant-extract mediated green approach for the synthesis of ZnONPs: characterization and evaluation of cytotoxic, antimicrobial and antioxidant potentials. J. Mol. Struct. 1189: 315–27. Http://www.sciencedirect.com/science/article/pii/s0022286019304612.

Iravani, S. 2014. Bacteria in nanoparticle synthesis: current status and future prospects. Int. Sch. Res. Not. 2014.

Jain, N., Bhargava, A., Tarafdar, J.C., Singh, S.K., Panwar, J. 2013. A biomimetic approach towards synthesis of zinc oxide nanoparticles. Appl. Microbiol. Biotechnol. 97(2): 859–69. Https://doi. org/10.1007/s00253-012-3934-2.

Jayaseelan, C., Rahuman, A.A., Kirthi, A.V., Marimuthu, S., Santhoshkumar, T., Bagavan, A., Gaurav, K., Karthik, L., Rao, K.V. 2012. Novel microbial route to synthesize zno nanoparticles using *Aeromonas hydrophila* and their activity against pathogenic bacteria and fungi. Spectrochim. Acta Part A Mol. Biomol. Spectrosc. 90: 78–84. Http://www.sciencedirect.com/science/article/pii/s1386142512000078.

Jeevanandam, J., Barhoum, A., Chan, Y.S., Dufresne, A., Danquah, M.K. 2018. Review on nanoparticles and nanostructured materials: history, sources, toxicity and regulations. Beilstein J. Nanotechnol. 9: 1050–74. Https://www.ncbi.nlm.nih.gov/pubmed/29719757.

Ji, M., Barnwell, C.V., Grunden, A.M. 2015. Characterization of recombinant glutathione reductase from the psychrophilic antarctic bacterium *Colwellia psychrerythraea*. Extremophiles 19(4): 863–74. Https://doi.org/10.1007/s00792-015-0762-1.

Joe, A., Park, S.H., Shim, K.D., Kim, D.J., Jhee, K.H., Lee, H.W., Heo, C.H., Kim, H.M., Jang, E.S. 2017. Antibacterial mechanism of ZnO nanoparticles under dark conditions. J. Ind. Eng. Chem. 45: 430–39. Http://www.sciencedirect.com/science/article/pii/s1226086x16303926.

Kalpana, V.N., Bala Anoop Sirish Kataru, Sravani, N., Vigneshwari, T., Panneerselvam, A., Devi Rajeswari, V. 2018. Biosynthesis of zinc oxide nanoparticles using culture filtrates of *Aspergillus niger*: antimicrobial textiles and dye degradation studies. Opennano 3(march): 48–55. Https:// doi.org/10.1016/j.onano.2018.06.001.

Karthikeyan, M., Jafar Ahamed, A., Karthikeyan, C., Vijaya Kumar, P. 2019. Enhancement of antibacterial and anticancer properties of pure and rem doped ZnO nanoparticles synthesized using *Gymnema sylvestre* leaves extract. Sn. Appl. Sci. 1(4): 355. Https://doi.org/10.1007/ s42452-019-0375-x.

Kaushik, M., Niranjan, R., Thangam, R., Madhan, B., Pandiyarasan, V., Ramachandran, C., Oh, D.H., Venkatasubbu, G.D. 2019. Investigations on the antimicrobial activity and wound healing potential of ZnO nanoparticles. Appl. Surf. Sci. 479: 1169–77. Http://www.sciencedirect.com/ science/article/pii/s0169433219305409.

Khalafi, T., Buzar, F., Ghanemi, K. 2019. Phycosynthesis and enhanced photocatalytic activity of zinc oxide nanoparticles toward organosulfur pollutants. Sci. Rep. 9(1): 6866. Https://doi. org/10.1038/s41598-019-43368-3.

Khan, I., Saeed, K., Khan, I. 2017. Nanoparticles: properties, applications and toxicities. Arab. J. Chem. Http://dx.doi.org/10.1016/j.arabjc.2017.05.011.

Kim, M.-H., Jeong, H.-J. 2015. Zinc oxide nanoparticles suppress LPS-induced NF-κB activation by inducing A20, a negative regulator of NF-κB, in RAW 264.7 macrophages. J. Nanosci. Nanotechnol. 15(9): 6509–15. Http://www.ingentaconnect.com/content/10.1166/ jnn.2015.10319.

Kirthi, A.V., Rahuman, A.A., Rajakumar, G., Marimuthu, S., Santhoshkumar, T., Jayaseelan, C., Velayutham, K. 2011. Acaricidal, pediculocidal and larvicidal activity of synthesized ZnO nanoparticles using wet chemical route against blood feeding parasites. Parasitol. Res. 109(2): 461–72. Https://doi.org/10.1007/s00436-011-2277-8.

Kuhn, D.A., Vanhecke, D., Michen, B., Blank, F., Gehr, P., Petri-Fink, A., Rothen-Rutishauser, B. 2014. Different endocytotic uptake mechanisms for nanoparticles in epithelial cells and macrophages. Beilstein J. Nanotechnol. 5: 1625–36. Http://www.ncbi.nlm.nih.gov/pmc/articles/pmc4222452/.

Kumar, A., Pandey, A.K., Singh, S.S., Shanker, R., Dhawan, A. 2011. Engineered ZnO and TiO$_2$ nanoparticles induce oxidative stress and dna damage leading to reduced viability of *Escherichia coli*. Free Radic. Biol. Med. 51(10): 1872–81. Http://www.sciencedirect.com/science/article/pii/ s0891584911005387.

Kumar, P.T.S., Lakshmanan, V.K., Anilkumar, T.V., Ramya, C., Reshmi, P., Unnikrishnan, A.G., Nair, S.V., Jayakumar, R. 2012. Flexible and microporous chitosan hydrogel/nano ZnO composite bandages for wound dressing: *in vitro* and *in vivo* evaluation. Acs. Appl. Mater. Interfaces 4(5): 2618–29. Https://doi.org/10.1021/am300292v.

Kundu, D., Hazra, C., Chatterjee, A., Chaudhari, A., Mishra, S. 2014. Extracellular biosynthesis of zinc oxide nanoparticles using *Rhodococcus pyridinivorans* NT2: multifunctional textile finishing, biosafety evaluation and *in vitro* drug delivery in colon carcinoma. J. Photochem. Photobiol. B Biol. 140: 194–204. Http://dx.doi.org/10.1016/j.jphotobiol.2014.08.001.

Liang, D.Y., Li, X.Q., Li, W.W., Fiorino, D., Qiao, Y., Sahbaie, P., Yeomans, D.C., Clark, J.D. 2010. Caspase-1 modulates incisional sensitization and inflammation. Anesthesiology 113(4): 945–56. Http://www.ncbi.nlm.nih.gov/pmc/articles/pmc2945456/.

Malaikozhundan, B., Vaseeharan, B., Vijayakumar, S., Thangaraj, M.P. 2017. *Bacillus thuringiensis* coated zinc oxide nanoparticle and its biopesticidal effects on the pulse beetle, callosobruchus maculatus. J. Photochem. Photobiol. B Biol. 174. https://www.sciencedirect.com/science/article/pii/s1011134417302245.

Manke, A., Wang, L., Rojanasakul, Y. 2013. Mechanisms of nanoparticle-induced oxidative stress and toxicity. Biomed Res. Int. 2013: 942916. Https://www.ncbi.nlm.nih.gov/pubmed/24027766.

Maret, W. 2013. Inhibitory zinc sites in enzymes. Biometals 26(2): 197–204. Https://doi.org/10.1007/s10534-013-9613-7.

Mirhendi, M., Emtiazi, G., Roghanian, R. 2013. Antibacterial activities of nano magnetite ZnO produced in aerobic and anaerobic condition by *Pseudomonas stutzeri*. Jundishapur J. Microbiol. 6(10): 1–6.

Mittal, A.K., Christi, Y., Banerjee, U.C. 2013. Synthesis of metallic nanoparticles using plant extracts. Biotechnol. Adv. 31(2): 346–56. Http://www.sciencedirect.com/science/article/pii/s0734975013000050.

Mittal, M., Siddiqui, M.R., Tran, K., Reddy, S.P., Malik, A.B. 2014. Reactive oxygen species in inflammation and tissue injury. Antioxid. Redox Signal. 20(7): 1126–67. Http://online.liebertpub.com/doi/abs/10.1089/ars.2012.5149.

Mohammad, G.R.K.S., Tabrizi, M.H., Ardalan, T., Yadamani, S., Safavi, E. 2019. Green synthesis of zinc oxide nanoparticles and evaluation of anti-angiogenesis, anti-inflammatory and cytotoxicity properties. J. Biosci. 44(2): 30. Https://doi.org/10.1007/s12038-019-9845-y.

Nagajyothi, P.C., Cha, S.J., Yang, I.J., Sreekanth, T.V.M., Kim, K.J., Shin, H.M. 2015. Antioxidant and anti-inflammatory activities of zinc oxide nanoparticles synthesized using *Polygala tenuifolia* root extract. J. Photochem. Photobiol. B Biol. 146: 10–17. Https://doi.org/10.1016/j.photobiol.2015.02.008.

Naveed Ul Haq, A., Nadhman, A., Ullah, I., Mustafa, G., Yasinzai, M., Khan, I. 2017. Synthesis approaches of zinc oxide nanoparticles: the dilemma of ecotoxicity. J. Nanomater. 2017. Https://doi.org/10.1155/2017/8510342.

Ogunyemi, S.O., Abdallah, Y., Zhang, M., Fouad, H., Hong, X., Ibrahim, E., Islam Masum, M.M., Hossain, A., Mo, J., Li, B. 2019. Green synthesis of zinc oxide nanoparticles using different plant extracts and their antibacterial activity against *Xanthomonas oryzae* pv. *oryzae*. Artif. Cells, Nanomedicine, Biotechnol. 47(1): 341–52. Https://doi.org/10.1080/21691401.2018.1557671.

Prasad, K., Jha, A.K. 2009. ZnO nanoparticles: synthesis and adsorption study. Nat. Sci. 01(02): 129–35. Http://www.scirp.org/journal/doi.aspx?Doi=10.4236/ns.2009.12016.

Raguvaran, R., Manuja, B.K., Chopra, M., Thakur, R., Anand, T., Kalia, A., Manuja, A. 2017. Sodium alginate and gum acacia hydrogels of ZnO nanoparticles show wound healing effect on fibroblast cells. Int. J. Biol. Macromol. 96: 185–91. Https://doi.org/10.1016/j.ijbiomac.2016.12.009.

Rajan, A., Cherian, E., Gurunathan, B. 2016. Biosynthesis of zinc oxide nanoparticles using *Aspergillus fumigatus* JCF and its antibacterial activity. Int. J. Mod. Sci. Technol. 1: 52–57.

Rajiv, P., Sivaraj Rajeshwari, Rajendran Venckatesh. 2013. Bio-fabrication of zinc oxide nanoparticles using leaf extract of *Parthenium hysterophorus* L. and its size-dependent antifungal activity against plant fungal pathogens. Spectrochim. Acta Part A Mol. Biomol. Spectrosc. 112: 384–87. Http://www.sciencedirect.com/science/article/pii/s1386142513004125.

Raliya, R., Tarafdar, J.C. 2013. ZnO nanoparticle biosynthesis and its effect on phosphorous-mobilizing enzyme secretion and gum contents in clusterbean (*Cyamopsis tetragonoloba* L.). Agric. Res. 2(1): 48–57. Https://doi.org/10.1007/s40003-012-0049-z.

Ramesh, M., Anbuvannan, M., Viruthagiri, G. 2015. Green synthesis of ZnO nanoparticles using *Solanum nigrum* leaf extract and their antibacterial activity. Spectrochim. Acta Part A

Mol. Biomol. Spectrosc. 136: 864–70. Http://www.sciencedirect.com/science/article/pii/s1386142514014632.

Rasmussen, J.W., Martinez, E., Louka, P., Wingett, D.G. 2010. Zinc oxide nanoparticles for selective destruction of tumor cells and potential for drug delivery applications. Expert Opin. Drug Deliv. 7(9): 1063–77. Https://www.ncbi.nlm.nih.gov/pubmed/20716019.

Rath, G., Hussain, T., Chauhan, G., Garg, T., Goyal, A.K. 2016. Development and characterization of cefazolin loaded zinc oxide nanoparticles composite gelatin nanofiber mats for postoperative surgical wounds. Mater. Sci. Eng. C 58: 242–53. Http://www.sciencedirect.com/science/article/pii/s0928493115303052.

Salem, W., Leitner, D.R., Zingl, F.G., Schratter, G., Prassl, R., Goessler, W., Reidl, J., Schild, S. 2015. Antibacterial activity of silver and zinc nanoparticles against *Vibrio cholerae* and enterotoxic *Escherichia coli*. Int. J. Med. Microbiol. 305(1): 85–95. Https://www.ncbi.nlm.nih.gov/pubmed/25466205.

Sarkar, J., Ghosh, M., Mukherjee, A., Chattopadhyay, D., Acharya, K. 2014. Biosynthesis and safety evaluation of ZnO nanoparticles. Bioprocess Biosyst. Eng. 37(2): 165–71. Https://doi.org/10.1007/s00449-013-0982-7.

Schwirn, K., Tietjen, L., Beer, I. 2014. Why are nanomaterials different and how can they be appropriately regulated under reach? Environ. Sci. Eur. 26(1): 4. Https://doi.org/10.1186/2190-4715-26-4.

Selvarajan, E., Mohanasrinivasan, V. 2013. Biosynthesis and characterization of ZnO nanoparticles using *Lactobacillus plantarum* VITES07. Mater. Lett. 112: 180–82. Http://www.sciencedirect.com/science/article/pii/s0167577x1301255x.

Sen, C.K., Khanna, S., Babior, B.M., Hunt, T.K., Ellison, E.C., Roy, S. 2002. Oxidant-induced vascular endothelial growth factor expression in human keratinocytes and cutaneous wound healing. J. Biol. Chem. 277(36): 33284–90. Http://www.jbc.org/content/277/36/33284.abstract.

Shah, M., Fawcett, D., Sharma, S., Tripathy, S.K., Poinern, G.E.J. 2015. Green synthesis of metallic nanoparticles via biological entities. Mater. (Basel, Switzerland) 8(11): 7278–7308. Https://www.ncbi.nlm.nih.gov/pubmed/28793638.

Shamsuzzaman, Ali, A., Assif, M.A., Mashrai, A.A.H., Khanam, H. 2014. Green synthesis of ZnO nanoparticles using *Bacillus subtilis* and their catalytic performance in the one-pot synthesis of steroidal thiophenes. Eur. Chem. Bull. 3. Http://dx.doi.org/10.17628/ecb.2014.3.939-945.

Shamsuzzaman, Mashrai, A., Khanam, H., Aljawfi, RN. 2017. Biological synthesis of ZnO nanoparticles using *C. albicans* and studying their catalytic performance in the synthesis of steroidal pyrazolines. Arab. J. Chem. 10: s1530–36. Http://dx.doi.org/10.1016/j.arabjc.2013.05.004.

Sharma, V., Anderson, D., Dhawan A. 2012. Zinc oxide nanoparticles induce oxidative DNA damage and ros-triggered mitochondria mediated apoptosis in human liver cells (hepG2). Apoptosis 17(8): 852–70. Https://doi.org/10.1007/s10495-012-0705-6.

Shen, C., James, S.A., de Jonge, M.D., Turney, T.W., Wright, P.F., Feltis B.N. 2013. Relating cytotoxicity, zinc ions, and reactive oxygen in ZnO nanoparticle-exposed human immune cells. Toxicol. Sci. 136(1): 120–30. Http://dx.doi.org/10.1093/toxsci/kft187.

Shi, H., Hudson, L.G., Liu, K.J. 2004. Oxidative stress and apoptosis in metal ion-induced carcinogenesis. Free Radic. Biol. Med. 37(5): 582–93. Http://www.sciencedirect.com/science/article/pii/s0891584904002734.

Shobha, N., Nanda, N., Giresha, A.S., Manjappa, P., Sophiya, P., Dharmappa, K.K., Nagabhushana, B.M. 2019. Synthesis and characterization of zinc oxide nanoparticles utilizing seed source of *Ricinus communis* and study of its antioxidant, antifungal and anticancer activity. Mater. Sci. Eng. C. 97: 842–50. Http://www.sciencedirect.com/science/article/pii/s0928493118307367.

Simkó, M., Fiedeler, U., Gazsó, A., Nentwich, M. 2011. The impact of nanoparticles on cellular functions. Nanotrust. Dossiers 007(november): 1–4.

Singh, A.K. 2010. Synthesis, characterization, electrical and sensing properties of ZnO nanoparticles. Adv. Powder Technol. 21(6): 609–13. Http://dx.doi.org/10.1016/j.apt.2010.02.002.

Singh, B.N., Rawat, A.K., Khan, W., Naqvi, A.H. and Singh, B.R. 2014. Biosynthesis of stable antioxidant ZnO nanoparticles by *Pseudomonas aeruginosa* rhamnolipids. Plos One 9(9). Https://doi.org/10.1371/journal.pone.0106937.

Sirelkhatim, A., Mahmud, S., Seeni, A., Kaus, N.H.M., Ann, L.C., Bakhori, S.K.M., Hasan, H., Mohamad, D. 2015. Review on zinc oxide nanoparticles: antibacterial activity and toxicity mechanism. Nano-micro Lett. 7. Http://dx.doi.org/10.1007/s40820-015-0040-x.

Song, W., Zhang, J., Guo, J., Zhang, J., Ding, F., Li, L., Sun, Z. 2010. Role of the dissolved zinc ion and reactive oxygen species in cytotoxicity of ZnO nanoparticles. Toxicol. Lett. 199(3): 389–97. Http://www.sciencedirect.com/science/article/pii/s0378427410017285.

Suresh, D., Nethravathi, P.C., Udayabhanu, Rajanaika, H., Nagabhushana, H., Sharma, S.C. 2015. Green synthesis of multifunctional zinc oxide (zno) nanoparticles using *Cassia fistula* plant extract and their photodegradative, antioxidant and antibacterial activities. Mater. Sci. Semicond. Process. 31: 446–54. Http://www.sciencedirect.com/science/article/pii/s1369800114007148.

Thurber A., Wingett, G., Rasmussen, J.W., Layne, J., Johnson, L., Tenne, D.A., Zhang, J., Hanna, C.B., Punnoose, A. 2012. Improving the selective cancer killing ability of ZnO nanoparticles using Fe doping. Nanotoxicology 6(4): 440–52. Https://doi.org/10.3109/17435390.2011.587031.

Tripathi, R.M., Bhadwal, A.S., Gupta, R.K., Singh, P., Shrivastav, A., Shrivastav, B.R. 2014. ZnO nanoflowers: novel biogenic synthesis and enhanced photocatalytic activity amity institute of nanotechnology. J. Photochem. Photobiol. B Biol. Http://dx.doi.org/10.1016/j.jphotobiol.2014.10.001.

Vasile, B.S., Oprea, O., Voicu, G., Ficai, A., Andronescu, A., Teodorescu, A., Holban, A. 2014. Synthesis and characterization of a novel controlled release zinc oxide/gentamicin–chitosan composite with potential applications in wounds care. Int. J. Pharm. 463(2): 161–69. Http://www.sciencedirect.com/science/article/pii/s0378517313010296.

Vinulan, R.D., Zheng, J. 2015. Serum protein adsorption and excretion pathways of metal nanoparticles. Nanomedicine 10(17): 2781–94. Https://doi.org/10.2217/nnm.15.97.

Walkey, C.D., Chan, C.W. 2012. Understanding and controlling the interaction of nanomaterials with proteins in a physiological environment. Chem. Soc. Rev. 41(7): 2780–99. Http://dx.doi.org/10.1039/c1cs15233e.

Warren, E.A.K., Payne, C.K. 2015. Cellular binding of nanoparticles disrupts the membrane potential. Rsc. Adv. 5(18): 13660–66. Http://dx.doi.org/10.1039/c4ra15727c.

Xie, Y., He, Y., Irwin, P.L., Jin, T., Shi, X. 2011. Antibacterial activity and mechanism of action of zinc oxide nanoparticles against *Campylobacter jejuni*. Appl. Environ. Microbiol. 77. Https:// www.Ncbi.nlm.nih.gov/pubmed/21296935.

Yan, H.X., Wu, H.P., Zhang, H.L., Ashton, C., Tong, C., Wu, H., Qian, Q.J., Wang, H.Y., Ying, Q.L. 2013. P53 promotes inflammation-associated hepatocarcinogenesis by inducing HMGB1 release. J. Hepatol. 59(4): 762–68. Http://www.ncbi.nlm.nih.gov/pmc/articles/pmc3805120/.

Yusof, M.H., Mahammad, R., Zaidan, U.H., Rahman, N.A.A. 2019. Microbial synthesis of zinc oxide nanoparticles and their potential application as an antimicrobial agent and a feed supplement in animal industry: a review. J. Anim. Sci. Biotechnol. 10(1): 57. Https://doi.org/10.1186/s40104-019-0368-z.

Zhang, J., Qin, X., Wang, B., Xu, G., Qin, Z., Wang, J., Wu, L., Ju, X., Bose, D.D., Qiu, F., Zhou, H., Zou, Z. 2017. Zinc oxide nanoparticles harness autophagy to induce cell death in lung epithelial cells. Cell Death Dis. 8(7): e2954–e2954. Https://www.ncbi.nlm.nih.gov/pubmed/28749469.

Chapter 10

Enhancement of Antimicrobial Activity by Biosynthesized Nano-sized Materials

Irena Maliszewska* and Ewelina Wanarska

Introduction

Infectious diseases are among the main problems of modern medicine and mortality throughout the world. The WHO and CDC have expressed serious concerns about the continuing increase in multi-drug resistance among microorganisms. The crisis associated with antibiotic resistance remains one of the most urgent problems in global public health. It is known that the major routes of resistance to antibiotics include enzymatic modification or degradation of the antibiotic molecule, the efflux of antibiotics from the bacterial cell through efflux pumps, and change of the target antibiotic, which prevents antibiotic binding and thus leads to loss of its activity (Kapoor et al. 2017). These different mechanisms show that different approaches to this problem are needed. However, the high prevalence of resistant microbes makes this problem extremely difficult to solve and requires innovative approaches.

One interesting solution is the use of the synergistic activity of antibiotics and non-antibiotics. Nanomaterials may offer a promising solution as they can not only combat bacteria themselves but also act as carriers for antibiotics and natural antimicrobial compounds. The combined use of nanoparticles with antimicrobial agents makes it possible to reduce the toxicity of both agents towards human cells because it decreases the required dosage and synergistically enhances their biocidal properties. For example, combining antibiotics with nanoparticles restores the ability

Division of Medicinal Chemistry and Microbiology, Faculty of Chemistry, Wroclaw University of Science and Technology, 50-370 Wroclaw, Wybrzeże Wyspiańskiego 27, Poland.
Email: ewelina.wanarska@pwr.edu.pl
* Corresponding author: irena.helena.maliszewska@pwr.edu.pl

to destroy bacteria that have acquired resistance to antibiotics. Moreover, it has been shown that nanoparticles tagged with antibiotics increase the concentration of antibiotics at the site of bacterial and antibiotic interactions, and facilitate the binding of antibiotics to bacteria.

In the context of the development of new antimicrobial therapeutics, antimicrobial photodynamic inactivation (PDI) is an attractive method of destroying pathogens. PDI is based on combining photosensitizers with light radiation in visible or near-infrared wavelengths in order to trigger cell death. The development of new antibacterial photodynamic therapy strategies is of great value for improving the antimicrobial efficacy of photosensitizers without introducing cytotoxicity, which is a great challenge for current leading efforts on antimicrobial PDI based on nanometric materials. Therefore, the design of photosensitizer-nanoparticles provides a new strategy for improving the efficiency of PDI and creates systems with excellent biocompatibility.

In this chapter, we summarize recent studies on the interaction between nanomaterials and antibiotics as well as other antibacterial agents to formulate new perspectives for future research. We focus on nanoparticles obtained by eco-friendly methods. In this context, biologically synthesized nanoparticles using various biological sources overcome several of the disadvantages of chemically synthesized nanoparticles. This green chemistry approach to the synthesis of nanoparticles has proved to be cost-effective and multifunctional, with high scalability and stability. The future of green synthesis is incredibly bright, given its low cost and high durability. We believe that the combination of biogenic nanoparticles and antimicrobial agents is a potential candidate for further research into antibiotic-resistant microorganisms.

Combinations of biogenic nanoparticles with antibiotics

Silver nanoparticles

In recent years, several studies have been carried out on the antibacterial efficacies of nanoparticle–antibiotic combinations. Investigation of the interactions of antibiotics with silver nanoparticles is the most common among studies dedicated to the examination of combined action of metallic nanoparticles with antibiotics (Haiyan et al. 2019). One of the earliest studies on this topic was published by Shahverdi et al. (2007). These authors impregnated standard antibiotic disks with 10 µl of silver nanoparticles synthesized by supernatant of *Klebsiella pneumoniae* at a final content of 10 µg per disk. The combined activity of silver nanoparticles with antibiotics from different groups (penicillin G, amoxicillin, carbenicillin, cephalexin, cefixime, gentamicin, amikacin, erythromycin, tetracycline, cotrimoxazole, clindamycin, nitrofurantoin, nalidixic acid, and vancomycin) was examined against Gram-positive (*Staphylococcus aureus*) and Gram-negative bacteria (*Escherichia coli*). The obtained results showed enhancement of antibacterial activity of penicillin G, amoxicillin, erythromycin, clindamycin and vancomycin in the presence of silver nanoparticles against both bacteria. The authors emphasized that the effect of silver nanoparticles on the activity of antibiotics against *E. coli* was lower than that against *S. aureus*. It was observed that fold increases in area ranged from 0.7 to 2.5 against *S. aureus* and from 0.4 to 1.3 against *E. coli*. Mechanisms of either synergistic or indifferent interactions were not proposed in this study. However, the highest enhanced effect of penicillins and differences in the interaction levels between Gram-positive and Gram-negative bacteria indicated the key role of the bacterial cell wall, which is further discussed in the following studies.

Fayaz et al. (2010) investigated combined antibacterial effects of silver nanoparticles synthesized by filtrate of *Trichoderma viride* with four different antibiotics (ampicillin, erythromycin, kanamycin and chloramphenicol) against Gram-positive (*S. aureus* and *Micrococcus luteus*) and Gram-negative (*E. coli* and *Salmonella typhi*) bacteria. It was shown that the antibacterial activity of all tested antibiotics increased in the presence of silver nanoparticles. However, in contrast to the study by Shahverdi et al. (2007) the increase was more prominent against Gram-negative bacteria than against Gram-positives. The highest enhancing effect was observed between silver nanoparticles and ampicillin. The high amplifying effect of ampicillin and silver nanoparticles was explained by cell wall lysis caused by ampicillin and increased cell wall penetration for silver nanoparticles (Fig. 10.1) (Fayaz et al. 2010).

Birla et al. (2009) studied the combined effect of silver nanoparticles produced by *Phoma glomerata* with five antibiotics (ampicillin, gentamycin, kanamycin, streptomycin and vancomycin) on the three human pathogens *S. aureus*, *E. coli* and *Pseudomonas aeruginosa*. In contrast to the studies by Shahverdi et al. (2007) and in agreement with results obtained by Fayaz et al. (2010), the antibacterial activity of antibiotics in the presence of silver nanoparticles has increased more significantly against Gram-negative bacteria for the majority of studied antibiotics (with the exception of streptomycin). The highest enhancing antibacterial effect was observed for vancomycin and ampicillin and the authors explained this effect by the increased

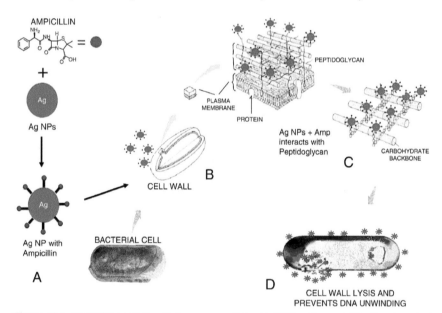

Figure 10.1. Synergistic activity of silver nanoparticles (AgNPs) with ampicillin (Amp) against bacteria. (A) Formation of core silver nanoparticles with ampicillin. (B) Interaction of AgNPs-Amp complex over the cell wall of bacteria. (C) AgNPs-Amp complex inhibits the formation of cross-links in the peptidoglycan layer leading to cell wall lysis. (D) AgNPs-Amp complex prevents the DNA from unwinding (reprinted from Fayaz et al. 2010; with permission from Elsevier).

penetration of silver nanoparticles through the cell membrane in the presence of these antibiotics.

The newly identified pathogenic species of the genus *Cryphonectria* were used for the synthesis of silver nanoparticles by Dar et al. (2013). These authors demonstrated the antibacterial activity of the obtained nanoparticles against *E. coli, S. typhi, S. aureus* and *Candida albicans*. Moreover, it was shown that these silver nanoparticles enhanced the antimicrobial activity of amphotericin.

The combined effectiveness of antibiotics and biogenic silver nanoparticles has also been proven by Zarina and Nanda (2014). These authors used biosynthesized silver nanoparticles by *Streptomyces albaduncus* (MTCC-924) to study the synergistic effect of antibiotics (ampicillin, azithromycin, cefotaxime, erythromycin, ofloxacin) against Gram-negative bacteria (*E. coli, Pseudomonas* sp., *K. pneumoniae*) and Gram-positive bacteria (*S. aureus, Micrococcus luteus, Streptococcus mutans*). It has been found that the antibacterial efficacy of various antibiotics is increased in combination with the silver nanoparticles.

Some authors have used silver nanoparticles synthesized by plant extracts to enhance the antimicrobial activity of antibiotics. For example, Jyoti et al. (2016) synthesized silver nanoparticles (AgNPs) by aqueous leaf extract of *Urtica dioica* (Linn.). *Urtica dioica* (stinging nettle) has been reported to possess antibacterial, antifungal, antiviral, and antioxidative activity (Gulcin et al. 2004). Antibacterial activity of these nanoparticles against Gram-positive bacteria (*Bacillus cereus, B. subtilis, S. aureus, Staphylococcus epidermidis*) and Gram-negative bacteria (*E. coli, K. pneumoniae, Serratia marcescens, Salmonella typhimurium*) was studied. In addition, the synergistic effect of AgNPs with various antibiotics was assessed against the above-mentioned pathogens. The results showed that AgNPs in combination with antibiotics have a better antibacterial effect than AgNPs alone and the maximum effect, with a 17.8-fold increase in the inhibition zone, was observed for amoxicillin and AgNPs against *S. marcescens.*

Literature studies have shown that among the various antibiotics whose antimicrobial activity has been enhanced by silver nanoparticles, aminoglycosides (gentamicin, tobramycin, kanamycin, and streptomycin) benefited the most from silver ions as adjuvants (Barras et al. 2018). Aminoglycosides are a group of bactericidal antibiotics that target the 30S ribosomal subunit and induce amino acid misincorporation. Aminoglycoside needs to be transported through the cytoplasmic membrane to reach its target. These transport systems are energized via proton motive force (PMF)-dependent pathways (Taber et al. 1987). Moreover, a so-called feed-forward loop model postulates the occurrence of a two-step process: Aminoglycosides pass through the cytoplasmic membrane before being combined with the membrane-associated ribosome (EDP-I), resulting in disruption of product translation and further destabilization of the membrane, thus allowing better access of aminoglycoside (EDP-II) (Hurwitz et al. 1981). Herisse et al. (2017) demonstrated that silver increases the aminoglycoside toxicity by acting independently of PMF when it bypasses the step of the EDP-I PMF-dependent aminoglycoside incorporation process. Silver circumvented the antagonistic effect of PMF scattering of carbonyl-m-chlorophenylhydrazone cyanide. Moreover, silver restored aminoglycoside uptake by strains showing a decreased level of PMF, such as mutants lacking complex I and II (Δnuo Δsdh) or the biosynthesis of Fe-S clusters ($\Delta iscUA$) (Herisse et al. 2017). In contrast, the silver-enhancing aminoglycoside toxicity activity remained dependent on translation, the translation of the EDP-II-dependent protein (Herisse et al. 2017).

Gold nanoparticles

Among all metal nanoparticles, gold nanoparticles (AuNPs) have been considered to be a highly useful platform for the efficient drug delivery/carrier system because of their easy surface functionalization, biocompatibility and low toxicity (Singh et al. 2018). The important advantages of nanoparticle-based antimicrobial drug delivery include improved solubility of poorly water-soluble drugs, prolonged drug half-life and systemic circulation time, and sustained and stimuli-responsive drug release, which eventually lowers administration frequency and dose (Gupta et al. 2016). In addition, minimized systemic side effects through targeted delivery of antimicrobial drugs as well as combined, synergistic, and resistance overcoming effects via co-delivery of multiple antimicrobial drugs can be achieved using nanoparticle carriers (Wang et al. 2017).

One of the first approaches to enhance the activity of antibiotics was shown by Fayaz et al. (2011). The gold nanoparticles were biologically synthesized using the non-pathogenic fungus *Trichoderma viride* at room temperature. These biogenic nanoparticles were bound with vancomycin, and antibacterial activity of these conjugates (VBGNP) was studied against some pathogens (Fig. 10.2). These nanoparticles were active against VRSA at a MIC of 8 µg/ml, which was explained by the possible non-specific binding of nanoparticles to cell surface peptides that were involved in cell wall synthesis. In addition, gold nanoparticles coated with vancomycin appeared to be active against *E. coli*, which is usually resistant to vancomycin because of its inability to penetrate the outer membrane of Gram-negative bacteria (Nikaido 1989). Owing to the presence of pits in the cell membrane, a possible mechanism of action against *E. coli* is the penetration of Gram-negative bacteria. Therefore, gold nanoparticles facilitate the binding of vancomycin to the surface of bacterial cells regardless of their structure in both Gram-positive and Gram-negative bacteria.

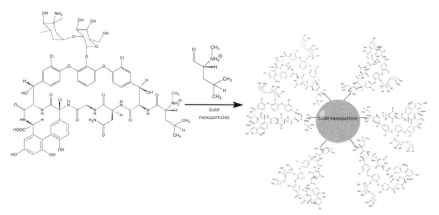

Figure 10.2. Structure of vancomycin and the conjugation method for biogenic gold nanoparticles with vancomycin (based on: Fayaz et al. 2011). Journal of Photochemistry and Photobiology B: Biology.

In another study, Roshmi et al. (2015) synthesized highly stable gold nanoparticles using *Bacillus* sp., isolated from soil, and their antibacterial activity functionalized with antibiotics was studied to combat bacteria that are resistant to many drugs. The nanoparticles obtained were functionalized with antibiotics ciprofloxacin, gentamycin, rifampicin, and vancomycin. It was shown that these biogenic gold nanoparticles have no antimicrobial activity, but act as a drug carrier, and antibiotic-bound drugs have very effective antibacterial activity against *S. epidermidis* and *Staphylococcus haemolyticus.*

An interesting biosynthesis of biocompatible gold nanoparticles by aqueous extract of the aerial parts of a pteridophyte *Adiantum philippense* by microwave irradiation and its surface functionalization with amoxicillin was presented by Kalita et al. (2016). The functionalization of amoxicillin on the biogenic gold nanoparticles (GNP-Amox) was carried out via electrostatic interaction of protonated amino group and thioether moiety-mediated attractive forces. The synthesized GNPs and GNP-Amox conjugates demonstrated an enhanced broad-spectrum bactericidal activity against both Gram-positive and Gram-negative bacteria. Furthermore, *in vitro* and *in vivo* assays of GNP-Amox revealed potent anti-MRSA activity and improved the survival rate.

Bimetallic silver–gold nanoparticles

Baker et al. (2017) emphasized the synthesis of silver and gold bimetallic nanoparticles from cell-free supernatant of *Pseudomonas veronii* strain AS41G, living in *Annona squamosa* L. Antibacterial activity of these nanoparticles against essential human pathogens was confirmed by diffusion test, and its synergistic effect with standard antibiotics showed 87.5% increased antibiotic activity of bacitracin against bacitracin-resistant *B. subtilis, E. coli* and *K. pneumoniae.* The authors are convinced that the mechanism of formation of conjugates between antibiotics and bimetallic nanoparticles could result from electrostatic interactions between negatively charged nanoparticles and a positively charged area of antibiotics, with the exception of those hydrophobic interactions and covalent bonding of nanoparticles with sulfhydryl groups (-SH) present in antibiotics.

Synergistic effect of silver nanoparticles and biosurfactants

An interesting synergistic effect of biogenic silver nanoparticles (AgNPs) and biosurfactant towards environmental bacteria and fungi was demonstrated by Chojniak et al. (2018). The AgNPs were synthesized in the culture supernatants of the biosurfactant producer *B. subtilis.* The following phytopathogens that were isolated from various parts of caraway, angelica and grapevine were used in this study: *Alternaria alternata, Boeremia strassesi, Colletotrichum dematium, Colletotrichum fuscum, Cylindrocarpon destructans, Diaporthe eres, Diplocereus hypericinum, Fusarium equiseti, Fusarium oxysporum, Phyllosticta plantaginis, Rhizoctonia solani, Sclerotinia sclerotiorum, and Fusarium avenaceum.* The authors showed that the presence of biosurfactant significantly increased the stability of the biogenic AgNPs and enhanced their antimicrobial activity. It was suggested that the biocidal activity of biologically synthesized AgNPs was enhanced through their stabilization in the presence of biosurfactants in cell-free supernatants used in the experiment (Kiran et al. 2010). It was also found that lipopeptides extracted from the supernatant have a strong antimicrobial effect on bacteria, including effects on planktonic growth, and the processes of biofilm formation and dislodging (Moryl et al. 2015).

Combinations of biogenic nanoparticles with plant extracts

Enhancing the activity of medicinal plant extracts

Phytomedicine (use of medicinal herbs for treatment) for oral disorders such as dental caries and periodontal disease has been widely practiced in traditional Indian, Egyptian, Greek, and Chinese medicine (Gourhan 1975, Falodun 2010, Pan et al. 2014). Many polyphenolic and flavonoid compounds isolated from plants are known for their healing and nutritional properties and have been used since the dawn of time (Li and Xu 2008, Abdollahzadeh et al. 2011, Shahindokht and Doostkam 2019). Most of these phytomedicinal compounds, especially phenolics, are poorly adsorbed by living cells (Manach et al. 2004). In order to solve this problem, many strategies have been developed to enhance the biological activity of phytomedicines. The selected strategies involving the use of biologically synthesized nanoparticles are described below.

Green tea (*Camellia sinensis*) contains a number of bioactive chemicals. Tea catechins and polyphenols are effective scavengers of reactive oxygen species (ROS) *in vitro* and may also function indirectly as antioxidants through their effects on transcription factors and enzyme activities (Higdon and Frei 2003). Studies conducted over the last 20 years have shown that the green tea polyphenolic catechins, particularly epigallocatechin gallate and epicatechin gallate, can inhibit the growth of a wide range of Gram-positive and Gram-negative bacteria species with moderate potency (Taylor et al. 2005). Evidence is emerging that these molecules may be useful in the control of common oral infections, such as dental caries and periodontal disease (Sakanaka et al. 1996, Sakanaka and Okada 2004, Hirasawa et al. 2002, Hirasawa and Takada 2004). The use of catechin as an antibacterial agent is becoming increasingly common, whereas unstable and easy oxidation has limited its application. To overcome this problem, Li et al. (2015) synthesized catechin-Cu nanoparticles, and the particle size, surface charge of the materials, antibacterial efficiency, changes in cellular morphology, as well as the degree of contact between the material and the bacteria were studied. The results showed that these catechin-Cu nanoparticles at the concentration of 20 ppm and 40 ppm provided rapid and effective killing of *S. aureus* and *E. coli* within 3 h. These authors demonstrated that catechin-Cu nanoparticles enhanced the antibacterial efficiency of catechin. Furthermore, it was believed that disruption of the bacterial cell membrane is a probable mode of action of the nanoparticles because it paves the way to bacterial cells leading to damage of the membrane protein and lipid bilayer. The damage could be the synergistic antibacterial effect of catechin and copper (II) (as supported by the bacterial efficacy comparison of catechin and catechin–copper nanoparticles). Leakage of intracellular molecules due to membrane disruption may cause shrinkage of the cell membrane, ultimately leading to cellular lysis.

For over a thousand years, the leaves of tulsi (*Ocimum sanctum*) are believed to have medicinal properties, mainly due to the presence of essential oils and phytonutrients. Tulsi is considered an excellent antibiotic, bactericidal, antifungal and disinfectant, which after consumption increases the resistance of the human body to various bacterial, fungal and viral infections (Prasannabalaji 2012, Yamani et al. 2016). The bactericidal activity of tulsi leaves was also confirmed by Mallikarjun et al. (2016). These authors obtained information on the antimicrobial efficacy of tulsi, particularly against three periodontal pathogens, namely *Aggregatibacter actinomycetemcomitans*, *Prevotella intermedia* and *Porphyromonas gingivalis*; these microbes are more commonly associated with initiation and progression of various periodontal diseases, especially aggressive periodontitis. Results of *in vitro* experiments showed that tulsi at a concentration of 5% and 10% can effectively inhibit the growth of *A. actinomycetemcomitans*, a rate comparable to that of doxycycline. Recently, leaf extracts of tulsi and quercetin (flavonoid present in tulsi) were used as precursors to study the role of biomolecules present in tulsi in the formation of silver nanoparticles (AgNP) (Jain and Mehata 2017). In addition, it was shown that AgNPs synthesized using both tulsi extract and quercetin had increased antibacterial activity as compared to pure tulsi extract, aqueous quercetin and $AgNO_3$.

Rajendran et al. (2013) studied the antimicrobial activity of ethanolic, methanolic, petroleum ether, and aqueous extracts of leaves of *O. sanctum* (Lamiaceae). It was found that the best antimicrobial effect was observed for methanolic extracts against *Bacillus subtilis*, *S. aureus*, *E. coli*, *P. aeruginosa*, *Aspergillus niger* and *Penicillium* spp. Then, the obtained methanolic extracts were loaded into sodium alginate chitosan nanoparticles (OSN), through a cation-induced, controlled gelation method. The obtained particles were deposited on cotton fabric. Compared to extracts and nanoparticles only, OSN demonstrated better and longer-lasting antimicrobial activity than the unloaded formulation, producing cotton fabrics with excellent antimicrobial activity (Rajendran et al. 2013).

Another example of the use of biologically synthesized nanoparticles to enhance antimicrobial activity was described by Sun et al. (2014). The authors used antibacterial traditional Chinese medicines (TCMs): *Polygonum cuspidatum* Sieb. et (*P.C.*), *Fagopyrum dibotrys* (D. Don), *Sanguisorba officinalis* L., *Agrimonia pilosa* Ledeb., *Hedyotis diffusa* Willd, *Rheum palmatum* L., and *Geranium wilfordii* containing various reductive components to synthesize TCM-silver nanoparticles. These authors obtained a series of TCM-mediated silver nanoparticles (TCM-AgNPs), in which TCM acted not only as an antibacterial agent but also as a stabilizer. It was found that both TCM extract and AgNPs significantly promoted the antibacterial potency of the various TCM-AgNPs. Moreover, these TCM-AgNPs were highly effective in inhibiting growth of *P. aeruginosa*, *S. epidermidis*, and *S. aureus*. Taking *Polygonum cuspidatum*-AgNPs as an example, the antibacterial effect on *P. aeruginosa* was enhanced by about 30,000 times when compared with the *P. cuspidatum* extract, suggesting that the AgNPs were a predominant contributor to the antibacterial activity. Meanwhile, the MIC_{50} for *Polygonum cuspidatum*-AgNPs was 20-fold higher than that for bare AgNPs, indicating a synergistic effect between TCM extract and AgNPs on antibacterial activity. The authors concluded that both AgNP and TCM have their respective advantages and limitations; therefore, the combination of natural antibacterial drugs and AgNP seemed to be justified for the development of stable and highly effective anti-bacterial AgNPs.

Phytosomes (often known as herbosomes) are nano-sized delivery systems that are prepared through the attachment of individual ingredients of herbal extracts to phosphatidylcholine, resulting in a formulation having a higher solubility and hence better absorption leading to promoted pharmacokinetic and pharmacodynamic properties compared to the conventional herbal extracts (Prajkta et al. 2017). Recently, some authors have described the use of these nanometric systems to increase antimicrobial activity. *Calendula officinalis* (Linn.) (calendula or marigold), which belongs to *Astracea* (*Compositae*) family, has been widely used for ornamental and medicinal purposes as folk therapy (Khaid and Silva 2012). Calendula has been used for its anti-tumoral, anti-inflammatory, wound healing, antimicrobial and antioxidant activities (Bashir et al. 2006, Kuppast and Nayak 2006, Leach 2008, Muley et al. 2009, Efstratios et al. 2012).

Demir et al. (2014) showed a very interesting method of increasing the biological activity of marigold extract. These authors performed double encapsulation of the herbal extract and gold nanoparticles (AuNPs) using calendula and AuNP as AuNP-phytosomes. The AuNP-phytosomes showed extremely high stability among other preparations used for comparison. At the same time, the authors showed that these structures did not have toxic effects up to 400 mg mL^{-1}, but were characterized by a higher antioxidant capacity as well as wound healing properties. Moreover, it was found that formulated vesicles that were less than 100 nm could penetrate cells and accumulate around the nucleus.

Enhancing of food preservation

It is well known that meat and other food products are contaminated by bacteria and fungi (Rad et al. 2019, Sajid et al. 2019). Food-borne pathogens such as *Salmonella*, *Shigella*, *Escherichia*, *Listeria*, *Clostridium* and *Vibrio* and their toxins are health problems. Spices (especially in meat processing) are an important aspect of food preservation. Spices are added to improve the taste and color of food. In addition, some spices are documented as bactericidal or bacteriostatic and can be used as preservatives (Liu et al. 2017, Sakkas and Papadopoulou 2017).

One of the methods of enhancing food preservation is the combination of spices with nanoparticles. This method was used by Abd El-Aziz and Yousef (2018), who demonstrated antibacterial synergy between antimicrobial agents and bioactive plant extracts. The aim of their research was to estimate the antimicrobial activity of four spices (carnations, rosemary, thyme and black pepper) and the biosynthesized nanoparticles themselves and in a combination of each spice with silver nanoparticles. The combination demonstrated 8 synergistic and 8 additive effects for inhibition of growth of the tested microorganisms (*P. aeruginosa*, *S. typhimurium*, *E. coli*, *B. subtilis*, *Candida parapsilosis* and *Trichosporon domesticum*). The microbicidal effects were 7 synergistic and 10 additive effects.

Another approach to prolonging food durability is the use of coatings and protective films. Food coatings and films are currently used in many food products, including fruits, vegetables, meats, chocolates, candies, bakery products and French fries (Morillon et al. 2002, Ahvenainen 2003, Rhim 2004). The basic functional properties of edible coatings and films depend on the characteristics of the film-forming materials used for their preparation. At present, the main film-forming materials used in the production of edible films are polysaccharides, proteins and lipids (Morillon et al. 2002, Cagri 2004, Cha and Chinnan 2004). Generally, lipid-based films are good moisture barriers, but they offer little resistance to gas transfer and have poor mechanical strength (Cha and Chinnan 2004). In contrast, biopolymer-based films are often good oxygen and carbon dioxide barriers, but they offer little protection against moisture migration (Jochenweiss and McClements 2006). The use of protective nanolayers and appropriate packaging has become a subject of great interest in the field of food nanotechnology due to their potential to extend the shelf life of many food products (Ahvenainen 2003). An interesting example of the use of nanotechnology are studies carried out by Fayaz et al. (2009). These authors synthesized silver nanoparticles using *T. viride*, and their incorporation into sodium alginate for vegetable and fruit preservation has been demonstrated. The silver nanoparticle incorporated into sodium alginate thin film shows good antibacterial activity against bacteria (*E. coli* and *S. aureus*). In addition, it was found that this film increased the persistence of carrot and pear compared to the control with respect to weight loss and soluble protein content.

Chitosan-based nanoparticles

Chitosan is a natural polymer having a good film-forming ability. Because of its high versatility, this polymer has been extensively evaluated for uses in food conservation (Britto et al. 2005, Bhoir et al. 2019) and biomedical applications (Singh and Ray 2000, Rabea et al. 2003, Vunain et al. 2017, Zhao et al. 2018). Owing to its biocompatibility, low toxicity, and biodegradability, chitosan is considered as a Generally Recognized and Safe (GRAS) component and its chloride salt is reported in a monograph of the European Pharmacopoeia. Several models suggested that the antimicrobial activity of chitosan is a result of its cationic nature (Rabea et al. 2003, Goy and Assis 2014). The electrostatic interaction between positively charged $R\mathrm{-\!-\!}N(CH_3)_3^+$ sites and negatively charged microbial cell membranes is predicted to be responsible for cellular lysis and assumed as the main antimicrobial mechanism (Rabea et al. 2003). Charged chitosan can also interact with essential nutrients, thereby interfering with microbial growth (Jia et al. 2001). Two approaches to chitosan-based nanomaterials are being considered.

First, chitosan nanoparticles can be a drug carrier with wide development potential and have the advantage that the controlled release of the drug improves the solubility and stability of the drug, increases efficacy and reduces toxicity. Acyclovir (9-[(2-hydroxyethoxy)methyl])-9H-guanine), which is a synthetic nucleoside analog derived from guanosine, is the drug of choice for treating *Herpes simplex* virus infections. However, because of its short half-life and incomplete absorption, it must be taken repeatedly throughout the day to be effective. Hence, development of acyclovir-loaded chitosan nanoparticles was undertaken for effective ocular delivery of the drug with improved bioavailability (Hao and Deng 2008). These authors described the nanosphere colloidal suspension containing acyclovir as a potential ophthalmic drug delivery system. Acyclovir-loaded chitosan nanoparticles displayed more crystallinity than acyclovir. The *in vitro* diffusion profile of acyclovir from the nanoparticles showed a sustained release of the drug over a period of 24 h. The results demonstrated the effective use of acyclovir-loaded chitosan nanoparticles as a controlled release preparation for treatment of ocular viral infections.

In the other studies, Donalisio et al. (2018) developed an effective topical acyclovir formulation that has been able to increase the efficacy of currently available commercial products in the local treatment of *Herpes simplex* virus infection. A modified nanoemulsion method was used as an innovative experimental set to prepare chitosan nanospheres containing acyclovir. The proposed chitosan nanoparticle system has improved properties such as increased drug loading, sustained-release kinetics, no cytotoxicity, and improved *in vitro* antiviral efficacy. Biological studies have shown higher antiviral activity in acyclovir-charged chitosan nanospheres than in free acyclovir against both HSV-1 and HSV-2 strains.

Second, chitosan can be used in biosynthesis of metallic nanoparticles and increase its antimicrobial activity. Amarnatha et al. (2012) presented a green approach to synthesize palladium nanoparticles by reducing palladium chloride salts with nontoxic and biodegradable polymeric chitosan and grape polyphenols. The results showed the efficacy of the grape- and chitosan-impregnated palladium nanoparticles as an antibacterial agent against *S. aureus* and *E. coli*. An electron microscopy technique has revealed size-dependent interaction of palladium nanoparticle conjugates with bacteria by disrupting cell membranes and the leakage of cellular proteins due to the change of membrane penetrability.

Some publications reported that chitin/chitosan/silver nanoparticle composites have enhanced antimicrobial activity against bacteria, fungi and viruses (Mori et al. 2013, Nguyen et al. 2013, Cardelle-Cobas et al. 2016). Comparative studies showed that AgNPs–chitosan composite is much more effective against bacteria than pure chitosan (Wei et al. 2009, Nguyen et al. 2013, Cardelle-Cobas et al. 2016). In addition, Kumar-Krishnan et al. (2015) believed that chitosan stabilizes nanoparticles and prevents their agglomeration and providing a positive surface charge strengthens their binding with negative charges present in the bacterial cell wall.

Enhancement of antimicrobial photodynamic inactivation

Photosensitizers that are necessary for the photodynamic inactivation of microorganisms are relatively harmless molecules in the dark, but when excited by light they generate ROS, such as singlet oxygen (Kochevar and Redmond 2000, Dai et al. 2009). Excited photosensitizers and ROS can kill cells by damaging biomolecules, such as proteins, lipids and nucleic acids. This approach seems to kill antibiotic-resistant strains as effectively as their antibiotic-sensitive counterparts (Maisch et al. 2011, Vera et al. 2012, Nakonieczna et al. 2019). However, PDI has some shortcomings. For example, most photosensitizers are highly hydrophobic and tend to accumulate in an aqueous medium. Many photosensitizers exhibit low to moderate quantum yield in ROS generation and can not be excited with light of a long wavelength, such as red or infrared light, which has a greater depth of tissue penetration than blue or green light.

Several types of nanoparticle systems have been tested to improve the solubility of the photosensitizer, photochemistry, photophysics, and targeting (Hamblin 2016). There are three different ways to combine nanoparticles and antimicrobial photodynamic therapies for antimicrobial applications: (1) non-covalent encapsulation or incorporation of photosensitizers in nanosystems, (2) covalent binding of photosensitizers to the surfaces of the nanoparticles, and (3) coadministration of photosensitizers and nanoparticles. Among the many nanomaterials used to enhance the effectiveness of this therapy, there are few examples showing the use of biologically synthesized nanostructures. For example, naturally occurring polymers such as chitosan can be used as novel starting material for the preparation of nontoxic nanoparticles with photobactericidal action (Chen et al. 2012, Shrestha and Kishen 2012, Shrestha and Kishen 2014).

Maliszewska et al. (2014) reported successful implementation of PDI of *S. epidermidis* using methylene blue (MB) in combination with biogenic gold nanoparticles synthesized by reduction of Au^{+3} in the presence of cell-free filtrate of *Trichoderma koningii* (GNP). A Xe lamp (550–780 nm) or a He-Ne laser (632 nm) was used as a light source to study the effect of MB alone, GNP alone, and the MB-GNP mixture on the viability of the tested bacterial cells. Lethal photosensitization of *S. epidermidis* with the MB-GNP mixture was achieved after 5 and 10 min exposure to laser or Xe lamp, respectively.

A successful photodynamic inactivation of planktonic and biofilm cells of *Candida albicans* using rose Bengal in combination with biogenic gold nanoparticles synthesized by the cell-free filtrate of *Penicillium funiculosum* BL1 strain was also described by Maliszewska et al. (2017). It has been observed that the rose Bengal–gold nanoparticles mixture exhibits a significant antifungal activity already in the absence of any light source and gives an enhanced antifungal response when using Xe lamp for photosensitization.

An enhancement of the phototoxic effect on *P. aeruginosa* was achieved by combination of tetrasulfonated hydroxyaluminum phthalocyanine ($AlPcS_4$) and bimetallic gold/silver nanoparticles (Au/Ag-NPs) synthesized by the cell-free filtrate of *Aureobasidium pullulans* (Maliszewska et al. 2018). Particularly high efficiency in killing bacterial cells was obtained for light intensity 105 mW/cm^2, after 20, 30, and 40 min of irradiation corresponding to 126, 189, and 252 J/cm^2 energy fluences. For $AlPcS_4$ + Au/Ag-NPs treatment, the viable count reduction was equal to 99.90%, 99.96%, and 99.975%, respectively. These results were significantly better than those achieved for irradiated separated assays of $AlPcS_4$ and Au/Ag-NPs.

Conclusion and future perspectives

Despite the various methods proposed to combat microbial resistance and the high prevalence of multi-drug-resistant bacteria, it is necessary to look for new approaches to cope with these problems. Nanoparticles have attracted great interest in recent years and have been widely described in the literature. The unique physicochemical properties of the surface, depending on size and shape, make nanoparticles more interactive and reactive to certain chemical species than their bulk scale counterparts. In addition, a high biological and chemical activity makes metal nanoparticles promising antibacterial agents.

Biologically synthesized nanoparticles deserve special attention as an alternative approach to conventional production processes that tend to use toxic chemicals and solvents. By combining antibiotics with biogenic metallic nanoparticles, it is possible to effectively restore some effective antibiotics, for example, penicillins, and also to overcome emerging resistance to presently highly effective antibiotics, such as vancomycin. It seems that the combined use of nanoparticles with antibiotics or other antimicrobial agents makes it possible to reduce the toxicity of both agents towards human cells. Studies on the interaction between metallic nanoparticles and antimicrobial agents are mainly devoted to conventional antibiotics. We are convinced that combinations of nanoparticles with antimicrobial antiseptics and especially with natural antimicrobial agents require more attention.

The known method of destroying pathogens is photodynamic inactivation. It seems that PDI has many more advantages than antibiotic therapies. It works in a short time and in a limited space, deactivating only microorganisms that are in the place of infection without adversely affecting the physiological flora. Numerous studies have shown that incubation of bacterial cells with sublethal doses of PDI does not cause photodamage resistance. However, an important limitation of the PDI method is its lower bactericidal effectiveness in relation to microorganisms growing in biofilms and the fact that the efficacy demonstrated in *in vitro* tests rarely translates into animal models. Moreover, after the effective elimination of microorganisms from the place of infection, their regrowth and the recurring development of infection is observed within 24 h after treatment. The use of biogenic metal nanoparticles is an interesting protocol to enhance the effectiveness of PDI and gives the opportunity to solve the problems of widespread use of this method.

In conclusion, we believe that the research results described represent a new and promising approach to overcoming the problems associated with widespread drug resistance in pathogens.

Acknowledgments

This work was financed by a statutory activity subsidy from the Polish Ministry of Science and Higher Education (PMSHE) for the Faculty of Chemistry of Wrocław University of Science and Technology.

References

Abd El-Aziz, D.M., Yousef, N.M.H. 2018. Enhancement of antimicrobial effect of some spices extract by using biosynthesized silver nanoparticles. Int. Food Res. J. 25(2): 589–596.

Abdollahzadeh, S.H., Mashouf, R., Mortazavi, H., Moghaddam, M., Roozbahani, N., Vahedi, M. 2011. Antibacterial and antifungal activities of *Punica granatum* peel extracts against oral pathogens. J. Dent. 8: 1–6.

Ahvenainen, R. 2003. Novel Food Packaging Techniques. CRC Press: Boca Raton, FL.

Amarnatha, K., Kumar, J., Reddy, T., Mahesh, V., Ayyappan, S.R., Nellore, J. 2012. Synthesis and characterization of chitosan and grape polyphenols stabilized palladium nanoparticles and their antibacterial activity. Colloids Surface B 92: 254–261.

Baker, S., Pasha, A., Satish, S. 2017. Biogenic nanoparticles bearing antibacterial activity and their synergistic effect with broad spectrum antibiotics: emerging strategy to combat drug resistant pathogens. Saudi Pharm. J. 25: 44–51.

Barras, F., Aussel, L., Ezraty, B. 2018. Silver and antibiotic, new facts to an old story. Antibiotics 7(79): 1–10

Bashir, S., Janbaz, K.H., Jabeen, Q., Gilani, A.H. 2006. Studies on spasmogenic and spasmolitic activities of *Calendula officinalis* flower. Phytother. Res. 20: 906–910.

Bhoir, S.H., Jhaveri, M., Chawla, S.P. 2019. Evaluation and predictive modeling of the effect of chitosan and gamma irradiation on quality of stored chilled chicken meat. J. Food Proc. Eng. DOI: 10.1111/jfpe.13254.

Birla, S.S., Tiwari, V.V., Gade, A.K., Ingle, A.P., Yadav, A.P., Rai, M.K. 2009. Fabrication of silver nanoparticles by *Phoma glomerata* and its combined effect against *Escherichia coli, Pseudomonas aeruginosa* and *Staphylococcus aureus*. Lett. Appl. Microbiol. 48(2): 173–179.

Britto, D., Campana-Filho, S.P., Assis, O.B.G. 2005. Mechanical properties of N,N,N-trimethylchitosan chloride films. Polimeros 5: 129–132.

Cagri, A., Ustunol, Z., Ryser, E.T. 2004. Antimicrobial edible films and coatings. J. Food Prot. 67: 833–848.

Cardelle-Cobas, A., Lima, D.S., Gullon, B., Brito, L.M., Rodrigues, K.A.F., Quelemes, P.V., Ramos-Jesus, J., Arcanjo, D.D.R., Plácido, A., Batziou, K., Quaresma, P., Eaton, P., Delerue-Matos, C., Carvalho, F.A.A., Silva, D.A., Pintado, M., Leite, J.R. 2016. Chitosan-based silver nanoparticles: A study of the antibacterial, antileishmanial and cytotoxic effects. J. Bioact. Compat. Polym. 32(4): 397–410.

Cha, D.S., Chinnan, M.S. 2004. Biopolymer-based antimicrobial packaging: review. Crit. Rev. Food. Sci. Nutr. 44: 223–237.

Chen, C.P., Chen, C.T., Tsai, T. 2012. Chitosan nanoparticles for antimicrobial photodynamic inactivation: characterization and *in vitro* investigation. Photochem. Photobiol. Sci. 88(3): 570–576.

Chojniak, J., Libera, M., Król, E., Płaza, G. 2018. A nonspecific synergistic effect of biogenic silver nanoparticles and biosurfactant towards environmental bacteria and fungi. Ecotoxicology 27: 352–359.

Dai, T., Huang, Y.Y., Hamblin, M.R. 2009. Photodynamic therapy for localized infections-state of the art. Photodiagn. Photodyn. 6: 170–188.

Dar, M.A., Ingle, A., Rai, M. 2013. Enhanced antimicrobial activity of silver nanoparticles synthesized by *Cryphonectria* sp. evaluated singly and in combination with antibiotics. Nanomedicine: NBM 9: 105–110.

Demir, B., Barlas, F.B., Guler, E., Gumus, P.Z., Can, M., Yavuz, M., Coskunolbef, H., Timur, S. 2014. Gold nanoparticle loaded phytosomal systems: synthesis, characterization and *in vitro* investigations. RSC Adv. 4: 34687–34695.

Donalisio, M., Leone, F., Civra, A., Spagnolo, R., Ozer, O., Lembo, D., Cavalli, R. 2018. Acyclovir-loaded chitosan nanospheres from nano-emulsion templating for the topical treatment of herpesviruses infections. Pharmaceutics 10(2): 46.

Efstratios, E., Hussain, A.I., Nigam, P.S., Moore, J.E., Ayub, M.A., Rao, J.R. 2012. Antimicrobial activity of *Calendula officinalis* petal extracts against fungi, as well as gram-negative and gram-positive clinical pathogens. Complement Ther. Clin. Pract. 18(3): 173–176.

Falodun, A. 2010. Herbal medicine in Africa-distribution, standardization and prospects. Res. J. Phytochemistry 4(3): 154–161.

Fayaz, M.A., Balaji, K., Girilal, M., Kalaichelvan, P.T., Venkatesan, R. 2009. Mycobased synthesis of silver nanoparticles and their incorporation into sodium alginate films for vegetable and fruit preservation. J. Agric. Food Chem. 57(14): 6246–6252.

Fayaz, A.M., Balaji, K., Girilal, M., Yadav, R., Kalaichelvan, P.T., Venketesan, R. 2010. Biogenic synthesis of silver nanoparticles and their synergistic effect with antibiotics: a study against Gram-positive and Gram-negative bacteria. Nanomedicine 6(1): 103–109.

Fayaz, A.M., Girilal, M., Mandy, S.A., Somsundar, S.S., Venkatesan, R., Kalaichelvan, P.T. 2011. Vancomycin bound biogenic gold nanoparticles: a different perspective for development of anti VRSA agents. Process Biochem. 46: 636–641.

Gourhan, L.A. 1975. The flowers found with Shanidar IV, a Neanderthal burial in Iraq. Sci. 190(4214): 562–564.

Goy, R.C., Assis, O.B.G. 2014. Antimicrobial analysis of films processed from chitosan and N,N,N-trimethylchitosan. Braz. J. Chem. Eng. 31: 643–648.

Gulcin, I., Kufrevioglu, O.I., Oktay, M., Buyukokuroglu, M.E. 2004. Antioxidant, antimicrobial, antiulcer and analgesic activities of nettle (Urtica dioica). J. Ethnopharmacol. 90: 205–215.

Gupta, A., Landis, R.F., Rotello, V.M. 2016. Nanoparticle-based antimicrobials: surface functionality is critical. Faculty Rev. 364: 1–10.

Haiyan, Y., Haoyu, S., Chunsheng, Y., Zhifen, L. 2019. Combination of sulfonamides, silver antimicrobial agents and quorum sensing inhibitors as a preferred approach for improving antimicrobial efficacy against *Bacillus subtilis*. Ecotoxicol. Environm. Safety 181: 43–48.

Hamblin, M.R. 2016. Antimicrobial photodynamic inactivation: a bright new technique to kill resistant microbes. Curr. Opin. Microbiol. 33: 67–73.

Hao, P.P., Deng, S.H. 2008. Preparation and detection of acyclovir loaded chitosan nanoparticles. Chin. Med. J. 5(1): 28–29.

Herisse, M., Duverger, Y., Martin-Verstraete, I., Barras, F., Ezraty, B. 2017. Silver potentiates aminoglycoside toxicity by enhancing their uptake. Mol. Microbiol. 105: 115–126.

Higdon, J.V., Frei, B. 2003. Tea catechins and polyphenols: health effects, metabolism and antioxidant functions. Crit. Rev Food. Sci. Nutr. 43(1): 89–143.

Hirasawa, M., Takada, K. 2004. Multiple effects of green tea catechin on the natifungal activity of antimycotics against *Candida albicans*. J. Antimicrob. Chemother. 53: 225–229.

Hirasawa, M., Takada, K., Makimura, M., Otake, S. 2002. Improvement of periodontal status by green tea catechins using a local delivery system: a clinical pilot study. J. Periodontal. Res. 37: 433–438.

Hurwitz, C., Braun, C.B., Rosano, C.L. 1981. Role of ribosome recycling in uptake of dihydrostreptomycin by sensitive and resistant *Escherichia coli*. Biochim. Biophys. Acta.-Nucleic Acids and Protein Synthesis 652: 168–176.

Jain, S., Mehata, S.M. 2017. Medicinal plant leaf extract and pure flavonoid mediated green synthesis of silver nanoparticles and their enhanced antibacterial property. Sci. Rep. 7: 15867.

Jia, Z., Shen, D., Xu, W. 2001. Synthesis and antibacterial activities of quaternary ammonium salt of chitosan. Carbohydr. Res. 333: 1–6.

Jochenweiss, P.T., McClements, D.J. 2006. Functional materials in food nanotechnology. J. Food Sci. 71: 107–116.

Jyoti, K., Baunthiyal, M., Singh, A. 2016. Characterization of silver nanoparticles synthesized using *Urtica dioica* Linn. leaves and their synergistic effects with antibiotics. J. Radiat. Res. Appl. Sc. 9: 217–227.

Kalita, S., Kandimalla, R., Sharma, K.K., Kataki, A.C., Deka, M., Kotoky, J. 2016. Amoxicillin functionalized gold nanoparticles reverts MRSA resistance. Mater. Sci. Eng. C 61: 720–727.

Kapoor, G., Saigal, S., Elongavan, A.E. 2017. Action and resistance mechanisms of antibiotics: A guide for clinicians. J. Anaesthesiol. Clin. Pharmacol. 33(3): 300–305.

Khaid, A.K., Silva, J.A.T. 2012. Biology of *Calendula officinalis* Linn. focus on pharmacology biological activities and agronomic practices. Med. Aromat. Plant. Sci. Biotechnol. 6(1): 12–27.

Kiran, G.S., Sabu, A., Selvin, J. 2010. Synthesis of silver nanoparticles by glicololid biosurfactant produced from marine *Brevibacterium casei* MSA 19. J. Biotechnol. 148(4): 221–225.

Kochevar, I.E., Redmond, R.W. 2000. Photosensitized production of singlet oxygen. Methods Enzymol. 319: 20–28.

Kumar-Krishnan, S., Prokhorov, E., Hernandez-Iturriaga, M., Mota-Morales, J.D., Vázquez-Lepe, M., Kovalenko, Y., Sanchez, I.C., Luna-Bárcenas, G. 2015. Chitosan/silver nanocomposites: synergistic antibacterial action of silver nanoparticles and silver ions. Eur. Polym. J. 67: 242–251.

Kuppast, I.J., Nayak, P.V. 2006. Wound healing activity of *Cardia dichotoma* Forst f. fruits. Nat. Prod. Rad. 5(2): 99–102.

Leach, M.J. 2008. *Calendula officinalis* and wound healing: a systematic review. Wounds 20(8): 1–7.

Li, H., Chen, Q., Zhao, J., Urmila, K. 2015. Enhancing the antimicrobial activity of natural extraction using the synthetic ultrasmall metal nanoparticles. Sci. Rep. 5: 11033.

Li, M., Xu, Z. 2008. Quercetin in a lotus leaves extract may be responsible for antibacterial activity. Arch. Pharmacal. Res. 31: 640–644.

Liu, Q., Meng, X., Li, Y., Zhao, C.N., Tang, G.Y., Li, H.B. 2017. Antibacterial and antifungal activities of spices. Int. J. Mol. Sci. 18(6): 1283.

Maisch, T., Hackbarth, S., Regensburger, J., Felgentrager, A., Baumler, W., Landthaler, M., Roder, B. 2011. Photodynamic inactivation of multi-resistant bacteria (PIB)—a new approach to treat superficial infections in the 21st century. J. Dtsch. Dermatol. Ges. 9: 360–366.

Maliszewska, I., Kałas, W., Wysokińska, E., Tylus, W., Pietrzyk, N., Popko, K., Palewska, K. 2018. Enhancement of photo-bactericidal effect of tetrasulfonated hydroxyaluminum phthalocyanine on *Pseudomonas aeruginosa*. Lasers Med. Sci. 33(1): 79–88.

Maliszewska, I., Leśniewska, A., Olesiak-Bańska, J., Matczyszyn, K., Samoć, M. 2014. Biogenic gold nanoparticles enhance methylene blue-induced phototoxic effect on *Staphylococcus epidermidis*. J. Nanopart. Res. 16(6): 1–16.

Maliszewska, I., Lisiak, B., Popko, K., Matczyszyn, K. 2017. Enhancement of the efficacy of photodynamic iactivation of *Candida albicans* with the use of biogenic gold nanoparticles. Photochem. Photobiol. 93(4): 1081–1090.

Mallikarjun, S., Rao, A., Rajesh, G., Shenoy, R., Pai, M. 2016. Antimicrobial efficacy of Tulsi leaf (*Ocimum sanctum*) extract on periodontal pathogens: an *in vitro* study. J. Indian. Soc. Periodontol. 20(2): 145–150.

Manach, C., Scalbert, A., Morand, C., Rémésy, C., Jiménez, L. 2004. Polyphenols: food sources and bioavailability. Am. J. Clin. Nutr. 79(5): 727–747.

Mori, Y., Ono, T., Miyahira, Y., Nguyen, V.Q., Matsui, T., Ishihara, M. 2013. Antiviral activity of silver nanoparticle/chitosan composites against H1N1 influenza A virus. Nanoscale Res. Lett. 8: 93.

Morillon, V., Debeaufort, F., Blond, G., Capelle, M., Voilley, A. 2002. Factors affecting the moisture permeability of lipid-based edible films: a review. Crit. Rev. Food Sci. Nutr. 42: 67–89.

Moryl, M., Spętana, M., Dziubek, K., Paraszkiewicz, K., Różalska, S., Płaza, G.A., Różalski, A. 2015. Antimicrobial, antiadhesive and antibiofilm potential of lipopeptides synthetized by *Bacillus subtilis* on uropathogenic bacteria. Acta Biochim. Pol. 62(4): 725–732.

Muley, B.P., Khadabadi, S.S., Banarase, N.B. 2009. Phytochemical constituents and pharmacological activities of *Calendula officinalis* Linn (Asteraceae): a review. Trop. J. Pharm. Res. 8(5): 455–465.

Nakonieczna, J., Wozniak, A., Pieranski, M., Rapacka-Zdonczyk, A., Ogonowska, P., Grinholc, M. 2019. Photoinactivation of ESKAPE pathogens: overview of novel therapeutic strategy. Fut. Med. Chem. 11(5): 443–446.

Nguyen, V.Q., Ishihara, M., Mori, Y., Nakamura, S., Kishimoto, S., Fujita, M., Hattori, H., Kanatani, Y., Ono, T., Miyahira, Y., Matsui, T. 2013. Preparation of size-controlled silver nanoparticles and chitosan-based composites and their anti-microbial activities. Biomed. Mater. Eng. 23: 473–483.

Nikaido, H. 1989. Outer membrane barrier as a mechanism of antimicrobial resistance. Antimicrob. Agents Chemother. 33: 1831–1836.

Pan, S.Y., Litscher, G., Gao, S.H., Zhou, S.F., Yu, Z.L., Chen, H.C., Zhang, S.F., Tang, M.K., Sun, J.N., Ko, K.M. 2014. Historical perspective of traditional indigenous medical practices: the current renaissance and conservation of herbal resources. Evid.-Based Complementary Altern. Med. 2014: 525340.

Prajkta, C., Vinal, P., Vineeta, J., Rani, A. 2017. A review on therapeutic applications of phytosomes. J. Drug. Deliv. Sci. Technol. 7(5): 17–21.

Prasannabalaji, N. 2012. Antibacterial activities of some Indian traditional plant extracts. Asian Pac. J. Trop. Dis. 2: 291–295.

Rabea, E.I., Badawy, M., Stevens, C.V., Smagghe, G., Steurbaut, W. 2003. Chitosan as antimicrobial agent: applications and mode of action. Biomacromolecules 4: 1457–1465.

Rad, A.H., Javadi, M., Kafil, H.S., Pirouzian, H.R., Khaleghi, M. 2019. The safety perspective of probiotic and non-probiotic yoghurts: a review. Food Qual. Safe. 3: 9–14.

Rajendran, R., Radhai, R., Kotresh, T.M., Csiszar, E. 2013. Development of antimicrobial cotton fabrics using herb loaded nanoparticles. Carbohydr. Polym. 91(2): 613–617.

Rhim, J.W. 2004. Increase in water vapor barrier property of biopolymer-based edible films and coatings by compositing with lipid materials. J. Food Sci. Biotechnol. 13: 528–535.

Roshmi, T., Soumya, K.R., Jyothis, M., Radhakrishnan, E.K. 2015. Effect of biofabricated gold nanoparticle-based antibiotic conjugates on minimum inhibitory concentration of bacterial isolates of clinical origin. Gold Bull. 48: 63–71.

Sajid, M., Mehmood, S., Yuan, Y., Yue, T. 2019. Mycotoxin patulin in food matrices: occurrence and its biological degradation strategies. Drug Met. Rev. 51(1): 105–120.

Sakanaka, S., Aizawa, M., Kim, M., Yamamoto, T. 1996. Inhibitory effects of green tea polyphenols on growth and cellular adherence of an oral bacterium, *Porphyromonas gingivalis*. Biosci. Biotechnol. Biochem. 60: 745–749.

Sakanaka, S., Okada, Y. 2004. Inhibitory effects of green tea polyphenols on the production of a virulence factor of the periodontal-disease-causing anaerobic bacterium *Porphyromonas gingivalis*. J. Agric. Food Chem. 52: 1688–1692.

Sakkas, H., Papadopoulou, C. 2017. Antimicrobial activity of basil, oregano, and thyme essential oils. J. Microbiol. Biotechnol. 27(3): 429–438.

Shahindokht, B.-J., Doostkam, A. 2019. Comparative evaluation of bioactive compounds of various cultivars of pomegranate (*Punica granatum*) in different world regions. AIMS Agric. Food 4(1): 41–55.

Shahverdi, A.R., Fakhimi, A., Shahverdi, H.R., Minaian, S. 2007. Synthesis and effect of silver nanoparticles on the antibacterial activity of different antibiotics against *Staphylococcus aureus* and *Escherichia coli*. Nanomedicine 3(2): 168–171.

Shrestha, A., Kishen, A. 2012. Polycationic chitosan-conjugated photosensitizer for antibacterial photodynamic therapy. Photochem. Photobiol. 88(3): 577–583.

Shrestha, A., Kishen, A. 2014. Antibacterial efficacy of photosensitizer functionalized biopolymeric nanoparticles in the presence of tissue inhibitors in root canal. J. Endod. 40(4): 566–570.

Singh, P., Pandit, S., Mokkapati, V.R.S.S.S., Garg, A., Ravikumar, V., Mijakovic, I. 2018. Gold nanoparticles in diagnostics and therapeutics for human cancer international. Journal Int. J. Mol. Sci. 19: 1979.

Singh, D.K., Ray, A.R. 2000. Biomedical applications of chitin, chitosan, and their derivatives. Macromol. Chem. Phys. C. 40: 69–83.

Sun, W., Qu, D., Chen, Y., Liu, C., Zhou, J. 2014. Enhanced stability and antibacterial efficacy of a traditional Chinese medicine-mediated silver nanoparticle delivery system. Int. J. Nanomedicine 9: 5491–5502.

Taber, H.W., Mueller, J.P., Miller, P.F., Arrow, A.M.Y.S. 1987. Bacterial uptake of aminoglycoside antibiotics. Microbiol. Rev. 51: 439–457.

Taylor, P.W., Hamilton-Miller, J.M.T., Stapleton, P.D. 2005. Antimicrobial properties of green tea catechins. Food Sci. Technol. Bull. 2: 71–81.

Vera, D.M., Haynes, M.H., Ball, A.R., Dai, T., Astrakas, C., Kelso, M.J., Hamblin, M.R., Tegos, G.P. 2012. Strategies to potentiate antimicrobial photoinactivation by overcoming resistant phenotypes (dagger). J. Photochem. Photobiol. B 13: 39–50.

Vunain, E., Mishra, A.K., Mamba, B.B. 2017. Fundamentals of chitosan for biomedical applications. Chitosan Based Biomaterials 1: 3–30.

Wang, L., Hu, C., Shao, L. 2017. The antimicrobial activity of nanoparticles: present situation and prospects for the future. Int. J. Nanomed. 12: 1227–1249.

Wei, D., Sun, W., Qian, W., Ye, Y., Ma, X. 2009. The synthesis of chitosan-based silver nanoparticles and their antibacterial activity. Carbohydr. Res. 344: 2375–2382.

Yamani, H.A., Pang, E.C., Mantri, N., Deighton, M.A. 2016. Antimicrobial activity of Tulsi (*Ocimum tenuiflorum*) essential oil and their major constituents against three species of bacteria. Front. Microbiol. 7: 681.

Zarina, A., Nanda, A. 2014. Combined efficacy of antibiotics and biosynthesised silver nanoparticles from *Streptomyces*. Int. J. Pharmtech. Res. 6(6): 1862–1869.

Zhao, D., Yu, S., Sun, B., Guo, S., Zhao, K. 2018. Biomedical applications of chitosan and its derivative nanoparticles. Polymers 10(4): 462.

Chapter 11

Characterization of Bacteria-mediated Metal Nanoparticles and Their Biological Applications

Jayakodi Santhoshkumar,[1] *Shanmugam Rajeshkumar*[2,]* and
Venkat Kumar Shanmugam[1]

Introduction

Nanoparticles are microscopic particles that are 1–100 nm in size and differ in shape. (Rajeshkumar et al. 2013). The commonly used nanoparticles are silver, gold, and copper, which are widely available. Nanoparticles have a range of uses in various fields based on their unique characteristics. For example, silver nanoparticles play a substantial role in electrochemical biosensors (Narayanan and Sakthivel 2010), cosmetics, textiles and medicine. They have applications in the medical field because of their potential antimicrobial activity against Gram-positive and Gram-negative microorganisms (Ponnanikajamideen et al. 2016). Nanoparticles are found to be useful against resistant strains of bacteria that are pathogenic to human beings (Rajeshkumar et al. 2016). Research has so far focused predominantly on the synthesis of multiple nanoparticles (Paulkumar et al. 2013).

Nanoparticles can be synthesized by physical, chemical and biological methods. Biological techniques are useful for the synthesis and reduction of metal ions (Santhoshkumar et al. 2017). Physical and chemical methods, in comparison, involve high pressure, high temperature, and toxicity (Garmasheva et al. 2016). Chemical methods include photochemical and electrochemical reactions and require high energy,

[1] Nano-Therapy Lab, Department of Biotechnology, School of Bio-Sciences and Technology, VIT, Vellore 632014, Tamil Nadu, India.
[2] Nanobiomedicine Lab, Department of Pharmacology, Saveetha Dental College and Hospitals, SIMATS, Saveetha University, Chennai 600077, Tamil Nadu, India.
* Corresponding author: ssrajeshkumar@hotmail.com

pressure, and temperature. The synthesis of nanoparticles using physical methods involving lasers, sprays and physical vapor deposition is expensive. Biological synthesis can be carried out using bacteria, fungi, actinomycetes, algae, and enzymes (Wadhwani et al. 2017). It can be rapidly carried out at room temperature (Nagababu et al. 2019). Biological methods are readily available, non-toxic, and cost-effective. In the green synthesis process, the plants are chosen based on their medicinal properties and diagnostic and therapeutic applications (Malarkodi et al. 2013).

The synthesized nanoparticles can be primarily characterized by UV-spectroscopy. To study the size, shape, and morphology of the nanoparticles, transmission electron microscopy (TEM) and scanning electron microscopy (SEM) analyses are made. X-ray diffraction (XRD) is used to find the crystalline structure of the nanoparticles. Fourier Transform Infrared Spectroscopy (FTIR) is used to reveal biomolecules and a functional group present in the nanoparticles for the biosynthesis. In this chapter, all the above characterizations of nanoparticles synthesized by bacteria are presented in Tables 11.1–11.6.

The present chapter discusses the bacterially mediated biosynthesis of silver (Ag), gold (Au), titanium oxide (TiO_2), copper (Cu), selenium (Se) and iron (Fe) nanoparticles shown in Fig. 11.1 and their mechanism of synthesis followed by advances in characterization techniques and their application in different domains.

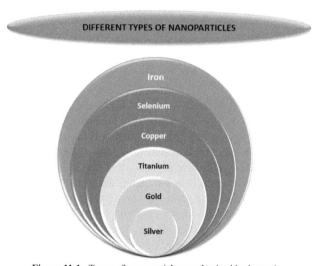

Figure 11.1. Types of nanoparticles synthesized by bacteria.

Characterization of nanoparticles

There are two methods used for the synthesis of nanoparticles: the top-down approach and the bottom-up approach. The top-down approach manages the material size decrease of particles using physical and chemical procedures. The size and shape of the nanoparticles are analyzed by SEM and TEM. The size range is 2–100 nm. SEM, TEM, XRD, and FTIR samples are prepared by centrifuging the material at 10,000 rpm and drying it in hot air oven; the samples are scraped from the plate and analyzed for the production of peripheral-cell nanoparticles. The methods of analysis focus on specific characteristics (Singaravelu et al. 2007). Colour change due to the formation of nanoparticles is observed using the UV-visible spectra. The optical density values of extra cellular are taken from 200 to 800 nm at different time intervals and recorded (Duhan et al. 2017). UV is used to record the firmness of the particle. Then broad surface plasmon resonance (SPR) is recorded due to the increase in the intensity and indicates the shape and size of the metal nanoparticles. The peak value denotes the presence of nanoparticles. The shape and size of the metal nanoparticles are measured by using SEM, and energy-dispersive X-ray (EDX) spectra can also be measured using SEM (Kumar and Ghosh 2016). The shapes may differ, or nanoparticles may acquire a new shape when organic compounds or protein are present on them (Siva Kumar et al. 2014). Morphological analysis is carried out by TEM. The TEM results can be confirmed with XRD results. The shape of nanoparticles obtained may be spherical, rod, polydisperse, or pseudo-spherical. The XRD pattern is shown in 2θ values ranging from 10 to 80. The intense peaks show the nanoparticle size. The source of the EDX analysis of nanoparticles is found, indicating the presence of basic nanoparticles in weight percentage. The document shows the main discharge energies for silver (Ag), and these are similar to the ridges in the range 3 kV, so it is certain that silver is accurately recognized. FTIR shows the functional group present in the nanoparticle that helps in the reduction from toxic to non-toxic. It shows the functional groups such as alkanes, aldehydes, ketones, phenol, and primary amines. The presence of organic compounds such as flavonoids or a terpenoid may be responsible for the synthesis of different nanoparticles (Rehana et al. 2017).

Multiple applications

Metal and metal oxide nanomaterials with antibacterial and antifungal activities have been produced for a wide range of applications. Among them, silver nanoparticles show higher activity, suggesting that they are more toxic to bacteria than other nanoparticles. For example, Nabila and Kannabiran (2018) evaluated the influence of silver nanoparticles on *E. coli* and reported that silver nanoparticles enter and damage the DNA of *E. coli*. The zone of inhibition indicates the inhibition of the test strain. Such an evaluation is performed by calculating the formula for radical scavenging activity. Cytotoxicity studies were also performed under the application part of the research (Baek and An 2011). Antibacterial activity was evaluated against *E. coli, Trichoderma, Staphylococcus aureus, Legionella pneumophila, Klebsiella, Pseudomonas aeruginosa, Bacillus cereus, Proteus vulgaris, Salmonella typhi, Vibrio cholerae,* and *Bacillus subtilis*. Antifungal activity was tested against fungi *Aspergillus niger* and *Candida albicans*.

Using green science, specialists can create biocompatible nanoparticles that have fewer impacts on human well-being and the environment. ZnO and Ag NPs have shown promise in this regard and merit further study (Sumbal et al. 2019). The characterized CrTiO$_2$ nanoparticles exhibited good biofilm inhibition against Gram-positive bacterium *Streptococcus mutans*, Gram-negative bacterium *Proteus vulgaris* and fungus *Candida albicans*. Photocatalysis demonstrated that the CrTiO$_2$ nanoparticles were effectively explored for the degradation of dyes. The outcomes suggest that CrTiO$_2$ nanoparticles are a phenomenal bactericidal, fungicidal and photocatalytic agent that can be regularly used for biomedical and mechanical applications (Rekha et al. 2019).

Silver antibiotics may even kill resistant microorganisms, for example, methicillin-resistant *Staphylococcus aureus*. Furthermore, microorganisms are not prepared to form a barrier against silver as anti-toxins and are protected by its concentration (Sondi and Salopek-Sondi 2004). The vast majority of specialists applied auxiliary metabolites for the blend of antimicrobial gold and silver nanoparticles, though its efficacy has been demonstrated. Along with other biogenic metal nanoparticles, such as zinc oxide, copper and selenium could be used for antimicrobial applications. From this toxicity assessment of metal nanoparticles is significant across the therapeutic application (Amini 2019).

Nanoparticles enhance the permeability of the bacterial cell membrane for better uptake of antibiotics. The conjugates also kill the bacteria effectively (Jelinkova et al. 2019). Bacterial cell lysis occurs subsequent to the use of amoxicillin because the third and last stage of bacterial cell wall union is inhibited (Chopra 2007). The extracellular biosynthesis of nanoparticles is highly profitable because it allows the nanoparticles to be easily purified during the union shown in Figs. 11.2 and 11.3. From this cell is filtered through the extracellular biosynthesis of silver nanoparticles via ERI-3 *Streptomyces* sp. (Zonooz and Salouti 2011, Kasana et al. 2017). Intracellular biosynthesis of copper and copper oxide NPs has been performed using *Serratia marcescens,* while *Salmonella typhimurium*, *Thermoanaerobacter* sp. X513, *E. coli*, *Pseudomonas aeruginosa,* and *Pseudomonas fluorescens* were applied for extracellular processes (Nakhaeepour et al. 2019). The Ag^{2+} HEPES solution is removed from the cells of the human embryonic kidney cell (HEK-293), the HeLa cell, and the cells in the studies on a new human cell line (SiHa) (Subramanian and Shanmugam 2016). Several other bacteria such as *Pseudomonas stutzeri*, *Pseudomonas* sp., *Streptomyces* sp., *Lactobacillus* sp., *Morganella morganii*, and *Morganella psychrotolerans* were also considered for CuONP synthesis in which no indication for their mode of production was presented (Ramanathan et al. 2013).

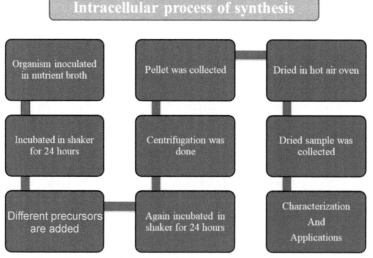

Figure 11.2. Intracellular synthesis of nanoparticles by different bacteria.

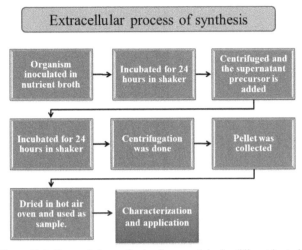

Figure 11.3. Extracellular synthesis of nanoparticles by different bacteria.

Bacteria-mediated synthesis and applications of copper nanoparticles

Pseudomonas fluorescens

In TEM analysis, the nanoparticle was found to be 49 nm, with hexagonal and spherical shape. The UV-vis peak absorption spectrum was obtained between 550 and 650 nm, which indicates the Cu particles. EDX analysis obtained a result as copper and oxygen are present in the nanoparticle at 11% and 24%, respectively. Selected area electron diffraction (SAED) analysis revealed that the copper was present as $Cu(OH)_2$ (Shantkriti and Rani 2014).

Morganella morganii

UV-vis spectroscopy was recorded and from that SPR was also obtained at the range maintained from 590 to 630 absorbance and the peak was observed at 610 nm, indicating the presence of copper nanoparticles. Nanoparticles synthesized by this organism were 3–10 nm in size, and they were polydisperse in shape. SAED investigation demonstrated the crystalline nature of nanoparticles. The protein surrounded the nanoparticles. The FTIR result confirmed that there is a presence of amide and protein group present in 1650 cm^{-1} and 1540 cm^{-1}. X-ray photoelectron spectroscopy (XPS) data showed the presence of C 1S, N 1S, O 1S and Cu 2p (Ramanathan et al. 2010).

Bacillus cereus **NCIM 2458, SWSD1**

The XRD analysis of *Bacillus cereus* NCIM2458 was 50 nm. TEM results revealed that the copper nanoparticles were spherical. FTIR and SEM analysis were not performed for this organism. Antimicrobial activity was tested against *E. coli, P. aeruginosa,*

Table 11.1. Copper nanoparticles.

Sample no.	Bacteria	XRD	TEM	SEM	FTIR	Remarks and application	Reference
1	*Pseudomonas fluorescens*	Not observed	49 nm spherical and hexagonal	Not observed	Not observed	UV: 550–650 nm EDX: copper, Oxygen SAED: $Cu(OH)_2$	Shantkriti and Rani 2014
2	*Morganella morganii*	Not observed	3–10 nm polydispersed	Not observed	Amide 1–1650 cm^{-1} amide 2–1540 cm^{-1}	UV: 610 nm SAED:15–20 nm SPR: 610 XPS: C 1S, N 1S, O 1S	Ramanathan et al. 2010
3	*Bacillus cereus* NCIM 2458 SWSD1	50 nm	Spherical 26–97 nm	Spherical slightly higher than TEM	The amine is stretching and bending for the presence of protein	Antibacterial activity against Gram-positive and Gram-negative organism SPR: yellow turn to green UV: 570–630 nm, 300–370 nm EDX: copper and protein	Tiwari et al. 2016
4	*Serratia*	Presence of copper and copper oxides	10–30 nm polydispersed		580 and 522 cm^{-1} Cu–O stretch 610-IR active mode of Cu_2O 1550 and 1650 cm^{-1}	Extracellular, UV: 590 nm and 630 nm SAED: crystalline EDX: copper, oxygen, nitrogen, carbon XPS: C 1S, Cu 2p core level	Dharne et al. 2008
5	*Shewanella oneidensis*	Presence of copper	Approximately 5 nm, uniform particle size		Not observed	XPS: 932.1 eV And 952.9 eV Cu 2p3/2 and Cu 2p1/2 respectively atomic ratio of Cu:S was 0.94:1	Tian et al. 2016

B. subtilis, and *S. aureus.* Among them, *P. aeruginosa* formed zone of inhibition at 37 mg/ml CuNPs, which was the highest (Tiwari et al. 2016).

Bacillus cereus SWSD1

Nanoparticles synthesized by *B. cereus* SWSD1 were 11–33 nm in size and spherical. FTIR result showed the peak of amide and primary amine stretching and bending for the presence of protein. There was no peak of copper oxide. Oligodynamic activity against *B. cereus* showed 18 mm zone, which was recorded to be the highest. The UV absorbance was between 570–630 nm and 300–350 nm. EDX results showed the presence of copper and protein (Dharne et al. 2008).

Serratia sp.

The copper nanoparticles were synthesized extracellularly from *Serratia* sp. isolated from the midgut of *Stibara* sp.; UV absorbance was recorded as 590 and 630 nm. The nanoparticles were 10–30 nm and polydisperse when analyzed through TEM. SAED results showed that they were crystalline. XRD revealed the presence of copper and copper oxides. EDX revealed the presence of copper, oxygen, nitrogen, and carbon. XPS showed C 1S and Cu 2p at the ore level. FTIR results confirmed in 580 and 522 cm^{-1} Cu-O stretch, in 610-IR active mode of Cu_2O and 1550 cm^{-1} and 1650 cm^{-1} (amide 1 and amide 2) presence of protein on the surface of the copper oxide (Dharne et al. 2008).

Shewanella oneidensis

The TEM image displayed nanoparticles approximately 5 nm in size. XRD results corresponded with TEM results, which showed the presence of copper nanoparticles. SEM, FTIR, EDS and analyses were not carried out. XPS showed a ratio of 0.94:1 for Cu:S (Zhou et al. 2016).

Gold nanoparticles, synthesis and characterization

Rhodopseudomonas capsulata

TEM analysis showed that the particle was 10–20 nm and spherical. XRS results showed that the particles were spherical. SAED result revealed that the particle was single-crystalline. SPR peak value was recorded at 540 nm-band1 and 900 nm-band2. UV peak absorbance value was recorded at 580 nm. The XRD recorded peak value was 0.039–0.068 nm (He et al. 2007).

E. coli

TEM analysis indicated that particles were spherical and triangle, and 0.5–1.8 μm in size. FTIR showed 1767 cm^{-1}, 1718 cm^{-1} carbonyl group, and 1606 cm^{-1} -C-C group were present in the particle. UV-SPR peak absorbance value was recorded at 580 nm.

Table 11.2. Gold nanoparticles.

Sample no.	Bacteria (reducing agent)	XRD	TEM	SEM	FTIR	Remarks and application	Reference
1	*Rhodopseudomonas capsulata*	0.039–0.068 nm	10–20 nm spherical	Not observed	Not observed	Extracellular, UV: 580 nm, EDX: spherical, SAED-single crystalline, SPR: 540-band1, 900 nm-band2	He et al. 2007
2	*E. coli*	Not observed	Spherical, triangle 0.5–1.8 μm	Not observed	1767,1718-carbonyl group, 1606-C-C group	UV: SPR-580 nm, SAED: polycrystalline AFM: length 440 nm, thickness 14 nm	Du et al. 2007
3	*Bacillus marisflavi*	14 nm	Extremely small Nanocrystalline, spherical, 14 nm	Spherical 14 nm	3121-N-H-bending in amines, 2889-C-H in aldehydes, 1639,1523-N-H-amines, 1332-C-N-N-proteins	UV: 560 nm, DLS: 13.5 nm	Nadaf and Kanase 2016
4	*Deinococcus radiodurans*	36.28 nm	39.5 nm spherical, pseudo-spherical, triangular, irregular	39.5 nm spherical pseudo-spherical, triangular, irregular	3322-OH, 3302, 3404-OH-NH, 1543-amide1, 2 polypeptides, proteins, 1073-P-O-C	DLS: spherical, triangular, irregular-43.75 nm, antibacterial activity against *E. coli*, *S. aureus*, EDX: C, N, O, P, S	Li et al. 2016
5	*Staphylococcus epidermis*	21 nm	21 nm, tetragonal	Not observed	Not observed	UV: 664 nm	Srinath and Ravishankar Rai 2015

SAED result showed that the particle was polycrystalline. Atomic force microscopy (AFM) measured the length as 440 nm and thickness as 14 nm (Du et al. 2007).

Bacillus marisflavi

TEM analysis of particles showed them to be extremely small nanocrystalline, spherical, and 14 nm in size. FTIR was performed and found that 3121 cm^{-1} -N-H-bending in amines, 2889 cm^{-1} -C-H in aldehydes, 1639 cm^{-1}, 1523 cm^{-1} -N-H–amines, and 1332 cm^{-1} -C-N- proteins were present in the particles. UV peak absorbance value was recorded at 560 nm. By DLS the peak value was observed at 13.5 nm (Nadaf and Kanase 2016).

Deinococcus radiodurans

TEM analysis indicated that particle size was 39.5 nm and the shape was spherical, pseudo-spherical, triangular, and irregular. FTIR was performed and established the possible molecules and groups present in the nanoparticle as 3322 cm^{-1} -OH, 3302 cm^{-1}, 3404 cm^{-1} -OH-NH, 1543 cm^{-1} -amide 1, 2polypeptides, proteins, 1073-P-O-C. DLS found that the nanoparticles were spherical, triangular, and irregular and 43.75 nm in size. Antibacterial activity was evaluated against *E. coli* and *S. aureus*. EDX analysis confirmed that C, N, O, P, and S were present in the nanoparticle. XRD values were recorded as 36.28 nm (Li et al. 2016).

Staphylococcus epidermidis

TEM analysis indicated particle size of 21 nm and tetragonal shape. UV peak absorbance value was recorded at 664 nm (Srinath and Ravishankar Rai 2015).

Bacteria-assisted synthesis of iron oxide nanoparticles

Pseudomonas aeruginosa

TEM analysis showed that the particle size was 57 nm. FTIR identified the groups and molecules found in the particle based on peaks as 3360.13 cm^{-1} -O-H, 2921.32 cm^{-1} -OH, 2820.45 cm^{-1} -CH, 1765.98 cm^{-1} C=C, 1632.67 cm^{-1} -C=O. UV peak absorbance value was recorded at 360 nm. EDX revealed the presence of zinc and oxygen in nanoparticles. DLS showed the size of nanoparticles at 81 nm. The XRD values calculated for the nanoparticles synthesized by this bacterium was 31.11°, 34.23°, 35.61°, 47.13°, 56.15°, 62.18°, 65.65°, 67.59°, 69.21°, and 72.18° (Creanga et al. 2011).

Staphylococcus aureus

SEM results revealed that 0.1562 mM of Zn (CH$_3$CO$_2$) embedded ZnO 646 nm and spherical in shape. Using FTIR, groups present in the particles were identified as 1399 cm^{-1} -C-C, 1650–1580 cm^{-1} -N-H bond, and 3385 cm^{-1} -N-H. UV peak absorbance

Table 11.3. Iron oxide nanoparticles.

Sample no.	Bacteria	XRD	TEM	SEM	FTIR	Remarks and application	Reference
1	*Pseudomonas aeruginosa*	31.11°, 34.23°, 35.61°, 47.13°, 56.15°, 62.18°, 65.65°, 67.59°, 69.21°, 72.18°	57 nm	Not observed	3360.13 cm⁻¹ -O-H, 2921.32 cm⁻¹ -OH, 2820.45 cm⁻¹ -CH, 1765.98 cm⁻¹, 1632.67 cm⁻¹ -C=O	UV: 360 nm EDX: zinc and oxygen DLS: 81 nm	Creanga et al. 2011
2	*Staphylococcus aureus*	XRD-crystalline nature of ZnO nanoparticles	20–90 nm irregular, spherical	100 nm, spherical, irregular	Not observed	EDS: presence of clay leaching studies	Shi et al. 2016
3	*Lactobacillus sporogens*	11 nm	5–15 nm	Hexagonal	Not observed	Not observed	Vignesh et al. 2015
4	*Bacillus licheniformis*	Not observed	200 with nanoparticles 40 in width and 400 in length	Nanoflowers	O-H, N-H, C-O (carbonyl) stretching in the amide1 and amide-2 linkage of protein, C-N stretching bond	Not observed	Sundaram et al. 2012
5	*Streptomyces MS2*	Not observed	10–20 nm irregular with a narrow shape	30–50 nm spherical	3260–3270 cm⁻¹ O-H 400, 600 cm⁻¹ -C=O, O-H bonds, 449, 430 cm⁻¹ octahedral metal stretching, 1350–1370 cm⁻¹ -C-N, 1224–1227 cm⁻¹ -C-H-OH of carboxylic groups and N-H bond of amide2	EDX: C, N, P UV: 370–400 nm ferric reductase activity against *Legionella pneumophila*	Sherine et al. 2016

value was recorded at 426 nm. XRD 2θ value was calculated and it was found that 2θ = 100° was face-centered cubic (Shi et al. 2016).

Lactobacillus sporogens

TEM analysis showed particle size 5–15 nm. SEM analysis showed hexagonal shape. XRD value was calculated and recorded as 11 nm (Vignesh et al. 2015).

B. licheniformis

TEM analysis revealed that the particle was 40 in width and 400 in length. SEM analysis showed that nanoparticles were like nanoflowers. FTIR showed O-H, N-H, C-O (carbonyl) stretching in the amide 1 and amide 2 linkage of protein, C-N stretching bond (Sundaram et al. 2012).

Streptomyces **MS2**

TEM analysis demonstrated the particle size as 10–20 nm and a narrow, irregular shape. SEM analysis also confirmed silver nanoparticles 30–50 nm in size and spherical in shape. FTIR demonstrated groups present in the nanoparticles were 3260–3270 cm^{-1} -O-H 400, 600 cm^{-1} -C=O, O-H bonds, 449, 430 cm^{-1} octahedral metal stretching, 1350–1370 cm^{-1} -C-N, 1224–1227 cm^{-1} -C-H-OH of carboxylic groups, and N-H relationship of amide 2. EDX showed the presence of C, N, and P in the silver particle. UV peak absorbance was recorded between 370 and 400 nm (Sherine et al. 2016).

Biosynthesis and characterization of selenium nanoparticles

Lactobacillus acidophilus

TEM results showed that this bacterium synthesized selenium nanoparticles 15–50 nm in size and spherical. The calculated crystallite size of the nanoparticles was 23–31 nm by XRD. UV peak absorption values were recorded between 280 and 353 nm (Radhika and Gayathri 2015).

Klebsiella pneumoniae

TEM result showed that the particle was 245 nm. SEM, FTIR, and XRD were not performed in this study. UV peak absorption was recorded at 218 and 248 nm. EDS absorption peak bands were 1.5, 11.2, and 12.5 KeV (Fesharaki et al. 2010).

Staphylococcus aureus

TEM results showed nanoparticles 40–60 nm in size and spherical. DLS revealed that the particles had hydrodynamic diameters. The bacterial assay was conducted against *S. aureus*. SEM, FTIR, and XRD were not performed (Presentato et al. 2018).

Table 11.4. Selenium nanoparticles.

Sample no.	Bacteria	XRD	TEM	SEM	FTIR	Remarks and application	Reference
1	*Lactobacillus acidophilus*	23–32 nm	15–50 nm spherical	Not observed	Not observed	UV: 280–353 nm	Radhika and Gayathri 2015
2	*Klebsiella pneumoniae*	Not observed	245 nm	Not observed	Not observed	UV: 218 and 248 nm EDS: optical absorption bands-1.5, 11.2, 12.5 KeV	Fesharaki et al. 2010
3	*Staphylococcus aureus*	Not observed	40–60 nm spherical	Not observed	Not observed	DLS: hydrodynamic diameters of 100 nm, four bacterial assays against *S. aureus*	Presentato et al. 2018
4	*Azoarcus*	Not observed	88 ± 40 nm spherical	Not observed	Not observed	EDX: spherical, SAED: amorphous form	Fernández-Llamosas et al. 2016
5	*Bacillus* JAPSK2	Not observed	21.9 nm spherical	Not observed	Not observed	UV: 222 nm Antimicrobial assay against *E. coli*, *Klebsiella* sp., *S. aureus*	Singh et al. 2014
6	*Lactobacillus plantarum*	Not observed	60–80 nm spherical	Not observed	3430 cm^{-1} -O-H, 3417 cm^{-1} -N-H, 747, 783, 791 cm^{-1} -C-H 1645 and 1651 cm^{-1} -C-C	UV: 300 nm antimicrobial activity against *Aspergillus niger* and *Candida albicans*	Radhika and Gayathri 2015
7	*Lactobacillus rhamnosus*	Not observed	10–20 nm	Not observed	Not observed	UV: 300 nm antimicrobial activity against *Aspergillus niger* and *Candida albicans*	Radhika and Gayathri 2015
8	*Rhodococcus aetherivorans* BCPI	Not observed	71–78 nm	Not observed	Not observed	EDX: silicon DLS: presence of macromolecules	Presentato et al. 2018

Azoarcus

TEM image revealed that the particle size was 88 ± 40 nm and the shape was spherical. EDX results showed the presence of dense electron particles. SAED pattern showed the amorphous form of a nanoparticle. SEM, FTIR, and XRD analyses were not done (Fernández-Llamosas et al. 2016).

Bacillus JAPSK2

UV spectral analysis peak absorbance was at 222 nm. XRD values were calculated as 31.002, 46.761, 56.700 and 75.488. AFM was observed to be 69.9 nm. Antimicrobial assay was done against *E. coli, Klebsiella, Staphylococcus,* and *Pseudomonas* sp. *Pseudomonas* sp. results were good (Singh et al. 2014).

L. plantarum

TEM analysis showed particle spherical and 60–80 nm in size. FTIR revealed that 3430 cm^{-1} -O-H, 3417 cm^{-1} -N-H, 791 cm^{-1} -C-H 1645 and 1651 cm^{-1} -C-C were present in the nanoparticle. UV peak absorbance was recorded at 300 nm. Antimicrobial activity was seen against *Aspergillus niger* and *Candida albicans* (Radhika and Gayathri 2015).

L. rhamnosus

TEM analysis showed particle of size 10–20 nm and spherical shape. UV peak absorbance was recorded at 300 nm (Radhika and Gayathri 2015).

Rhodococcus aetherivorans BCPI

TEM analysis showed that the size of nanoparticles was 71–78 nm, and the shape was spherical. EDX results demonstrated that selenium was present. DLS confirmed the presence of macromolecules (Presentato et al. 2018).

Silver nanoparticle synthesis and characterization

Rhodopseudomonas

TEM analysis found the silver nanoparticle to be 6–10 nm in size. The FTIR showed 3300–2500 cm^{-1} O-H-carboxylic, 2260–2210 cm^{-1} nitriles, 1650–1580 cm^{-1} primary amines, and 1370–1350 cm^{-1} -alkanes groups to be present. UV absorbance at peak value was recorded to be at 450 nm (Manisha et al. 2014).

Pseudomonas aeruginosa

The XRD peak values were recorded at 38.13°, 46.31°, 64.47°, and 77.15°. These peak values denote the presence of silver nanoparticles. SEM analysis recorded particles to be of spherical shape and 45 nm size. FTIR showed the presence of

Table 11.5. Silver nanoparticles.

Sample no.	Bacteria	XRD	TEM	SEM	FTIR	Remarks and application	Reference
1	*Rhodopseudomonas*	Not observed	6–10 nm	Not observed	3300–2500 cm^{-1} -O-H-carboxylic 2260–2210-nitriles 1650–1580-primary amines 1370–1350-alkanes	UV: 450 nm antibacterial activity against *E. coli, Bacillus substiles, Klebsiella pneumoniae*	Manisha et al. 2014
2	*Pseudomonas aeruginosa*	38.13°, 46.31°, 64.47° and 77.15°	Not observed	Spherical 45 nm	1270.55 cm^{-1}, 1300–1270 cm^{-1}, -symmetric nitrate stretching, 805.89 cm^{-1} -N-H stretching of organic compounds, 3306.14 cm^{-1} and 1640.78 cm^{-1} -O-H hydrogen stretch and secondary amide C=O stretching	ICP-MS: 20–200 nm. UV: 442 nm	Kumar et al. 2018
3	*Klebsiella pneumoniae*	38.2°, 44.2°, 64.3°, 77.7°	Not observed	Spherical 35 ± 9.4 nm and 32.3 ± 5.5 nm	1657 and 1539 cm^{-1} -vibrations of amide 1(C=O) and amide 2 vibrations of proteins, 1235 and 1062 cm^{-1} -C-N stretching vibrations of N-H bending of peptide linkage	UV: 425 nm and 520 nm extracellular	Siva Kumar et al. 2014

4	*Brevibacillus bortelensis*	32°; 46°; 57°; 76°	Not observed	Cubic 5–15 nm	1345 cm⁻¹ the presence of nitrate 3626/cm-O-h Stretching H-bonded alcohols and phenols 3072/cm-O-H-carboxylic acids 1651/cm-N-H primary bond amines.	Not observed	Kumar and Ghosh 2016
5	*Lactobacillus* sp.	Not observed	10–40 nm spherical	Not observed	Not observed	Antimicrobial activity against Gram-positive bacteria *S. aureus*, *S. epidermidis, Bacillus cereus*. Gram-negative bactéria *Pseudomonas aeruginosa, Proteus vulgaris, E. coli* GC/MS–250°C–280°C	Garmasheva et al. 2016
6	*Chryseobacterium artocarpi*	28.5 may indicate the presence of the flex Rubin complex	Not observed	Spherical 49 nm	3454 cm⁻¹ -OH 1643 cm⁻¹ -C=O 1365 cm⁻¹ -C-O 1217 cm⁻¹ -C-H 3429, 1521, 1502, 1382, 1263, 3429, 1382, 1263, 1037 cm⁻¹ -O-H, C-O, alkyl C-H	UV: 300–700 nm, UV: 420 nm EDX: 3 KeV Anticancer activity	Sathishkumar et al. 2015
7	*Bacillus cereus*	Not observed	Spherical	Spherical 20–40 nm	The 3510 cm⁻¹ -OH stretch of carboxylic acid 1636 cm⁻¹ -N-H bending of primary amines	UV: 432 nm EDX-3KV SPR: C, N, O, antibacterial activity against *E. coli, Staphylococcus aureus, Salmonella typhi, Klebsiella pneumoniae*	Sunkar and Nachiyar 2012

Table 11.5 contd. ...

... *Table 11.5 contd.*

Sample no.	Bacteria	XRD	TEM	SEM	FTIR	Remarks and application	Reference
8	*Nocardiopsis* sp. MBRC-1	38.44°, 44.38°, 56.77°, 64.38°, 77.50°	30–90 nm spherical	Not observed	3440, 2923, 2853, 1655, 1460, 1685 cm⁻¹ 3440 cm⁻¹ -H-bonded 2923 and 2853- alkaline group	UV: 420 nm EDX: silver region-3eKV SPR-Cl and O Antimicrobial activity against *E. coli, Bacillus subtilis, S. aureus*	Manivasagan et al. 2013
9	*Ochrobactrum anthropi*	Not observed	Spherical 38-85 nm	Spherical	Not observed	UV: 450 nm Antibacterial against *Salmonella typhi, Vibrio chlorae, S. aureus*	Thomas et al. 2014
10	*Bacillus pumilus, Bacillus licheniformis, Bacillus persicus*	Not observed	Monodispersed triangular, hexagonal, spherical	Not observed	3373 cm⁻¹, 2359 cm⁻¹, 1650 cm⁻¹, 1600–1000 cm⁻¹	UV: 450 nm DLS: 80 nm Charge carried-18.5 Mv EDX: O, N, C Antimicrobial activity against *Aeruginosa, K. pneumoniae*	Elbeshehy et al. 2015
11	*Acinetobacter calcoaceticus*	38.1° 44.3° 64.4° 77.2°	10–60 nm polydisperse	Dispersed	Not observed	UV: 440 nm SPR: 410–440 nm metal nanoparticle spherical-Ag NPs EDS: 3 KeV Antibacterial activity against *P. aeruginosa* and *A. baumanii*	Singh et al. 2013

No.	Organism						Reference
12	*Brevibacterium frigoritolerans*	20°–80°	50–150 nm hydrodynamic	Not observed	Not observed	Antimicrobial activity against *E. coli, B. cereus, Candida albicans, S. aureus* UV: 200–800 nm (420 nm) SPR: 420 nm-Ag	Singh et al. 2015
13	*Stenotrophomonas maltophilia*	93 nm	Not observed	Cubic capped	3376 cm^{-1} -CHO, alkyl 2167–2851 cm^{-1} -C=O 1644 cm^{-1} -COO	UV: 250–600 nm (420 nm) SPR: 428 nm-Ag	Oves et al. 2013
14	*Streptomyces hygroscopicus*	38°, 45°, 65°	20–30 nm spherical	Not observed	Not observed	UV: 420 nm–425 nm SPR: spherical NPs EDXA: Si, O, C, Cl Antimicrobial activity against *E. coli, S. aureus*	Sadhasivam et al. 2010
15	*Nocardiopsis valliformis*	Not observed	5–50 nm spherical and polydisperse	Not observed	$3437 \text{ cm}^{-1}, 1639 \text{ cm}^{-1},$ 1405, 1384, 1352 cm^{-1}	UV: 423 nm zeta potential –17.1 mV Antibacterial activity against *E. coli, S. aureus*	Rathod et al. 2016
16	*Bacillus JAPSK2*	75.488 nm	Not observed	Not observed	Not observed	AFM: 29.9nm UV: 393 nm Antimicrobial activity against *E. coli, Klebsiella sp., S. aureus*	Singh et al. 2014

1270.55 cm^{-1} C-H -symmetric nitrate stretching, 1300–1270 cm^{-1}, -symmetric nitrate stretching, 805.89 cm^{-1} -N-H stretching of organic compounds, 3306.14 cm^{-1} and 1640.78 cm^{-1} -O-H hydrogen stretch and secondary amide C=O stretching (Kumar et al. 2018).

Klebsiella pneumoniae

The XRD peak values were recorded as 38.2°, 44.2°, 64.3°, and 77.7°. SEM analysis showed particles spherical in shape and 35 ± 9.4 nm and 32.3 ± 5.5 nm in size. FTIR analysis revealed 1657 and 1539 cm^{-1}—vibrations of amide 1 (C=O) and amide 2 vibrations of proteins, 1235 and 1062 cm^{-1} -C-N stretching vibrations of N-H bending of peptide linkage. UV peak value absorbance was recorded to be 425 nm and 520 nm. The nanoparticles were synthesized extracellularly (Siva Kumar et al. 2014).

Brevibacillus borstelensis

The XRD peak values were recorded to be 32°, 46°, 57°, and 76°. SEM analysis found particles to be cubic and 5–15 nm in size. FTIR analysis showed 1345 cm^{-1} presence of nitrate 3626 cm^{-1} O-H stretching H-bonded alcohols and phenols 3072 cm^{-1} O-H-carboxylic acids 1651 cm^{-1} N-H bond primary amines (Kumar and Ghosh 2016).

Lactobacillus sp.

The particles were found by TEM analysis to be 10–40 nm in size and spherical.

Chryseobacterium artocarpi

XRD results revealed the peak value to be 28.5, and this indicates the presence of the flexirubin complex. SEM analysis found the particle to be spherical and 49 nm in size. FTIR results revealed 3454 cm^{-1} –OH, 1643 cm^{-1} C=O, 1365 cm^{-1} C-O 1217, cm^{-1} C-H, 3429 cm^{-1}, 1037 cm^{-1} O-H, C-O, alkyl C-H. The UV absorbance value was recorded between 300 and 700 nm and the peak value was 420 nm. EDX revealed the result to be 3 KeV (Sathishkumar et al. 2015)·

Bacillus cereus

The particles were found to be spherical in TEM analysis. In SEM analysis, it was found that the nanoparticles were spherical and 20–40 nm in size. FTIR found 3510 cm^{-1} OH stretch of carboxylic acid, 1636 cm^{-1} N-H bending of primary amines. UV peak value for absorbance was recorded at 432 nm. EDX showed the result 3 KV. SPR showed C, N, O, were present (Sunkar and Nachiyar 2012).

Nocardiopsis sp. **MBRC-1**

The XRD values recorded were 38.44°, 44.38°, 56.77°, 64.38°, and 77.50°, which indicated the presence of silver nanoparticles synthesized by this bacterium. The particles were found to be 30–90 nm and spherical by TEM analysis. FTIR analysis showed the presence of 3440 cm⁻¹, 2923 cm⁻¹, 2853 cm⁻¹, 1655 cm⁻¹, 1460 cm⁻¹, 1685 cm⁻¹, 3440 cm⁻¹ H-bonded, 2923 cm⁻¹ and 2853 cm⁻¹ alkaline group. UV peak absorbance value was recorded to be 420 nm. EDX showed the silver region 3 eKV. SPR revealed that Cl and O were present (Manivasagan et al. 2013).

Ochrobactrum anthropi

TEM analysis recorded the particle to be spherical and 38–85 nm in size. UV peak absorbance value was recorded at 450 nm. Antibacterial activity was evaluated against *Salmonella typhi, Vibrio chlorae,* and *S. aureus* (Thomas et al. 2014).

Bacillus pumilus, Bacillus licheniformis, Bacillus persicus

TEM analysis showed particles to be monodispersed, triangular, hexagonal, spherical. FTIR revealed that 3373 cm⁻¹, 2359 cm⁻¹, 1650 cm⁻¹, 1600–1000 cm⁻¹ groups were present. UV absorbance peak value was recorded at 450 nm. DLS showed the peak value at 80 nm. The charge carried at 18.5 Mv. EDX value recorded that O, N, C were present (Elbeshehy et al. 2015).

Acinetobacter calcoaceticus

The XRD value was recorded as 38.1°, 44.3°, 64.4°, and 77.2°, which indicates the presence of nanoparticles. The particles were found by TEM analysis to be 10–60 nm and polydispersed. UV peak absorbance was recorded at 440 nm. SPR showed peak between 410 and 440 nm, which represents silver nanoparticles, and the shape was spherical. EDX results were found to be 3 KeV (Singh et al. 2013).

Brevibacterium frigoritolerans

XRD results were recorded from 20° to 80° and showed peak at (1 1 1) and (1 1 2), confirming the crystalline nature of the nanoparticles. TEM analysis showed that the particles were 50–150 nm in size and hydrodynamic in shape. UV peak absorbance value was recorded at 200–800 nm (420 nm). SPR peak value was noticed at 420 nm, which indicates the presence of silver nanoparticles (Singh et al. 2015).

Stenotrophomonas maltophilia

XRD peak values were recorded at 93 nm. TEM analysis of the particles revealed cubic shape. FTIR peak values revealed the presence of 3376 cm⁻¹ CHO, alkyl 2167–2851 cm⁻¹ C=O, and 1644 cm⁻¹ COO. UV peak value was recorded at 250–600 nm (420 nm). SPR peak was noted at 428 nm, indicating the presence of Ag nanoparticles (Oves et al. 2013).

Streptomyces hygroscopicus

XRD peak values were marked to be 38°, 45°, and 65°, denoting the presence of silver nanoparticles. TEM analysis showed spherical nanoparticles 20–30 nm in size. UV peak value was recorded between 420 and 425 nm. SPR results showed that the nanoparticles were spherical. EDX revealed that Si, O, C, and Cl were present (Sadhasivam et al. 2010).

Nacardiopsis valliformis

TEM results showed particles 5–50 nm in size and spherical. UV absorbance peak value was recorded at 423 nm. Zeta potential was found to be 17.1 mV (Rathod et al. 2016).

Bacillus sp. JAPSK2

XRD peak value was recorded to be 31.02, 46.76, and 75.48 nm. AFM was recorded to be 29.9 nm. UV peak absorbance value was recorded at 393 nm. Antimicrobial activity was assessed against *E. coli*, *Klebsiella* sp., and *S. aureus* (Singh et al. 2014).

Bacillus mycoides-mediated titanium oxide nanoparticles

TEM analysis revealed that the nanoparticles were spherical and 40–60 nm in size. FTIR results showed the presence of 3400 cm^{-1}, 1630 cm^{-1}, hydroxyl group 3431 cm^{-1} O-H stretching, and 2985 cm^{-1} symmetric stretch of CH_2 group (Bravo et al. 2014).

Table 11.6. Titanium oxide nanoparticles.

Sample no.	Bacteria	XRD	TEM	SEM	FTIR	Remarks and application	Reference
1	*Bacillus mycoides*	Not observed	40–60 nm spherical	Not observed	3400 cm^{-1}, 1630 cm^{-1} -hydroxyl group 3431 cm^{-1} -O-H stretching, 2985 cm^{-1} -symmetric stretch of CH_2	Not observed	Bravo et al. 2014

Conclusion

In this chapter, bacterially synthesized nanoparticles such as gold, silver, selenium, titanium, copper, and iron have been discussed. Biosynthesis of metal nanoparticles using bacteria has advantages over physical or chemical processes as it does not apply chemicals or toxic substances for synthesis, and it is also eco-friendly. Synthesis using bacteria can save time and is cost-effective. Hence, biosynthesis of nanoparticles on a large scale will be a significant improvement in the field of nanotechnology.

References

Amini, Seyed Mohammad. 2019. Preparation of antimicrobial metallic nanoparticles with bioactive compounds. Materials Science and Engineering C 103(May): 109809.

Baek, Yong Wook, Youn Joo An. 2011. Microbial toxicity of metal oxide nanoparticles (CuO, NiO, ZnO, and Sb_2O_3) to *Escherichia coli, Bacillus subtilis*, and *Streptococcus aureus*. Science of the Total Environment 409(8): 1603–8.

Bravo, Denisse Margarita, José Manuel Pérez-Donoso, Juan Pablo Monrás, Vicente María Durán-Toro, Luis Alberto Saona, Nicolás Alexis Órdenes-Aenishanslins. 2014. Use of titanium dioxide nanoparticles biosynthesized by *Bacillus mycoides* in quantum dot sensitized solar cells. Microbial Cell Factories 13(1): 1–10.

Chopra, Ian. 2007. The increasing use of silver-based products as antimicrobial agents: A useful development or a cause for concern? Journal of Antimicrobial Chemotherapy 59(4): 587–90.

Creanga, D., Poiata, A., Fifere, N., Airinei, A. Nadejde, C. 2011. Fluorescence of pyoverdine synthesized by pseudomonas under the effect of iron oxide nanoparticles. Romanian Biotechnological Letters 16(4): 6336–43.

Dharne, Mahesh S., Yogesh S. Shouche, Sanjay Singh, Rasesh Y. Parikh, Prasad, B.L.V., Syed Saif Hasan, Milind S. Patole. 2008. Bacterial synthesis of copper/copper oxide nanoparticles. Journal of Nanoscience and Nanotechnology 8(6): 3191–96.

DR, Manisha, Ramchander, M., Prashanthi, Y., Pratap, M.P.R. 2014. Phototrophic bacteria mediated synthesis, characterisation and antibacterial activity of silver nanoparticles. Nanosci. Nanotechnol. Int. J. 4(2)(January): 20–24.

Du, Liangwei, Hong Jiang, Xiaohua Liu, Erkang Wang. 2007. Biosynthesis of gold nanoparticles assisted by *Escherichia coli* DH5α and its application on direct electrochemistry of hemoglobin. Electrochemistry Communications 9(5): 1165–70.

Duhan, Joginder Singh, Ravinder Kumar, Naresh Kumar, Pawan Kaur, Kiran Nehra, Surekha Duhan. 2017. Nanotechnology: The new perspective in precision agriculture. Biotechnology Reports 15(May): 11–23.

Elbeshehy, Essam K.F., Ahmed M. Elazzazy, George Aggelis. 2015. Silver nanoparticles synthesis mediated by new isolates of *Bacillus* Spp., nanoparticle characterization and their activity against bean yellow mosaic virus and human pathogens. Frontiers in Microbiology 6(MAY): 1–13.

Fernández-Llamosas, Helga, Laura Castro, María Luisa Blázquez, Eduardo Díaz, Manuel Carmona. 2016. Biosynthesis of selenium nanoparticles by *Azoarcus* sp. CIB. Microbial Cell Factories 15(1): 1–10.

Fesharaki, Parisa Jafari, Pardis Nazari, Mojtaba Shakibaie, Sassan Rezaie, Maryam Banoee, Mohammad Abdollahi, Ahmad Reza Shahverdi. 2010. Biosynthesis of selenium nanoparticles using *Klebsiella pneumoniae* and their recovery by a simple sterilization process. Brazilian Journal of Microbiology 41(2): 461–66.

Garmasheva, Inna, Nadezhda Kovalenko, Sergey Voychuk, Andriy Ostapchuk, Olena Livins'ka, Ljubov Oleschenko. 2016. *Lactobacillus* species mediated synthesis of silver nanoparticles and their antibacterial activity against opportunistic pathogens *in vitro*. BioImpacts 6(4): 219–23.

He, Shiying, Zhirui Guo, Yu Zhang, Song Zhang, Jing Wang, Ning Gu. 2007. Biosynthesis of gold nanoparticles using the bacteria *Rhodopseudomonas capsulata*. Materials Letters 61(18): 3984–87.

Jelinkova, Pavlina, Aninda Mazumdar, Vishma Pratap Sur, Silvia Kociova, Kristyna Dolezelikova, Ana Maria Jimenez Jimenez, Zuzana Koudelkova, Pawan Kumar Mishra, Kristyna Smerkova, Zbynek Heger, Marketa Vaculovicova, Amitava Moulick, Vojtech Adam. 2019. Nanoparticle-drug conjugates treating bacterial infections. Journal of Controlled Release 307: 166–85.

Kasana, Ramesh Chand, Nav Raten Panwar, Ramesh Kumar Kaul, Praveen Kumar. 2017. Biosynthesis and effects of copper nanoparticles on plants. Environmental Chemistry Letters 15(2): 233–40.

Kumar, Amar and Ashok Ghosh. 2016. Biosynthesis and characterization of silver nanoparticles with bacterial isolate from gangetic-alluvial soil. International Journal of Biotechnology and Biochemistry 12(2): 95–102.

Kumar, Sathish Sundar Dhilip, Nicolette Nadene Houreld, Eve M. Kroukamp, and Heidi Abrahamse. 2018. Cellular imaging and bactericidal mechanism of green-synthesized silver nanoparticles against human pathogenic bacteria. Journal of Photochemistry and Photobiology B: Biology 178(September 2017): 259–69.

Li, Jiulong, Qinghao Li, Xiaoqiong Ma, Bing Tian, Tao Li, Jiangliu Yu, Shang Dai, Yulan Weng, Yuejin Hua. 2016. Biosynthesis of gold nanoparticles by the extreme bacterium deinococcus radiodurans and an evaluation of their antibacterial properties. International Journal of Nanomedicine 11: 5931–44.

Malarkodi, C., Chitra, K., Gnanajobitha, G., Paulkumar, K., Vanaja, M., Annadurai, G. 2013. Pelagia research library novel eco-friendly synthesis of titanium oxide nanoparticles by using *Planomicrobium* sp. and its antimicrobial evaluation. Der Pharmacia Sinica 4(3): 59–66.

Manisha, D.R., Ramchander, M., Prashanthi, Y., Pratap, M.P.R. 2014. Phototrophic bacteria mediated synthesis, characterisation and antibacterial activity of silver nanoparticles. Nanoscience and Technology: An International Journal 4(2)(January): 20–24.

Manivasagan, Panchanathan, Jayachandran Venkatesan, Kalimuthu Senthilkumar, Kannan Sivakumar, Se Kwon Kim. 2013. Biosynthesis, antimicrobial and cytotoxic effect of silver nanoparticles using a novel Nocardiopsis sp. MBRC-1. Biomed Res Int. 2013: 287638. Doi: 10.1155/2013/287638.

Nabila, Mohammed Ishaque, Krishnan Kannabiran. 2018. Biosynthesis, characterization and antibacterial activity of copper oxide nanoparticles (CuO NPs) from actinomycetes. Biocatalysis and Agricultural Biotechnology 15: 56–62.

Nadaf, Nilofar Yakub, Shivangi Shivraj Kanase. 2016. Biosynthesis of gold nanoparticles by Bacillus marisflavi and its potential in catalytic dye degradation. Arabian Journal of Chemistry. 12(8): 4806–4814

Nagababu, U., Govindh, B., Diwakar, B.S., Kiran Kumar, G., Anindita Chatterjee. 2019. Synthesis & characterization of biologically active gigantic swallow-wort mediated silver nanoparticles. Materials Today: Proceedings 18: 2102–2106.

Nakhaeepour, Zahra, Mansour Mashreghi, Maryam M. Matin, Ali NakhaeiPour, Mohammad Reza Housaindokht. 2019. Multifunctional CuO nanoparticles with cytotoxic effects on KYSE30 esophageal cancer cells, antimicrobial and heavy metal sensing activities. Life Sciences 234: 116758.

Narayanan, Kannan Badri, Natarajan Sakthivel. 2010. Biological synthesis of metal nanoparticles by microbes. Advances in Colloid and Interface Science 156(1-2): 1–13.

Oves, Mohammad, Mohammad Saghir Khan, Almas Zaidi, Arham S. Ahmed, Faheem Ahmed, Ejaz Ahmad, Asif Sherwani, Mohammad Owais, Ameer Azam. 2013. Antibacterial and cytotoxic efficacy of extracellular silver nanoparticles biofabricated from chromium reducing novel OS4 strain of *Stenotrophomonas maltophilia*. PLoS ONE 8(3): e59140. https://doi.org/10.1371/journal.pone.0059140.

Paulkumar, Kanniah, Shanmugam Rajeshkumar, Gnanadhas Gnanajobitha, Mahendran Vanaja, Chelladurai Malarkodi, Gurusamy Annadurai, Kanniah Paulkumar, Shanmugam Rajeshkumar, Gnanadhas Gnanajobitha, Mahendran Vanaja, Chelladurai Malarkodi, Gurusamy Annadurai. 2013. Biosynthesis of silver chloride nanoparticles using *Bacillus subtilis* MTCC 3053 and assessment of its antifungal activity. ISRN Nanomaterials 2013: 1–8.

Ponnanikajamideen, M., Rajeshkumar, S., Annadurai, G. 2016. *In vivo* antidiabetic and *in vitro* antioxidant and antimicrobial activity of aqueous leaves extract of *Chamaecostus cuspidatus*. Research Journal of Pharmacy and Technology 9(8): 1204.

Presentato, Alessandro, Elena Piacenza, Max Anikovskiy, Martina Cappelletti, Davide Zannoni, Raymond J. Turner. 2018. Biosynthesis of selenium-nanoparticles and -nanorods as a product of selenite bioconversion by the aerobic bacterium *Rhodococcus aetherivorans* BCP1. New Biotechnology 41: 1–8.

R, Radhika Rajasree, S., Gayathri, S. 2015. Extracellular biosynthesis of selenium nanoparticles using some species of Lactobacillus. 43: 766–75.

Rajeshkumar, Shanmugam, Chelladurai Malarkodi, Gnanadhas Gnanajobitha, Kanniah Paulkumar, Mahendran Vanaja, Chellapandian Kannan, Gurusamy Annadurai. 2013. Seaweed-mediated synthesis of gold nanoparticles using turbinaria conoides and its characterization. Journal of Nanostructure in Chemistry 3(1): 44.

Rajeshkumar, Shanmugam, Chelladurai Malarkodi, Mahendran Vanaja, Gurusamy Annadurai. 2016. Anticancer and enhanced antimicrobial activity of biosynthesizd silver nanoparticles against clinical pathogens. Journal of Molecular Structure 1116: 165–73.

Ramanathan, Rajesh, Suresh K. Bhargava, Vipul Bansal. 2010. Biological synthesis of copper/copper oxide. pp. 1–8. *In*: Rose Amal (ed.). CHEMECA 2011—Engineering A Better World, Sydney, Australia, 18–21 September.

Ramanathan, Rajesh, Matthew R. Field, Anthony P. O'Mullane, Peter M. Smooker, Suresh K. Bhargava, Vipul Bansal. 2013. Aqueous phase synthesis of copper nanoparticles: A link between heavy metal resistance and nanoparticle synthesis ability in bacterial systems. Nanoscale 5(6): 2300–2306.

Rathod, Dnyaneshwar, Patrycja Golinska, Magdalena Wypij, Hanna Dahm, Mahendra Rai. 2016. A New report of nocardiopsis valliformis strain OT1 from alkaline lonar crater of india and its use in synthesis of silver nanoparticles with special reference to evaluation of antibacterial activity and cytotoxicity. Medical Microbiology and Immunology 205(5): 435–47.

Rehana, Dilaveez, Mahendiran, D., Senthil Kumar, R., Kalilur Rahiman, A. 2017. Evaluation of antioxidant and anticancer activity of copper oxide nanoparticles synthesized using medicinally important plant extracts. Biomedicine and Pharmacotherapy 89: 1067–77.

Rekha, Ravichandran, Mani Divya, Marimuthu Govindarajan, Naiyf S. Alharbi, Shine Kadaikunnan, Jamal M. Khaled, Mohammed N. Al-Anbr, Roman Pavela, Baskaralingam Vaseeharan. 2019. Synthesis and characterization of crustin capped titanium dioxide nanoparticles: photocatalytic, antibacterial, antifungal and insecticidal activities. Journal of Photochemistry and Photobiology B: Biology 199: 111620.

Sadhasivam, Sathya, Parthasarathi Shanmugam, Kyu Sik Yun. 2010. Biosynthesis of silver nanoparticles by streptomyces hygroscopicus and antimicrobial activity against medically important pathogenic microorganisms. Colloids and Surfaces B: Biointerfaces 81(1): 358–62.

Santhoshkumar, J., Rajeshkumar, S. Venkat Kumar, S. 2017. Phyto-assisted synthesis, characterization and applications of gold nanoparticles—A review. Biochemistry and Biophysics Reports 11(June): 46–57.

Sathishkumar, Palanivel, Mahalingam Malathi, Rajarajeswaran Jayakumar, Abdull Rahim Mohd Yusoff, Wan Azlina Ahmad, Chidambaram Kulandaisamy Venil, Rajamanickam Usha. 2015. Synthesis of flexirubin-mediated silver nanoparticles using *Chryseobacterium artocarpi* CECT 8497 and investigation of its anticancer activity. Materials Science and Engineering: C 59: 228–34.

Shantkriti, S., Rani, P. 2014. Biological synthesis of copper nanoparticles using *Pseudomonas fluorescens*. Int. J. Curr. Microbiol. App. Sci. 3(9): 374–383.

Sherine, Jositta, Annie Sujatha, Maheshwaran Rathinam. 2016. Biological synthesis of iron oxide nanoparticles using Streptomyces sp. and its antibacterial activity. (8): 58–60.

Shi, Si Feng, Jing Fu Jia, Xiao Kui Guo, Ya Ping Zhao, De Sheng Chen, Yong Yuan Guo, Xian Long Zhang. 2016. Reduced *Staphylococcus aureus* biofilm formation in the presence of chitosan-coated iron oxide nanoparticles. International Journal of Nanomedicine 11: 6499–6506.

Singaravelu, G., Arockiamary, J.S., Ganesh Kumar, V., Govindaraju, K. 2007. A novel extracellular synthesis of monodisperse gold nanoparticles using marine alga, *Sargassum wightii* greville. Colloids and Surfaces B: Biointerfaces 57(1): 97–101.

Singh, Nidhi, Prasenjit Saha, Karthik Rajkumar, Jayanthi Abraham. 2014. Biosynthesis of silver and selenium nanoparticles by *Bacillus* sp. JAPSK2 and evaluation of antimicrobial activity. Der Pharmacia Lettre 6(1): 175–81.

Singh, Priyanka, Yeon Ju Kim, Hina Singh, Chao Wang, Kyu Hyon Hwang, Mohamed El Agamy Farh, Deok Chun Yang. 2015. Biosynthesis, characterization, and antimicrobial applications of silver nanoparticles. International Journal of Nanomedicine 10: 2567–77.

Singh, R., Wagh, P., Wadhwani, S., Kumbhar, A., Bellare, J., Chopade, B.A. 2013. Synthesis, optimization, and characterization of silver nanoparticles from *Acinetobacter calcoaceticus* and their enhanced antibacterial activity when combined with antibiotics. Int. J. Nanomedicine 8: 4277–4290. Doi: 10.2147/IJN.S48913.

Siva Kumar, K., Kumar, G., Prokhorov, E., Luna-Bárcenas, G., Buitron, G., Khanna, V.G., Sanchez, I.C. 2014. Exploitation of anaerobic enriched mixed bacteria (AEMB) for the silver and gold nanoparticles synthesis. Colloids and Surfaces A: Physicochemical and Engineering Aspects 462: 264–70.

Sondi, Ivan, Branka Salopek-Sondi. 2004. Silver nanoparticles as antimicrobial agent: a case study on *E. coli* as a model for Gram-negative bacteria. Journal of Colloid and Interface Science 275(1): 177–82.

Srinath, B.S., Ravishankar Rai, V. 2015. Rapid biosynthesis of gold nanoparticles by *Staphylococcus epidermidis*: Its characterisation and catalytic activity. Materials Letters 146: 23–25.

Subramanian, Prasanna, Kirubanandan Shanmugam. 2016. Extracellular and intracellular synthesis of silver nanoparticles. Asian Journal of Pharmaceutical and Clinical Research 9: 133–39.

Sumbal, Asifa Nadeem, Sania Naz, Joham Sarfraz Ali, Abdul Mannan, Muhammad Zia. 2019. Synthesis, characterization and biological activities of monometallic and bimetallic nanoparticles using *Mirabilis jalapa* leaf extract. Biotechnology Reports 22: e00338.

Sundaram, P. Alagu, Robin Augustine, M. Kannan. 2012. Extracellular biosynthesis of iron oxide nanoparticles by *Bacillus subtilis* strains isolated from rhizosphere soil. Biotechnology and Bioprocess Engineering 17(4): 835–40.

Sunkar, Swetha, Valli Nachiyar, C. 2012. Biogenesis of antibacterial silver nanoparticles using the endophytic bacterium Bacillus cereus isolated from *Garcinia xanthochymus*. Asian Pacific Journal of Tropical Biomedicine 2(12): 953–59.

Thomas, Roshmi, Anju Janardhanan, Rintu T. Varghese, Soniya, E.V., Jyothis Mathew, Radhakrishnan, E.K. 2014. Antibacterial properties of silver nanoparticles synthesized by marine *Ochrobactrum* sp. Brazilian Journal of Microbiology 45(4): 1221–27.

Tian, Li-Jiao, Yu-Cai Wang, Nan-Qing Zhou, Han-Qing Yu, Dao-Bo Li, Pan-Pan Li, Xing Zhang. 2016. Extracellular biosynthesis of copper sulfide nanoparticles by *Shewanella oneidensis* MR-1 as a photothermal agent. Enzyme and Microbial Technology 95: 230–235.

Tiwari, Mradul, Prateek Jain, Raghu Chandrashekhar Hariharapura, Kashinathan Narayanan, Udaya Bhat K., Nayanabhirama Udupa, Josyula Venkata Rao. 2016. Biosynthesis of copper Nanoparticles using copper-resistant *Bacillus cereus*, a soil isolate. Process Biochemistry 51(10): 1348–56.

Vignesh, Venkatasamy, Ganesan Sathiyanarayanan, Gnanasekar Sathishkumar, Karuppaiah Parthiban, Kamaraj Sathish-Kumar, Ramasamy Thirumurugan. 2015. Formulation of iron oxide nanoparticles using exopolysaccharide: Evaluation of their antibacterial and anticancer activities. RSC Advances 5(35): 27794–804.

Wadhwani, Sweety A., Shradhda B. Nadhe, Utkarsha U. Shedbalkar, Balu A. Chopade, Richa Singh. 2017. Lignin peroxidase mediated silver nanoparticle synthesis in *Acinetobacter* sp. AMB Express 7(1).

Zhou, Nan-qing, Li-jiao Tian, Yu-cai Wang, Dao-bo Li, Pan-pan Li, Xing Zhang, Han-qing Yu. 2016. Enzyme and microbial technology extracellular biosynthesis of copper sulfide nanoparticles by shewanella oneidensis MR-1 as a photothermal agent. Enzyme and Microbial Technology 95: 230–35.

Zonooz, N. Faghri, Salouti, M. 2011. Sharif university of technology extracellular biosynthesis of silver nanoparticles using cell filtrate of Streptomyces sp. ERI-3. Scientia Iranica 18(6): 1631–35.

Chapter 12

Synthesis of Nanomaterials Using Biosurfactants

Mechanisms and Applications

Paulo Ricardo Franco Marcelino,[1,*] *Fernanda Gonçalves Barbosa,*[1] *Mariete Barbosa Moreira,*[2] *Talita Martins Lacerda*[1] and *Silvio Silvério da Silva*[1,*]

Introduction

Nano is a unit prefix derived from the Greek word *nanos* used to refer to things with dimensions of 10^{-9} meter (0.000000001 m) (Plaza et al. 2014). In the 20th century, the use of this prefix became part of the scientific vocabulary, giving rise to nanoscience, which focuses on the development of materials and technologies for various applications. Although the terms "nanomaterials" and "nanotechnology" were first used in the mid-20th century by researchers such as Richard Feynman (1918–1988) and Norio Taniguchi (1912–1999), the use of nanoscale systems by humankind dates back to antiquity. Some reports indicate that the first nanomaterials were developed by the Mesopotamians in the 9th century to give a shiny effect to vessel surfaces. In Ancient Rome, gold and silver nanoparticles were also used to give special characteristics to glass-based materials (https://nanomedicina.webnode.pt/nanotecnologia-e-medicina/historia/).

The history of nanoscience had a breakthrough during the 1980s. In 1981, Gerd Binning and Heinrich Rohrer at the IBM laboratory in Zurich, Switzerland, created the Scanning Tunneling Microscope (STM), capable of seeing and manipulating atoms, which was never imagined before. In 1986, Eric Drexler was the first scientist

[1] Department of Biotechnology, Engineering School of Lorena (EEL), São Paulo University (USP), CEP 12602-810, Lorena, Brazil.
[2] Chemistry Institute, Universidade Estadual Paulista Júlio de Mesquita Filho (UNESP), CEP 14800-900, Araraquara, Brazil.
* Corresponding authors: silviosilverio@usp.br; paulorfm1@hotmail.com

who obtained a PhD in nanotechnology, popularizing it through his book *Engines of Creation*. In the same year, Gerd Binning invented the atomic force microscope (AFM), which allowed a scientist to see atoms and also to move them one by one when an extremely strong electrical tension was applied between the tip of the microscope and the sample. In 1989, the physicist Donald M. Eigler of IBM in California, USA, wrote the company's initials on a nickel surface using 35 xenon atoms, showing that structures could be built up from individual atoms or molecules (Mody and Lynch 2010).

In 1991, Professor Sumio Iijima, in Japan, reported the discovery of carbon nanotubes (Cadioli and Salla 2006, Fiolhais 2007). Since then, several studies involving nanotechnology have been carried out all over the world for varied applications, with the first nanoproducts (passive and active nanostructures) being commercialized in the 2000s (Nanowerk 2012, Faria-Tischer and Tischer 2012).

At that time a new and very promising branch of nanoscience emerged, *nanobiotechnology*. It corresponds to a frontier area of nanoscience that uses concepts of nanotechnology and biotechnology for the development of materials, devices and systems, besides studying and understanding diverse biological systems. Nanobiotechnology can be employed in strategic fields such as medicine, agriculture, civil engineering, electronics, computer science, industry (chemical, pharmaceutical, cosmetics, petrochemical) and environment (Srinivas 2016, Nagamune 2017, Rai 2013). Nanobiotechnology is also often related to green chemistry principles, aiming to develop a society based on sustainable processes and products. Thus, the use of biotransformation and fermentation processes and bio-based molecules in the synthesis and functionalization of nanoparticles has become a reality (Rai 2013).

The importance of nanometric systems for the progress of society is unquestionable, having an impact on agriculture, electronics, medicine, personal care, chemistry and other fields. This is directly reflected in the evolution of the use of the terms "nanoscience", "nano" and "nanobiotechnology" in scientific publications over the last 20 years (Fig. 12.1).

Companies such as ABC Nanotech (South Korea), ATDBio (United Kingdom), Alpha Nano Technology (China), Advance NanoTek (Australia), APNano–NIS

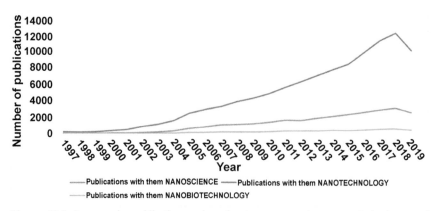

Figure 12.1. Increase in publications using the terms nanoscience, nanotechnology and nanobiotechnology in the last 20 years (www.sciencedirect.com). * 2019: the data obtained refer to the survey carried out in July 2019.

Corporation (USA), Antibodies Incorporated (USA), Bionano Genomics (USA), Biophan Technologies (USA), TEKNA (Canada), Cheap Tubes (USA), DAIS Analytic (USA), Nanovetores (Brazil), Intrinsiq Materials (USA), Lumiphore (USA), Nanoselect (Brazil), Magforce Nanotechnologies (Germany), NanoBioMagnetics Inc. (USA), F Group Nano LLC (USA), Nanotex LLC (USA), NovaCentrix (USA), Sarastro GmbH (Germany), nanoSPACE (Czech Republic), nanoComposix (USA), ULURU Inc. (USA) and Nanovet (Brazil) have been developing processes and products that use nanoscience, nanotechnology and nanobiotechnology (https://www.pesquisa-unificada.com/pesquisas/nanotecnolgia/empresas-de-nanotecnologia/).

Green nanomaterials: production mechanisms and biomolecules used

Nanoparticles, nanocrystals, nanotubes, nanofibers, nanowires, nanofilms, liposomes, and other materials found in the nanometer scale are today generally referred to as nanomaterials or nanostructured materials (Faria-Tischer and Tischer 2012). Several physicochemical routes are used for the synthesis of nanomaterials, as presented in Table 12.1. However, most of these methods are complex (multi-step) processes that use toxic solvents and generate toxic by-products and wastes that are harmful to human health and the environment. In addition, the physicochemical methods used in the synthesis of nanomaterials demand excessive energy and high-cost materials, making the final products often overpriced and of limited application. Therefore, it is necessary to find alternatives for the production of nanomaterials that are cheap, simple, and inert with respect to the health of living organisms and the environment (Ferreira and Rangel 2009). Green synthesis of nanomaterials may be a promising approach to achieve these objectives.

Table 12.1. Some chemical, physical and biological methods employed for nanoparticle and liposome synthesis (Yu et al. 2008, Ealias et al. 2017, He et al. 2007).

Chemical methods	Physical methods	Biological methods
• Colloidal methods • Sol-gel processing • Chemical precipitation • Redox reactions • Microemulsion methods • Electrochemical methods • Sonochemical decomposition reactions by effect acoustic cavitation • Chemical vapor condensation	• Vacuum deposition and vaporization • Hydrothermal synthesis • Flow injection synthesis • Spinning • Mechanical milling • Laser ablation • Sputtering • Spray pyrolysis • Physical vapor condensation • Supercritical fluid method • Electron beam lithography • Microwave-assisted synthesis	• Biogenic synthesis using microorganisms • Biogenic synthesis using biomolecules as the template • Biogenic synthesis using plant extracts
Liposomes		
• Hydration • Method of injection of ethanol or ether	• Extrusion through polycarbonate membranes with different porosities • French press • Use of homogenizer, microfluidizer or sonication	

Microorganisms, plants and biomolecules can be used for the green synthesis of nanomaterials (Narayanan and Sakthivel 2011). Reduction reactions of metals, such as the conversion of Ag^+ to Ag^0 in nanoparticle synthesis, can be conducted by plant extracts obtained from stems, bark, leaves and seeds, which contain reducing agents such as enzymes, proteins, amino acids, flavonoids, heterocyclic compounds and water-soluble metabolites. This method, often called bioreduction, is additionally advantageous considering that plants produce reducing biomolecules that adhere to the surface of the nanoparticles, ensuring stability, protection, high yield and low production cost. For this synthesis, in general, the extract is mixed with an aqueous solution of a metal salt, usually silver nitrate, and bioreduction occurs. It should be noted that physicochemical factors such as pH, temperature, concentration and nature of the plant extract, and concentration of the metal salt may affect the reaction rate as well as the amount and stability of the bionanoparticles (Mittal et al. 2013).

Microorganisms such as bacteria, yeasts and filamentous fungi can be used in nanoparticle bioreduction. The mechanisms that act on the formation of nanoparticles by microorganisms are diverse and may involve the internalization of some ions, which are reduced in the cytoplasmic (or periplasmic) environment, or the reduction outside the cell, either by the action of secreted metabolites or molecules on the outer surface of microorganisms (Gericke and Pinches 2006, Philip 2009, Chauhan et al. 2011, Baker et al. 2013). According to Plaza et al. (2014), one of the enzymes involved in the biosynthesis of metal nanoparticles is a nitrate reductase (NADPH-NADH dependent enzyme) that reduces metal ions (Me^{n+} to Me^0). Studies of golden and silver nanoparticle biosynthesis using respectively *Rhodopseudomonas capsulata* and *Bacillus licheniformis* bacteria proved the redox potential of nitrate reductase in bioreduction (He et al. 2007, Sadowski 2010). In addition to bioreduction, biomolecules are used in stabilization, nanoparticle functionalization and liposome production (Mittal et al. 2013) (Fig. 12.2).

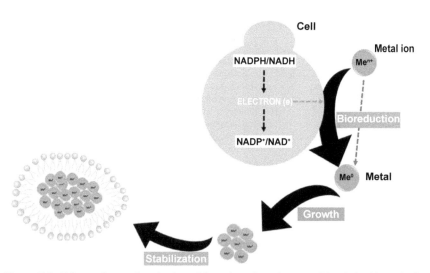

Figure 12.2. Scheme of general mechanism of formation of metal nanoparticles during biosynthesis (based on He et al. 2007, Mittal et al. 2013).

Carbohydrates, peptides, proteins, lipids and some polymers of organic acids are most commonly used biomolecules for biosynthesis of nanoparticles. These are described below.

Carbohydrates: mono- and oligosaccharides (glucose, sucrose, xylose and cellobiose), polysaccharides (starch, chitosan, xylan, vegetable and microbial cellulose, carboxymethylcellulose, hydroxypropylcellulose, gellan gum, xanthan gum, cyclodextrins, pullulan, alginate, pectin, mannan, levan, fucoidan, ghatti gum, kondagogu gum, arabinogalactan, lentinan, hyaluronic acid, arabic gum, katiragum, gumkaraya, k-carrageenan and algin) and polyols (xylitol) (Dhar et al. 2008, Venturini et al. 2008, Kora et al. 2010, Vu-Quang et al. 2011, Iconaru et al. 2012, Kora et al. 2012, Maity et al. 2012, Willis 2012, Kanmani and Lim 2013, Liu et al. 2013, Ahmed et al. 2014, Lai et al. 2014, Cheryl-Low et al. 2015, Faria-Tischer and Tischer 2012, Hussain et al. 2015, Jia et al. 2015, Li et al. 2015, Pooja et al. 2015, Anuradha et al. 2016, Narayanan et al. 2016, Rajeshkumar 2016, Rau et al. 2016, Mahdavinia et al. 2017, Pallavicini et al. 2017, Viana 2017, Wang et al. 2017, Anjum et al. 2019).

Amino acids, peptides and proteins: valine, leucine, isoleucine alanine, arginine, glutamine, lysine, aspartic acid, glutamic acid, proline, cysteine, threonine, methionine, histidine, phenylalanine, tyrosine, tryptophan, asparagine, glycine, serine. Proteins such as bovine serum albumin and human serum albumin also are used (Akbarzadeh et al. 2009, Zare et al. 2010, Ebrahiminezhad et al. 2012, Faria-Tischer and Tischer 2012, Maruyama et al. 2015, Jeong et al. 2018, Spicer et al. 2018).

Lipids, fatty acids and their derivatives: saturated and unsaturated fatty acids, phospholipids, cholesterol, fatty acid monoesters and ceramides (Nastruzzi et al. 1993, Patravale and Mandawgade 2008, Socaciu 2009, Singh et al. 2011a, Tapas and Oli 2014, Reva et al. 2015, Kurapati 2016, Ganesan and Choi 2016).

Polymers of organic acids and their derivatives: poly(L-lactic acid), poly(glycolic acid), poly(hydroxy-butyrate), poly(hydroxy-valerate), poly(malic acid), polygalacturonic acid and poly(L-glutamic acid) (Van Der Elst et al. 1999, Jan et al. 2011, Westwood et al. 2011, Bakó et al. 2016, Nagy et al. 2016).

The use of biomolecules in the preparation of nanomaterials is of great importance. In the medical field, for example, proteins, peptides, aptamers and specific antibodies are used to functionalize nanoparticles for the treatment of tumoral tissues (Oyelere et al. 2007, Yang et al. 2009, Nghiem et al. 2010, Medley et al. 2011). In the food industry, Siegrist et al. (2007) developed starch-coated β-carotene nanocapsules in order to improve the solubility of this pigment in aqueous medium and prolong its shelf life. Some active ingredients and adjuvants of cosmetic formulations also use particular liposomal systems (transferosomes, novasomes, marinosomes, ethosomes, sphingosomes, nanosomes, glycerosomes and oleosomes) based on fatty acids and lipids (Batista et al. 2007, Ashtiani et al. 2016). Recent research described the potential of nanocellulose and its derivatives in the development of new materials for applications as reinforcement composites, composition of pharmaceuticals and cosmetics, civil construction products, food and packaging components, and also in electronics and electrical industries, textiles, biomedicine and remediation of degraded environments (Ioelovich 2008, Kim et al. 2015, Shak et al. 2018).

Among the biomolecules used in the production of nanomaterials, biosurfactants can be highlighted.

Biosurfactants

Biosurfactants, also known as biological surfactants, are tensioactive or bioemulsifier molecules, synthesized by plants (saponins), microorganisms (biosurfactants) and higher organisms (bile salts) (Nitschke and Pastore 2002). Microbial biosurfactants are produced by bacteria, actinobacteria, yeasts and filamentous fungi. Besides their active surface and emulsifying properties, biosurfactants can also present antimicrobial, antitumor and larvicidal or insecticidal potentials (Nitschke and Pastore 2002, Nitschke and Costa 2007, Kim et al. 2011, Marcelino et al. 2017, Wu et al. 2017). The eco-friendly characteristics of these biomolecules, namely high biodegradability, low toxicity, environmental compatibility and the possibility of being produced from agroindustrial by-products, make biosurfactants promising substitutes for synthetic surfactants of petrochemical origin (Desai and Banat 1997, Nitschke and Pastore 2002, Mulligan et al. 2014). Some other advantages are mentioned below:

- **Higher surface and interfacial activities:** Biosurfactants present higher surface and interfacial activities than synthetic surfactants because they are more efficient in reducing the surface tension in lower concentrations, which means lower critical micelle concentration (CMC) (Cooper and Paddock 1984). The CMC is defined as the minimum concentration required to initiate the formation of micelles, and from that concentration there is no further variation of the surface tension (Becher 1965, Mulligan 2005).

- **Greater tolerance to temperature, pH and ionic strength:** Some biosurfactants have high thermal and pH stability. They can stand salt concentrations of up to 10% (w/v), while concentrations of 2–3% are sufficient to destabilize conventional surfactants (Bognolo 1999).

- **Biodegradability:** Biosurfactants are easily degradable in water and soil and can be metabolized by microorganisms for energy production (Mulligan and Gibbs 1993).

- **Low toxicity:** Biosurfactants are less likely to cause allergy and can be used in food, pharmaceutical and personal care products (Flasz et al. 1998).

- **Production based on renewable substrates:** Biosurfactants can be produced from renewable substrates and agroindustrial by-products. Some studies reported the use of sugarcane bagasse, hydrolysates, molasses, cassava meal, wheat bran, crude glycerol from biodiesel refineries and animal fat for this purpose (Deshpande and Daniels 1995, Nitschke et al. 2004, Nitschke and Pastore 2006, Raza et al. 2007, Accorsini et al. 2011, Marcelino et al. 2019). There are many low-cost substrates, such as soybean oil, which act not only as nutrients for microbial growth but also as a biosurfactant source (Guerra-Santos et al. 1986, Lee et al. 2008).

Classification of biosurfactants

Biosurfactants are classified according to their chemical structure and microbial origin. Most of them are neutral or anionic, with some few examples of cationic ones that have amino groups in their structure (Mulligan 2005). They possess a hydrophilic domain consisting of a carbohydrate, a carboxylic acid, an amino acid, a peptide or a hydroxyl function, and a hydrophobic domain corresponding to a hydrocarbon chain with one or more saturated or unsaturated fatty acids (Desai and Banat 1997, Rosenberg and Ron 1999). They are divided into molecules with low molar mass, which are efficient surface and interfacial tension reducers, and those with high molar mass, which are more effective as emulsion stabilizing agents (Rosenberg and Ron 1999).

According to the chemical composition, biosurfactants can be divided into six groups: glycolipids, phospholipids and fatty acids, lipopeptides, lipoproteins, polymeric surfactants and particulate surfactants (Table 12.2) (Desai and Desai 1993).

Currently, glycolipids (rhamnolipids, sophorolipids and mannosyl-erithritol lipids) and lipopeptides/lipoproteins (surfactin) (Fig. 12.3) are the most known and used biosurfactants.

These glycolipid and lipopeptide/lipoprotein biosurfactants are also the most widely used in nanotechnological systems, commonly applied in the development of liposomes, nanoparticles and other nanomaterials.

Table 12.2. Biosurfactant classification (Desai and Banat 1997, Rosenberg and Ron 1999, Nitschke and Costa 2007).

Groups	Subgroups	Types
Low molar mass	Glycolipids	Rhamnolipids
		Trehaloselipids
		Sophorolipids
		Mannosyl-erythritol lipids
	Fatty acids, neutral lipids and phospholipids	Fatty acids
		Neutral lipids
		Phospholipids
	Lipopeptides	Surfactin Viscosin Lichenysin Serrawettin
High molar mass	Polymeric surfactants	Emulsan
		Biodispersan
		Alasan
		Liposan
	Lipoproteins	Bioelan
	Particulate surfactants	Vesicles
		Whole microbial cells

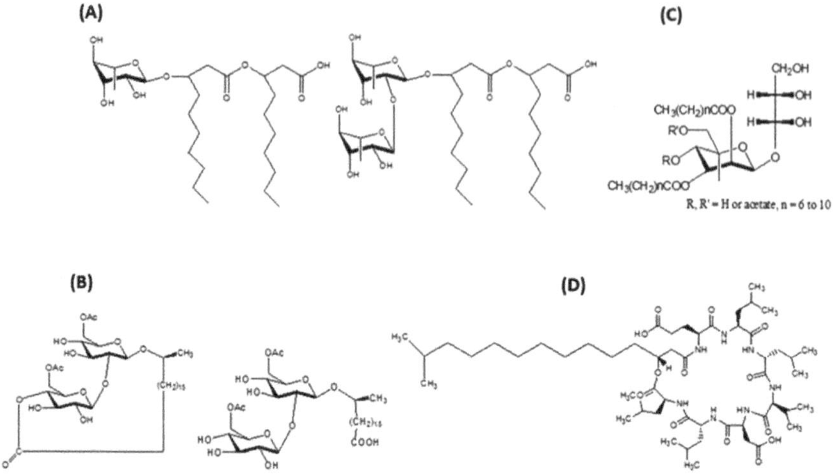

Figure 12.3. The most known and used biosurfactants: (A) rhamnolipids (mono- and dirhamnolipids), (B) sophorolipids (lactonic and acidic), (C) mannosyl-erithritol lipids and (D) surfactin.

Biosurfactant applications in nanostructured materials

The expansion of products containing nanostructured materials leads to concerns about toxicity, since humans and the environment are exposed to increasing levels of nanoparticles and liposomes. In this context, green synthesis has emerged as an important tool for the production of eco-friendly materials (Merroun et al. 2007), and biosurfactants have attracted significant interest from the scientific community as a non-toxic and biodegradable stabilizing agent (Kumar et al. 2014).

Rhamnolipids, sophorolipids and surfactin occupy a prominent position among the most frequently used biosurfactants for production of nanostructures, such as nanoparticles and liposomes. Nanomaterials produced with these biosurfactants have been widely studied by the scientific community and, owing to their interesting properties, are applied in pharmaceuticals (Thakur and Agrawal 2015), medical devices (Huang et al. 2011, Ge et al. 2014), biotechnological products (Tartaj et al. 2005), cosmetics (Katz et al. 2015) and agriculture (Rai and Ingle 2012). The potential application of biosurfactant-based nanomaterials was also investigated for drug delivery systems (Wu et al. 2017).

Rhamnolipids

Rhamnolipids are glycolipidic biosurfactants generally produced by bacteria of the genus *Pseudomonas* (*P. aeruginosa, P. putida, P. chlororaphis, P. fluorescens, P. nitroreducens* and *P. alcaligenes*) (Gunther et al. 2005, Oliveira et al. 2009, Nanganuru and Korrapati 2012, El-Amine Bendaha et al. 2012, Onwosi and Odibo 2012, Kaskatepe and Yildiz 2016). However, other bacteria or actinobacteria such as *Renibacterium salmoninarum*, *Cellulomonas cellulans*, *Nocardioides* sp., *Burkholderia thailandensis* and *Tetragenococcus koreensis* have also been reported as rhamnolipid producers (Arino et al. 1998, Christova et al. 2004, Lee et al. 2005,

Vasileva-Tonkova and Gesheva 2005, Dubeau et al. 2009). These biosurfactants are formed by one molecule (mono-rhamnolipid) or two molecules (di-rhamnolipid) of rhamnose (6-deoxy-L-mannose) linked to a fatty acid tail, generally β-hydroxydecanoic acid (C_{10}) (Monteiro et al. 2007, Cortés-Sanchés et al. 2013).

These glycolipids exhibit physicochemical and biological properties, such as tensioactivity, emulsification, antimicrobial and antitumoral potential and can, therefore, be applied in the most varied industrial sectors (Abdel-Mawgoud et al. 2010, Cortés-Sanchés et al. 2013). In nanobiotechnology, these molecules are generally used as capping or stabilizing agents for nanoparticles (Singh et al. 2009).

Farias et al. (2014) demonstrated a simple eco-friendly synthesis of spherical silver nanoparticles by microemulsion using rhamnolipid from *P. aeruginosa* as stabilizing agent. According to the authors, under experimental conditions, the biosurfactant prevented the formation of nanoparticle aggregates, stabilized biosynthesized nanoparticles and favored the reaction. Moreover, they observed that the structure of the biosurfactant is related to the morphology of the obtained nanoparticles. Rhamnolipids produced by *P. aeruginosa* AMBAS7 were more efficient than chemical surfactants in green synthesis of silver nanoparticles (AgNPs) (Pandian and Mariaamalraj 2014). Several synthetic methods using biosurfactants from bacteria as stabilizing agents were successful, as exemplified by the synthesis of AgNPs in reverse micelles, the synthesis of AgNPs using borohydrate reduction at different pH levels, the synthesis of spherical NiO nanoparticles by emulsion technique and the production of NiO nanorods by microemulsion technique (Xie et al. 2006, Palanisamy 2008, Palanisamy and Raichur 2009, Reddy et al. 2009a).

According to literature, rhamnolipids are able to form nanoparticles and considered good candidates for dermal delivery because of their non-toxicity to primary human fibroblast (Muller et al. 2016). Hazra et al. (2014a) demonstrated that biosurfactant-polymer hybrids are promising for biomedical applications. Biosurfactants from *P. aeruginosa* BSO1 were used in the synthesis of poly(methyl methacrylate)-(core)-biosurfactant (shell) nanoparticles that exhibited strong antibacterial activity against *Bacillus subtilis.* Also, a non-toxic concentration (0.005%) of a biosurfactant extracted from *P. aeruginosa* JS-11 culture was used for the synthesis of rhamnolipid-functionalized silver nanoparticles. The results indicate that these materials were more efficient against human breast adenocarcinoma (MCF-7) cells than normal peripheral blood mononuclear cells due to the generation of reactive oxygen species, causing membrane damage and apoptosis in MCF-7 cells (Dwivedi et al. 2015).

Biosurfactants have been used in the synthesis of self-assembly nanoparticles with membrane scaffold proteins. The results reported by Faas et al. (2017) suggest a strong relationship between the nanoparticle formation and the biosurfactant-membrane molar ratio and indicate that rhamnolipids are an important tool for biophysical and biochemical studies on membrane proteins.

Sophorolipids

Sophorolipids are glycolipidic biosurfactants produced by yeasts, mainly *Starmerella* (*Candida*) *bombicola* strains. According to Kulakovskaya and Kulakovskaya (2014), these biosurfactants are formed by a residue of sophorose (disaccharide consisting of two glucose residues linked by the β-1,2 glycosidic bond) and a saturated or unsaturated fatty acid (C_{16}–C_{22}). In some cases, these glycolipids exhibit different degree of acetylation and hydroxylation, respectively, in the glycidic and lipidic portions. Sophorolipids can be produced by microorganisms in the acidic form (with a carboxyl group) or in the lactonic form (with a lactone group).

Like rhamnolipids, sophorolipids present tensioactivity, emulsification, antimicrobial and antitumoral potential and can, therefore, be applied in the most varied industrial sectors. Generally, sophorolipids are used in cosmetic formulations of Korres, Bioderma, Germanie de Capuccini, Melvita, Naturopathica, Cattier and other companies, are certified by Ecocert and also appear in biodegradable detergent formulations in products of Ecover Company and Saraya Co., Ltd (Kulakovskaya and Kulakovskaya 2014). In nanobiotechnology, these molecules are generally used as capping or stabilizing agents of nanoparticles and for the development of liposomes for drug delivery systems.

A sophorolipid produced by *Candida bombicola* (polymorph *Starmerella bombicola*) was used for the production of iron oxide nanoparticles with diameters varying from 10 to 30 nm in water and in 0.01 M and 2 M KCl solutions. These nanoparticles can be used in biomedical applications because of their potential biocompatibility (Baccile et al. 2013).

Shikha and collaborators (2017) reported the synthesis of gold nanoparticles functionalized with sophorolipids and their antimicrobial potential against Gram-negative *Vibrio cholerae*, Gram-positive *Staphylococcus aureus* and fungi of *Candida albicans*. The results indicated that the nanoparticles were very active against *V. cholera* and may be used as potential drug in the treatment of cholera, a disease that afflicts underdeveloped countries with precarious basic sanitation.

Sophorolipids obtained from oleic acid were used as capping agent for cobalt nanoparticles. In this case, the sugar moiety of the sophorolipids was exposed to the external environment, making the nanoparticles stable and water-dispersible. These nanoparticles had potential for further attachment of bioactive molecules such as lectins or glycosidases to generate biocompatible surfaces for medical and diagnostic applications (Kasture et al. 2007).

Basak et al. (2014) used sophorolipids as stabilizer in the synthesis of zinc oxide nanoparticles. The antimicrobial effect of these biofunctionalized nanoparticles against *Salmonella enterica* and *C. albicans* was evaluated in concentrations varying from 1 to 5 mg/ml. *S. enterica* presented higher sensitivity to nanoparticles at concentration of 5 mg/ml than *C. albicans*. TEM analysis showed the interaction of nanostructures with thiol(-SH) groups of cell wall proteins. This interaction decreased the cell wall permeability, which led to cell lyses and permanent damage.

Vasudevan and Prabhune (2018) studied micelle formation by encapsulating curcumin together with acidic sophorolipid. The nanostructures obtained in these experiments were stable in various conditions and presented ellipsoid shape of approximately 100 nm. Additionally, bacterial fluorescence uptake studies showed the uptake of formed nanostructures into both Gram-positive and Gram-negative bacteria and quorum quenching activity against *P. aeruginosa*. This work also highlighted the fluorescence properties of the nanostructures, which allow them to be applied as biomarker agents for confocal imaging with potential theranostic applications.

Joshi-Navare and Prabhune (2013) reported the synergistic action of sophorolipids and antibiotics (cefaclor and tetracycline) against *Escherichia coli* and *S. aureus*.

Results showed the efficiency of sophorolipid-tetracycline and sophorolipid-cefaclor liposomes against both bacteria, with their growth inhibition observed in shorter time intervals than with pure antibiotics. Besides, an increase of 25% in the inhibition of bacterial growth for the sophorolipid-tetracycline liposomes and 48% for sophorolipid-cefaclor liposomes was observed. According to the authors, the cells treated with biosurfactant-antibiotic liposomes revealed bacterial cell membrane damage and pore formation (Fig. 12.4).

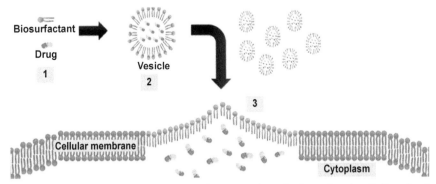

Figure 12.4. Schematic diagram showing the proposed mechanism of sophorolipid-mediated drug entry facilitation across cell membrane: (1) sophorolipid-drug combination, (2) development of vesicle/liposome, and (3) interaction of vesicle with cell membrane (based on Joshi-Navare and Prabhune 2013).

Surfactin

Surfactin is a lipopeptidic biosurfactant produced by bacteria of the genus *Bacillus* (*B. subtilis, B. cereus* and others). The cyclic structure of this biosurfactant contains seven amino acids linked to a β-hydroxy-fatty acid (13–15 carbon chains) by lactone bond, with the occurrence of isomeric or homologous compounds. Surfactin exhibits high tensioactive and emulsifier properties (Kluge et al. 1988, Kowall et al. 1998, Lang 2002, Barros et al. 2007, Shekhar et al. 2015).

Surfactin exhibits antibacterial and antifungal (Das et al. 2008), antimycoplasmal (Boettcher et al. 2010), antiviral (Seydlova et al. 2011, Sachdev and Cameotra 2013, Singla et al. 2014), anti-inflammatory (Byeon et al. 2008, Zhang et al. 2015), and thrombolytic (Kikuchi and Hasumi 2002, Singla et al. 2014) activities. Antitumoral effects of surfactin were also reported, according to a review by Wu and collaborators (2017). The molecule was efficient against Ehrlich ascites carcinoma, breast cancer, colon cancer, leukemia, hepatocellular carcinoma, cervical cancer, human oral epidermoid carcinoma, pancreatic cancer and rat melanoma. It is important to mention that the biological activities of a surfactin depend on the amino acid composition, the peptide sequence and the nature of its lipid part (Kowall et al. 1998).

Due to their physicochemical and biological properties, surfactins are good candidates as biosurfactants used in the development of nanomaterials. As in the case of glycolipids, nanomaterials produced from surfactins can be used in medical, food, environmental and other fields, as described below.

Surfactin nanomicelles with potential antitumoral applications were produced in aqueous environments. Results showed that micelles produced in distilled water were 100–200 nm spheres, while those produced in PBS buffer were shapeless objects of 100–400 nm. In assays against human cervical cancer cells (HeLa), the nanoparticles exhibited satisfactory effects. Surfactin nanomicelles inhibited the growth of tumoral cells tested in a time- and dosage-dependent method, so that their IC_{50} at 16, 24, and 48 h were 86.9, 73.1, and 50.2 μmol/l, respectively (Nozhat et al. 2012).

Doxorubicin is a drug used in the treatment of tumors because of its potential inhibition of the enzyme DNA gyrase (Almeida et al. 2005). Recently, surfactin–doxorubicin nanoparticles were synthesized by the solvent-emulsion method and tested against resistant tumoral cells. Nanoparticles obtained in these experiments induced higher cytotoxicity against doxorubicin-resistant human breast cancer MCF-7/ADR cells compared to free doxorubicin. Surfactin–doxorubicin nanoparticles exhibited enhanced cellular uptake and decreased cellular efflux, and *in vivo* animal experiment also showed that nanoparticles were more efficiently accumulated in tumors than free doxorubicin (Huang et al. 2018).

Liposomes obtained from surfactin–cationic surfactant (1,2-dioleoyl-*sn*-glycero-3-ethylphosphocholine—EDOPC) combinations were used as a delivery system of siRNA to cancer cells. Results indicated the production of liposomes smaller than 200 nm that exhibited enhanced cellular delivery of siRNA in HeLa cells when compared to surfactin-free liposomes. This result indicates that liposomes are potential biocompatible delivery systems of siRNA for enhanced cellular uptake and inhibition of target gene expression in tumoral cells (Shim et al. 2009).

Surfactin was also used for the development of self-(micro/nano)-emulsifying drug delivery systems with vitamin E, gelucire 44/14, labrasol and PEG 3000. In these experiments, physically stable formulations were generated with potential pharmaceutical applications (Kural and Gürsoy 2011).

Joe and collaborators (2012) studied nanoemulsion formulations based on some vegetable oils (sunflower, castor, coconut, groundnut and sesame) and surfactin. Sunflower oil–surfactin nanoemulsion showed better results in antimicrobial tests inhibiting the growth of *Salmonella typhi, Listeria monocytogenes, S. aureus, Rhizopus nigricans, Aspergillus niger* and *Penicillium* sp. This nanoemulsion also presented sporicidal activity against *Bacillus cereus* and *Bacillus circulans. In situ* tests of the sunflower oil–surfactin nanoemulsion on food products (raw chicken, apple juice, milk and mixed vegetable) resulted in a significant reduction in the native cultivable microorganisms, which indicated the potential of this emulsion for the development of antimicrobial formulations.

Poly(methyl-methacrylate) nanoparticles were surface-functionalized with surfactin and tested as multifunctional polymer nanoparticles with enhanced sorption properties. The nanostructures of 60 nm were obtained by sonochemical emulsion polymerization and were used for the removal of metallic ions and phenolic compounds from waste. Antimicrobial tests of this nanomaterial were also performed. In the adsorption experiments, ions formed chelate complexes or interacted electrostatically with nanoparticles, with the affinity adsorption following the order $Co^{2+} > Zn^{2+} > Ni^{2+} > Cr^{3+} > Fe^{2+} > Cu^{2+} > Cd^{2+} > Pb^{2+}$. Poly(methyl-methacrylate)–surfactin nanoparticles were good adsorbent for phenol and β-naphthol in the tests with phenolics. Finally, in antimicrobial tests, a bactericidal activity of nanomaterials (at concentration of 300 mg/ml) was observed against *E. coli*, with 80% efficiency (Kundu et al. 2016).

Surfactin–polyvinyl alcohol nanofibers produced by electrospinning exhibited antimicrobial activity against methicillin-resistant *S. aureus, L. monocytogenes, E. coli* and *P. aeruginosa*. In the adhesion tests of *L. monocytogenes* to the polystyrene films in the presence of surfactin–nanofibers, decreased cellular optical density at 595 nm (OD: 0.012 ± 0.001) was observed when compared with control sample (OD: 0.022 ± 0.002). Thus, these experiments suggested that surfactin–polyvinyl alcohol nanofibers may be used in wound dressings or in the coating of prosthetic devices to prevent biofilm formation and secondary infections (Ahire et al. 2017).

Aiming to obtain high performance surfactant and emulsifier blends for industrial applications, Onaizi et al. (2012) combined surfactin with the synthetic surfactant sodium dodecylbenzenesulfonate (SDOBS). The adsorption of the surfactant mixture from the liquid phase into the air–liquid interface resulted in the formation of a mixed interfacial monolayer with a mole fraction of the biosurfactant at 0.55, although its mole fraction in the liquid bulk was much lower (0.076). The formation of a mixed interfacial monolayer comprising SDOBS and surfactant was found to be a spontaneous process ($\Delta G = -2.42$ kJ/mol). This result shows the potential application of this mixture in industrial processes. The use of a biosurfactant–synthetic surfactant mixture has the advantage of reducing the use of a surfactant derived from petroleum, making the process or product eco-friendly and sustainable.

Studies by Reddy et al. (2009a, 2009b), Singh et al. (2011b) and Hazra et al. (2014b) showed the potential of surfactin as stabilizer, capping, anti-aggregation and anti-precipitation agent in inorganic nanoparticle formulations.

Conclusion

Green nanomaterials have emerged as a field of growing interest and have enormous potential as eco-friendly alternatives to their petrochemical counterparts. Nanobiotechnology-based systems can be produced from a large class of biomolecules using chemical, physical and biological tools. Among the biological methods, one can highlight bioreduction reactions and nanoparticle production and functionalization mediated by enzymes, proteins, amino acids, and other agents. Biosurfactants (mainly rhamnolipids, sophorolipids and surfactin) appear in this context as efficient non-toxic and biodegradable stabilizing agents for the synthesis of nanostructures. A series of interesting studies in recent years demonstrate that this strategy will soon be fully consolidated.

Acknowledgments

The authors acknowledge financial support from the Coordination for the Improvement of Higher Education Personnel (CAPES), the National Council for Scientific and Technological Development (CNPq) and the São Paulo Research Foundation (FAPESP, projects 2015/06238-4, 2016/10636-8, 2016/14852-7, 2017/16062-6).

References

Abdel-Mawgoud, A.M., Lépine, F., Déziel, E. 2010. Rhamnolipids: diversity of structures, microbial origins and roles. Appl. Microbiol. Biotechnol. 86: 1323–1336. doi:10.1007/s00253-010-2498-2.

Accorsini, F.R., Mutton, M.J.R., Lemos, E.G.M., Benincasa, M. 2011. Biosurfactants production by yeasts using soybean oil and glycerol as low cost substrate. Braz. J. Microbiol. 43: 116–125.

Ahire, J.J., Robertson, D.D., Van Reenen, A.J., Dicks, L.M.T. 2017. Surfactin-loaded polyvinyl alcohol (PVA) nanofibers alters adhesion of *Listeria monocytogenes* to polystyrene. Mater Sci. Eng. C. 77: 27–33.

Ahmed, K.B.A., Kalla, D., Uppuluri, K.B., Anbazhagan, V. 2014. Green synthesis of silver and gold nanoparticles employing levan, a biopolymer from *Acetobacter xylinum* NCIM 2526, as a reducing agent and capping agent. Carbohydr. Polym. 112: 539–545.

Akbarzadeh, A., Zare, D., Farhangi, A., Mehrabi, M.R., Norouzian, D., Tangestaninejad, S., Moghadam, M., Bararpour, N. 2009. Synthesis and characterization of gold nanoparticles by tryptophane. Am. J. Appl. Sci. 6: 691–695.

Almeida, V.L., Leitão, A., Reina, L.C.B., Montanari, C.A., Donnici, C.L., Lopes, M.T.P. 2005. Câncer e agentes antineoplásicos ciclo-celular específicos e ciclo-celular não específicos que interagem com o DNA: uma introdução. Química Nova. 28: 118–129.

Anjum, A., Chung, P.Y., Ng, S.F. 2019. PLGA/xylitol nanoparticles enhance antibiofilm activity via penetration into biofilm extracellular polymeric substances. RSC Adv. 9: 14198–14208.

Anu, M.E., Saravanakumar, M.P. 2017. A review on the classification, characterisation, synthesis of nanoparticles and their application. IOP Conf. Ser.: Mater. Sci. Eng. 263: 032019.

Anuradha, K., Bangal, P., Madhavendra, S.S. 2016. Macromolecular arabinogalactan polysaccharide mediated synthesis of silver nanoparticles, characterization and evaluation. Macromol. Res. 24: 152–162.

Arino, S., Marchal, R., Vandecasteele, J.P. 1998. Production of new extracellular glycolipids by a strain of *Cellulomonas cellulans (Oerskovia xanthineolytica)* and their structural characterization. Can. J. Microbiol. 44: 238–243.

Ashtiani, H.R.A., Bishe, P., Lashgari, N.A., Nilforoushzadeh, M.A., Zare, S. 2016. Liposomes in cosmetics. J. Skin Stem Cell. doi: 10.5812/jssc.65815.

Baccile, N., Noiville, R., Stievano, L., Bogaert, I.V. 2013. Sophorolipids-functionalized iron oxide nanoparticles. Phys. Chem. Chem. Phys. 15: 1606–1620. doi:10.1039/c2cp41977g.

Baker, S., Rakshith, D., Kavitha, K.S., Santosh, P., Kavitha, H.U., Rao, Y., Satish, S. 2013. Plants: emerging as nanofactories towards facile route in synthesis of nanoparticles. Bioimpacts. 3: 111–117.

Bakó, J., Kerényi, F., Hrubi, E., Varga, I., Daróczi, L., Dienes, B., Csernoch, L., Gáll, J., Hegedüs, C. 2016. Poly-γ-glutamic acid nanoparticles based visible light-curable hydrogel for biomedical application. J. Nanomater. 2016: ID 7350516. doi: 10.1155/2016/7350516.

Basak, G., Das, D., Das, N. 2014. Dual role of acidic diacetate sophorolipid as biostabilizer for ZnO nanoparticle synthesis and biofunctionalizing agent against *Salmonella enterica* and *Candida albicans*. J. Microbiol. Biotechnol. 24: 87–96.

Barros, F.F.C., Quadros, C.P., Maróstica Júnior, M.R., Pastore, G.M. 2007. Surfactina: propriedades químicas, tecnológicas e funcionais para aplicações em alimentos. Química Nova. 30: 409–414. https://dx.doi.org/10.1590/S0100-40422007000200031.

Batista, C.M., de Carvalho, C.M.B., Magalhães, N.S.S. 2007. Lipossomas e suas aplicações terapêuticas: Estado da arte. Rev. Bras. Cienc. Farm. 43: 167–179.

Becher, P. 1965. Emulsions: theory and practice. J. Chem. Educ. 42: 692. https://doi.org/10.1021/ed042p692.2.

Boettcher, C., Kell, H., Holzwarth, J.F., Vater, J. 2010. Flexible loops of thread-like micelles are formed upon interaction of L-alpha-dimyristoyl-phosphatidylcholine with the biosurfactant surfactin as revealed by cryo-electron tomography. Biophys. Chem. 149: 22–27. doi: 10.1016/j.bpc.2010.03.006.

Bognolo, G. 1999. Biosurfactants as emulsifying agents for hydrocarbons. Colloids Surf., A. 152: 41–52.

Byeon, S.E., Lee, Y.G., Kim, B.H., Shen, T., Lee, S.Y., Park, H.J. 2008. Surfactin blocks NO production in lipopolysaccharide-activated macrophages by inhibiting NF-kappaB activation. J. Microbiol. Biotechnol. 18: 1984–1989. doi: 10.4014/jmb.0800.189.

Cadioli, L.P., Salla, L.D. 2006. Nanotecnologia: um estudo sobre seu histórico, definição e principais aplicações desta inovadora tecnologia. Disponível em <http://www.unianhanguera.edu.br/programasinst/Revistas/revistas2006/rev_exata s/11.pdf>.

Chauhan, A., Zubair, S., Tufail, S., Sherwani, A., Sajid, M., Raman, S.C., Azam, A., Owais, M. 2011. Fungus-mediated biological synthesis of gold nanoparticles: potential in detection of liver cancer. Int. J. Nanomed. 6: 2305–2319.

Cheryl-Low, Y.L., Theam, K.L., Lee, H.V. 2015. Alginate-derived solid acid catalyst for esterification of low-cost palm fatty acid distillate. Energy Convers. Manag. 106: 932–940. https://doi.org/10.1016/j.enconman.2015.10.018.

Christova, N., Tuleva, B., Lalchev, Z., Jordanova, A., Jordanov, B. 2004. Rhamnolipid biosurfactants produced by *Renibacterium salmoninarum* 27BN during growth on n-hexadecane. Zeitschrift für Naturforschung C, DOI: https://doi.org/10.1515/znc-2004-1-215.

Cooper, D.G., Paddock, D.A. 1984. Production of a biosurfactant from *Torulopsis bombicola*. Appl. Environ. Microbiol. 47: 173–176.

Cortes-Sanchez, A.J., Hernandez-Sanchez, H., Jaramillo-Flores, M.E. 2013. Biological activity of glycolipids produced by microorganisms: new trends and possible therapeutic alternatives. Microbiol. Res. 168: 22–32.

Das, P., Mukherjee, S., Sen, R. 2008. Antimicrobial potential of a lipopeptide biosurfactant derived from a marine *Bacillus circulans*. J. Appl. Microbiol. 104: 1675–1684. doi: 10.1111/j.1365-2672.2007.03701.

Desai, J.D., Banat, I.M. 1997. Microbial production of surfactants and their commercial potential. Microbiol. Mol. Biol. Rev. 61: 47–64.

Desai, J.D., Desai, A.J. 1993. Production of Biosurfactants. Biosurfactants: Production, Properties, Application. New York: CRC Press.

Deshpande, M., Daniels, L. 1995. Evaluation of sophorolipid biosurfactant production by *Candida bombicola* using animal fat. Bioresour. Technol. 54: 143–150.doi: 10.1016/0960-8524(95)00116-6.

Dhar, S., Maheswara, R.E., Shiras, A., Pokharkar, V., Prasad, B.L. 2008. Natural gum reduced/stabilized gold nanoparticles for drug delivery formulations. Chemistry 14: 10244–10250. https://doi.org/10.1002/chem.200801093.

Dubeau, D., Déziel, E., Woods, D.E., Lépine, F. 2009. *Burkholderia thailandensis* harbors two identical rhl gene clusters responsible for the biosynthesis of rhamnolipids. BMC Microbiol. 9: 263.

Dwivedi, S., Saquib, Q., Al-Khedhairy, A.A., Ahmad, J., Siddiqui, M.A., Musarrata, J. 2015. Rhamnolipids functionalized AgNPs-induced oxidative stress and modulation of toxicity pathway genes in cultured MCF-7 cells. Colloids Surf. B. 132: 290–298.

Ealias, A.M., Saravanakumar, M. 2017. A review on the classification, characterisation, synthesis of nanoparticles and their application. IOP Conf. Ser. Mater. Sci. Eng. 263: 032019.

Ebrahiminezhad, A., Ghasemi, Y., Rasoul-Amini, S., Barar, J., Davaran, S. 2012. Impact of amino-acid coating on the synthesis and characteristics of iron-oxide nanoparticles (IONs). Bull. Korean Chem. Soc. 33: 3957–3962.

El-Amine Bendaha, M., Mebrek, S., Naimi, M., Tifrit A., Belaouni, H.A., Abbouni, B. 2012. Isolation and comparison of rhamnolipids production in *Pseudomonas aeruginosa* P.B:2 and *Pseudomonas fluorescens* P.V.:10, Open Access Scientific Reports. 1: 544.

Faas, R., Pohle, A., Mob, K., Henkel, M., Hausmann, R. 2017. Self-assembly of nanoscale particles with biosurfactants and membrane scaffold proteins. Biotechnol. Rep. 16: 1–4.

Farias, C.B.B., Silva, A.F., Rufino, R.D., Luna, J.M., Souza, J.E.G., Sarubbo, L.A. 2014. Synthesis of silver nanoparticles using a biosurfactant produced in low-cost medium as stabilizing agente. Electron. J. Biotechnol. 17: 122–125.

Faria-Tischer, P.C.S., Tischer, C.A. 2012. Nanobiotecnologia: plataforma tecnológica para biomateriais e aplicação biológica de nanoestruturas. BBR – Biochemistry and Biotechnology Reports. 1: 32. Disponível em: <www.uel.br/revistas/ uel/index.php/bbr/article/download/13190/11502>.

Ferreira, H.S., Rangel, M.C. 2009. Nanotecnologia: aspectos gerais e potencial de aplicação em catálise. Química Nova. 32: 1860–1870.

Fiolhais, C. Breve História da nanotecnologia. 2007. <http://nautilus.fis.uc.pt/cec/arquivo/Carlos%20 Fiolhais/Textos%20divulga%e7%e3o_2/tecnologia/Breve%20Hist%f3ria%20da%20 nanotecnologia.pdf>.

Flasz, A., Rocha, C.A., Mosquera, B., Sajo, C. 1998. A comparative study of the toxicity of a synthetic surfactant and one produced by *Pseudomonas aeruginosa* ATCC 55925. Med. Sci. Res. 26: 181–185.

Ganesan, P., Choi, D.K. 2016. Current application of phytocompound-based nanocosmeceuticals for beauty and skin therapy. Int. J. Nanomed. 11: 1987–2007. doi: 10.2147/IJN.S104701.

Ge, L., Li, Q., Wang, M., Ouyang, J., Li, X., Xing, M.M.Q. 2014. Nanosilver particles in medical applications: synthesis, performance and toxicity. Int. J. Nanomed. 9: 2399–2407.

Gericke, M., Pinches, A. 2006. Biological synthesis of metal nanoparticles. Hydrometallurgy 83: 132–40.

Guerra-Santos, L.H., Kappeli, O., Fiechter, A. 1986. Dependence of *Pseudomonas aeruginosa* continuous culture biosurfactant production on nutritional and environmental factors. Appl. Microbiol. Biotechnol. doi: 10.1007/BF00250320.

Gunther, N.W., Nunez, A., Fett, W., Solaiman, D.K. 2005. Production of rhamnolipids by *Pseudomonas chlororaphis*, a nonpathogenic bacterium. Appl. Environ. Microbiol. doi: 10.1128/ AEM.71.5.2288-2293.2005.

Hazra, C., Bari, S., Kundu, D., Chaudhari, A., Mishra, S., Chatterjee, A. 2014a. Ultrasound-assisted/ biosurfactant-templated size-tunable synthesis of nano-calcium sulfate with controllable crystal morphology. Ultrason. Sonochem. 21: 1117–1131.

Hazra, C., Kundua, D., Chatterjee, A., Chaudharia, A., Mishra, S. 2014b. Poly(methyl methacrylate) (core)–biosurfactant (shell) nanoparticles:Size controlled sub-100 nm synthesis, characterization, antibacterialactivity, cytotoxicity and sustained drug release behavior. Colloids Surf. A. 449: 96–113.

He, S., Guo, Z., Zhang, Y., Zhang, S., Wang, J., Gu, N. 2007. Biosynthesis of gold nanoparticles using the bacteria *Rhodopseudomonas capsulata*. Materials Letters 61: 3984–3987.

https://nanomedicina.webnode.pt/nanotecnologia-e-medicina/historia/.

https://www.pesquisa-unificada.com/pesquisas/nanotecnolgia/empresas-de-nanotecnologia/.

https://sciencedirect.com.

https://www.nanowerk.com/nanotechnology/ten_things_you_should_know_7. hp.

Huang, W., Lang, Y., Hakeem, A., Lei, Y., Gan, L., Yang, X. 2018. Surfactin-based nanoparticles loaded with doxorubicin to overcome multidrug resistance in cancers. Int. J. Nanomed. 21: 1723–1736. doi:10.2147/IJN.S157368.

Huang, Z., Jiang, X., Guo, D., Gu, D., Gu, N. 2011. Controllable synthesis and biomedical applications of silver nanomaterials. J. Nanosc. Nanotechnol. 11: 9395–9408.

Hussain, M.A., Shah, A., Jantan, I., Shah, M.R., Tahir, M.N., Ahmad, R., Bukhari, S.N.A. 2015. Hydroxypropylcellulose as a novel green reservoir for the synthesis, stabilization, and storage of silver nanoparticles. Int. J. Nanomed. 16: 2079–2088. doi: 10.2147/IJN.S75874.

Iconaru, S.L., Prodan, A.M., Motelica-Heino, M., Sizaret, S., Predoi, D. 2012. Synthesis and characterization of polysaccharide-maghemite composite nanoparticles and their antibacterial properties. Nanoscale Res. Lett. 7: 576.

Ioelovich, M. 2008. Cellulose as a nanostructured polymer: a short review. BioResources. 3: 1403–1418.

Jan, J.S., Chuang, T.H., Chen, P.J., Teng, H. 2011. Layer-by-layer polypeptide macromolecular assemblies-mediated synthesis of mesoporous silica and gold nanoparticle/mesoporous silica tubular nanostructures. Langmuir. 27: 2834–2843.

Jeong, W.J., Bu, J., Kubiatowicz, L.J., Chen, S.S., Kim, Y., Hong, S. 2018. Peptide-nanoparticle conjugates: a next generation of diagnostic and therapeutic platforms? Nano Converg. 5. doi: 10.1186/s40580-018-0170-1.

Jia, X., Liu, Q., Zou, S., Xu, X., Zhang, L. 2015. Construction of selenium nanoparticles/β-glucan composites for enhancement of the antitumor activity. Carbohydr. Polym. 117: 434–442. https://doi.org/10.1016/j.carbpol.2014.09.088.

Joe, M.M., Bradeeba, K., Parthasarathi, R., Sivakumaar, P.K., Chauhan, P.S., Tipayno, S., Benson, A., Sa, T. 2012. Development of surfactin based nanoemulsion formulation from selected cooking oils: evaluation for antimicrobial activity against selected food associated microorganisms. J. Taiwan Inst. Chem. Eng. 42: 172–180. https://doi.org/10.1016/j.jtice.2011.08.008.

Joshi-Navare, K., Prabhune, A. 2013. A biosurfactant-sophorolipid acts in synergy with antibiotics to enhance their efficiency. Biomed. Res. Int. ID 512495. doi:10.1155/2013/512495.

Kanmani, P., Lim, S.T. 2013. Synthesis and characterization of pullulan-mediated silver nanoparticles and its antimicrobial activities. Carbohydr. Polym. 97: 421–428. https://doi.org/10.1016/j.carbpol.2013.04.048.

Kaskatepe, B., Yildiz, S. 2016. Rhamnolipid biosurfactants produced by *Pseudomonas* species. Braz. Arch. Biol. Technol. 59: e16160786.

Kasture, M., Singh, S., Patel, P., Joy, P.A., Prabhune, A.A., Ramana, C.V., Prasad, B.L.V. 2007. Multiutility sophorolipids as nanoparticle capping agents: Synthesis of stable and water dispersible Co nanoparticles. Langmuir. 23: 11409–11412. doi:10.1021/la702931j.

Katz, L.M., Dewan, K., Bronaugh, R.L. 2015. Nanotechnology in cosmetic. Food. Chem. Toxicol. 85: 127–137.

Kikuchi, T., Hasumi, K. 2002. Enhancement of plasminogen activation by surfactin C: augmentation of fibrinolysis *in vitro* and *in vivo*. Biochim. Biophys. Acta. 1596: 234–245. doi: 10.1016/S0167-4838(02)00221-2.

Kim, S.K., Kim, Y.C., Lee, S., Kim, J.C., Yun, M.Y., Kim, I.S. 2011. Insecticidal activity of rhamnolipids isolated from *Pseudomonas* sp. EP-3 against Green Peach Aphid (*Myzus persicae*). J. Agric. Food Chem. 59: 934–938. doi: 10.1021/jf104027x.

Kim, J.H., Shim, B.S., Kim, H.S., Lee, Y.J., Min, S.K., Jang, D. 2015. Review of nanocellulose for sustainable future materials. Int. J. of Precis. Eng. and Manuf. Green Tech. 2: 197–213. https://doi.org/10.1007/s40684-015-0024-9.

Kluge, B., Vater, J., Sainikow, J., Eckart, K. 1989. Studies on the biosynthesis of surfactin, a lipopetide antibiotic from *Bacillus subtilis* ATCC 21332. FEBS Lett. 231: 107–110.

Kora, A., Beedu, S., Jayaraman, A. 2012. Size-controlled green synthesis of silver nanoparticles mediated by gum ghatti (*Anogeissus latifolia*) and its biological activity. Org. Med. Chem. Lett. 2: 17.

Kora, A.J., Sashidhar, R.B., Arunachalam, J. 2010. Gum kondagogu (*Cochlospermum gossypium*): A template for the green synthesis and stabilization of silver nanoparticles with antibacterial application. Carbohydr. Polym. 82: 670–679.

Kowall, M., Vater, J., Kluge, B., Stein, T., Franke, P., Ziessow, D. 1998. Separation and characterization of surfactin isoforms produced by *Bacillus subtilis* OKB 105. J. Colloid. Interface Sci. 204: 1–8.

Kulakovskaya, E., Kulakovskaya, T. 2014. Extracellular Glycolipids of Yeasts. Biodiversity, Biochemistry, and Prospects. Academic Press. doi: 0.1016/B978-0-12-420069-2.00004-2.

Kumar, V., Kumari, A., Kumar, D., Yadav, S.K. 2014. Biosurfactant stabilized anticancer biomolecule-loaded poly(d,l-lactide) nanoparticles. Colloids Surf. B. 117: 505–511.

Kundu, D., Hazra, C., Chatterjee, A., Chaudhari, A., Mishra, S., Kharat, A., Kharat, K. 2016. Surfactin-functionalized poly-(methyl-methacrylate) as an eco-friendly nano-adsorbent: from size-controlled scalable fabrication to adsorptive removal of inorganic and organic pollutants. RSC Advances. 6: 80438–80454.

Kural, F.H., Gürsoy, R.N. 2011. Formulation and characterization of surfactin-containing self-microemulsifying drug delivery systems (SF-SMEDDS). Hacettepe Univ. J. Faculty Pharmacy 30: 171–186.

Kurapati, S. 2016. The current role of nanomaterials in cosmetics. J. Chem. Pharm. Res. 8: 906–914.

Lai, C., Zeng, G.M., Huang, D.L., Zhao, M.H., Wei, Z., Huang, C., Xu, P., Li, N.J., Zhang, C., Chen, M., Li, X., Lai, M., He, Y. 2014. Synthesis of gold-cellobiose nanocomposites for colorimetric measurement of cellobiase activity. Spectrochim. Acta Part A Mol. Biomol. Spectrosc. https://doi.org/10.1016/j.saa.2014.04.091.

Lang, S. Biological amphiphiles (microbial biosurfactants). Curr. Opin. 2002. Colloid Inter. Sci. 7: 12–20.

Lee, M., Kim, M.K., Vancanneyt, M., Swings, J., Kim, S.H., Kang, M.S., Lee, S.T. 2005. *Tetragenococcus koreensis* sp. nov., a novel rhamnolipid-producing bacterium. Int. J. Syst. Evol. Micr.doi: 10.1099/ijs.0.63448-0.

Lee, S.C., Lee, S.J., Kim, S.H., Park, I.H., Lee, Y.S., Chung, S.Y., Choi, Y.L. 2008. Characterization of new biosurfactant produced by *Klebsiella* sp. Y6-1 isolated from waste soybean oil. Bioresour. Technol. 99: 2288–2292. doi: 10.1016/j.biortech.2007.05.020.

Li, Y., Li, G., Li, W., Yang, F., Liu, H. 2015. Greenly Synthesized Gold–Alginate Nanocomposites Catalyst for Reducing Decoloration of Azo-Dyes. Nano. https://doi.org/10.1142/S1793292015501088.

Liu, C.P., Lin, F.S., Chien, C.T., Tseng, S.Y., Luo, C.W., Chen, C.H., Chen, J.K., Tseng, F.G., Hwu, Y., Lo, L.W. 2013. *In-situ* formation and assembly of gold nanoparticles by gum Arabic as efficient photothermal agent for killing cancer cells. Macromol. Biosci, https://doi.org/10.1002/mabi.201300162.

Mahdavinia, G.R., Mosallanezhad, A., Soleymani, M., Sabzi, M. 2017. Magnetic- and pH-responsive κ-carrageenan/chitosan complexes for controlled release of methotrexate anticancer drug. Int. J. Biol. Macromol. https://doi.org/10.1016/j.ijbiomac.2017.01.012.

Maity, S., Sen, I.K., Islam, S.S. 2012. Green synthesis of gold nanoparticles using gum polysaccharide of *Cochlospermum religiosum* (katira gum) and study of catalytic activity. Phys. E Low-Dimens. Syst. Nanostruct.https://doi.org/10.1016/j.physe.2012.07.020.

Marcelino, P.R., da Silva, V.L., Rodrigues Philippini, R., Von Zuben, C.J., Contiero, J., dos Santos, J.C., da Silva, S.S. 2017. Biosurfactants produced by *Scheffersomyces stipitis* cultured in sugarcane bagasse hydrolysate as new green larvicides for the control of *Aedes aegypti*, a vector of neglected tropical diseases. PLoS One.12: e0187125.

Marcelino, P.R.F., Peres, G.F.D., Teran-Hilares, R., Pagnocca, F.C., Rosa, C.A., Lacerda, T.M., Dos Santos, J.C., Da Silva, S.S. 2019. Biosurfactants production by yeasts using sugarcane bagasse hemicellulosic hydrolysate as new sustainable alternative for lignocellulosic biorefineries. Ind. Crops Prod. 129: 212–223.

Maruyama, T., Fujimoto, Y., Maekawa, T. 2015. Synthesis of gold nanoparticles using various amino acids. J. Colloid Interf. Sci. 447: 254–257. doi: 10.1016/j.jcis.2014.12.046.

Medley, C.D., Bamrungsap, S., Tan, W., Smith, J.E. 2011. Aptamer-conjugated nanoparticles for cancer cell detection. Anal. Chem. 83: 727–734.

Merroun, M., Rossberg, A., Hennig, C., Scheinost, A.C., Selenska, S. 2007. Spectroscopic characterization of gold nanoparticles formed by cells and S-layer proteins of *Bacillus sphaericus* JG-A12. Material Sci. Eng. 27: 188–192.

Mittal, A.K., Chisti, Y., Banerjee, U.C. 2013. Synthesis of metallic nanoparticles using plant extracts. Biotechnol. Adv. 31: 346–356. doi:10.1016/j.biotechadv.2013.01.003.

Mody, C.C.M., Lynch, M. 2010. Test objects and other epistemic things: a history of a nanoscale object. Br. J. Hist. Sci. 43: 423–458.

Monteiro, S.A., Sassaki, G.L., de Souza, L.M., Meira, J.A., de Araújo, J.M., Mitchell, D.A., Ramos, L.P., Krieger, N. 2007. Molecular and structural characterization of the biosurfactant produced by *Pseudomonas aeruginosa* DAUPE 614. Chem. Phys. Lipids. 147: 1–13. https://doi.org/10.1016/j.chemphyslip.2007.02.001.

Müller, F., Hönzke, S., Luthardt, W., Wong, E.L., Unbehauen, M., Bauer, J., Haag, R., Hedtrich, S., Rühl, E., Rademann, J. 2016. Rhamnolipids form drug-loaded nanoparticles for dermal drug delivery. Eur J. Pharm. Biopharm. 116: 31–37.

Mulligan, C.N. 2005. Environmental applications for biosurfactants. Environ. Pollut. 133: 183–198.

Mulligan, C.N., Sharma, S.K., Mudhoo, A., Makhijani, K. 2014. Green chemistry and biosurfactant research. pp. 1–30. *In*: Mulligan, C.N., Sharma, S.K., Mudhoo, A. (eds.). Biosurfactants: Research Trends and Applications. CRC Press, Boca Raton.

Mulligan, C.N., Gibbs, B.F. 1993. Biosurfactants: Production, Properties, Applications; Kosaric, N., ed.; Marcel Decker: New York, 1993, cap.13.

Nagamune, T. 2017. Biomolecular engineering for nanobio/bionanotechnology. Nano Converg. 4: 9. doi:10.1186/s40580-017-0103-4.

Nagy, L.N., Polyák, A., Mihály, J., Szécsényi, Á., Szigyártó, I.C., Czégény, Z., Jakab, E., Németh, P., Magda, B., Szabó, P., Veres, Z., Jemnitz, K., Bertóti, I., Jóba, R.P., Trencsényi, G.Y., Balogh, L., Bóta, A. 2016. Silica zirconia poly(malic acid) nanoparticles: promising nanocarriers for theranostic applications. J. Mater. Chem. B. 4: 4420–4429.

Nanganuru, H.Y., Korrapati, N. 2012. Studies on the production of rhamnolipids by pseudomonas putida. International Journal of Research in Computer Science 2: 19.

Narayanan, G., Aguda, R., Hartman, M., Chung, C.C., Boy, R., Gupta, B.S., Tonelli, A.E. 2016. Fabrication and Characterization of Poly-(ε-caprolactone)/α-Cyclodextrin Pseudorotaxane Nanofibers. Biomacromolecules, https://doi.org/10.1021/acs.biomac.5b01379.

Narayanan, K.B., Sakthivel, N. 2011. Green synthesis of biogenic metal nanoparticles by terrestrial and aquatic phototrophic and heterotrophic eukaryotes and biocompatible agents. Adv. Colloid Interfac. Science 169: 59–79. https://doi.org/10.1016/j.cis.2011.08.004.

Nastruzzi, C., Esposito, E., Menegatti, E., Walde, P. 1993. Use ano stability of liposomes in dermatological preparations. J. Appl. Cosmetol. 11: 77–91.

Nghiem, T.H.L., La, T.H., Vu, X.H., Chu, V.H., Nguyen, T.H., Le, Q.H., Fort, E., Do, Q.H., Tran, H.N. 2010. Synthesis, capping and binding of colloidal gold nanoparticles to proteins. Advances in Natural Sciences 1: 1–5. doi:10.1088/2043-6254/1/2/025009.

Nitschke, M., Costa, S. 2007. Biosurfactants in food industry. Trends Food Sci. Tech. 18: 252–259.

Nitschke, M., Pastore, G.M. 2006. Production and properties of a surfactant obtained from *Bacillus subtilis* grown on cassava wastewater. Bioresour. Technol. 97: 336–341. doi: 10.1016/j.biortech.2005.02.044.

Nitschke, M., Ferraz, C., Pastore, G.M. 2004. Selection of microorganisms for biosurfactant production using agroindustrial wastes. Braz. J. Microbiol. 35: 81–85. doi: 10.1590/S1517-83822004000100013.

Nitschke, M., Pastore, G.M. 2002. Biossurfactantes: propriedades e aplicações. Química Nova. 25: 772–776.

Nozhat, Z., Asadi, A., Zahri, S. 2012. Properties of surfactin C-15 nanopeptide and its cytotoxic effect on human cervix cancer (HeLa) cell line. J. Nanomater. ID 526580. doi: 10.1155/2012/526580.

Oliveira, F.J.S., Vazquez, L., De Campos, N.P., De Franca, F.P. 2009. Production of rhamnolipids by a *Pseudomonas alcaligenes* strain. Process Biochemistry 44: 383–389.

Onaizi, S.A., Nasser, M.S., Twaiq, F.A. 2012. Micellization and interfacial behavior of a synthetic surfactant–biosurfactant mixture. Colloid Surface A. 415: 388–393. https://doi.org/10.1016/j.colsurfa.2012.09.014.

Onwosi, C.O., Odibo, F.J.C. 2012. Effects of carbon and nitrogen sources on rhamnolipid biosurfactant production by *Pseudomonas nitroreducens* isolated from soil. World J. Microbiol. Biotechnol. 28: 937–942. doi: 10.1007/s11274-011-0891-3.

Oyelere, A.K., Chen, P.C., Huang, X., El-Sayed, I.H., El-Sayed, M.A. 2007. Peptide-conjugated gold nanorods for nuclear targeting. Bioconjugate Chem. 18: 1490–1497. https://doi.org/10.1021/bc070132i.

Palanisamy, P. 2008. Biosurfactant mediated synthesis of NiO nanorods, Mater. Lett. 62: 743–746.

Palanisamy, P., Raichur, A.M. 2009. Synthesis of spherical NiO nanoparticles through a novel biosurfactant mediated emulsion technique, Mater. Sci. Eng C. 29: 199–204.

Pallavicini, P., Arciola, C.R., Bertoglio, F., Curtosi, S., Dacarro, G., D'Agostino, A., Ferrari, F., Merli, D., Milanese, C., Rossi, S., Taglietti, A., Tenci, M., Visai, L. 2017. Silver nanoparticles synthesized and coated with pectin: An ideal compromise for anti-bacterial and anti-biofilm action combined with wound-healing properties. J. Colloid Interf. Science 498: 271–28. https://doi.org/10.1016/j.jcis.2017.03.062.

Pandian, S.S., Mariaamalraj, A.A. 2014. Influence of biosurfactant in biosynthesis of silver nanoparticles by *Pseudomonas aeruginosa* AMBAS7. Int. J. Innov. Res. Sci. Eng. Technol. 3: 2319–8753.

Patravale, V.B., Mandawgade, S.D. 2008. Novel cosmetic delivery systems: an application update. Int. J. Cosmetic. Sci. 30: 19–33. doi: 10.1111/j.1468-2494.2008.00416.x.

Philip, D. 2009. Biosynthesis of Au, Ag and Au-Ag nanoparticles using edible mushroom extract. Spectrochim. Acta A. 73: 374–81.

Płaza, G.A., Chojniak, J., Banat, I.M. 2014. Biosurfactant mediated biosynthesis of selected metallic nanoparticles. Int. J. Mol. Sci. 15: 13720–13737.

Pooja, D., Panyaram, S., Kulhari, H., Reddy, B., Rachamalla, S.S., Sistla, R. 2015. Natural polysaccharide functionalized gold nanoparticles as biocompatible drug delivery carrier. Int. J. Biol. Macromol. 80: 48–56. https://doi.org/10.1016/j.ijbiomac.2015.06.022.

Rai, M. 2013. Nanobiotecnologia verde: biossínteses de nanopartículas metálicas e suas aplicações como nanoantimicrobianos. Ciência e Cultura. 65. http://dx.doi.org/10.21800/S0009-67252013000300014.

Rai, M., Ingle, A. 2012. Role of nanotechnology in agriculture with special reference to management of insect pest. App. Microbiol. Biotechnol. 94: 287–293.

Rajeshkumar, S. 2016. Phytochemical constituents of fucoidan (*Padina tetrastromatica*) and its assisted AgNPs for enhanced antibacterial activity. IET Nanobiotechnol. 11: 292–299. doi: 10.1049/iet-nbt.2016.0099.

Rau, L.R., Tsao, S.W., Liaw, J.W., Tsai, S.W. 2016. Selective targeting and restrictive damage for nonspecific cells by pulsed laser-activated hyaluronan-gold nanoparticles. Biomacromolecules. 17: 2514–2521.

Raza, Z.A., Khan, M.S., Khalid, Z.M. 2007. Physicochemical and surfaceactive properties of biosurfactant produced using molasses by a *Pseudomonas aeruginosa* mutant. J. Environ. Sci. Health A Tox. Hazard. Subst. Environ. Eng. 42: 73–80. doi: 10.1080/10934520601015784.

Reddy, A.S., Chen, C.Y., Baker, S.C., Chen, C.C., Jean, J.S., Fan, C.W., Chen, H.R., Wang, J.C. 2009a. Synthesis of silver nanoparticles using surfactin: A biosurfactant as stabilizing agent. Mater. Lett. 63: 1227–1230. https://doi.org/10.1016/j.matlet.2009.02.028.

Reddy, A.S., Chen, C.Y., Chen, C.C., Jean, J.S., Fan, C.W., Chen, H.R., Chen, J.W., Nimje, V.R. 2009b. Synthesis of gold nanoparticles via an environmentally benign route using a biosurfactant. J. Nanosci. Nanotechnol. 9: 6693–6699. doi: https://doi.org/10.1166/jnn.2009.1347.

Reva, T., Vaseem, A.A., Satyaprakash, S., Md. khalid, J.A. 2015. Liposomes: The novel approach in cosmaceuticals. World J. Pharm. Pharm. Sci. 4: 1616–1640.

Rosenberg, E., Ron, E.Z. 1999. High-and low-molecular-mass microbial surfactants. Appl. Microbiol. Biotechnol. 52: 54–162.

Sachdev, D.P., Cameotra, S.S. 2013. Biosurfactants in agriculture. Appl. Microbiol. Biotechnol. 97: 1005–1016. doi: 10.1007/s00253-012-4641-8.

Sadowski, Z. 2010. Biosynthesis and application of silver and gold nanoparticles. pp. 258–276. *In*: Perez, D.P. (ed.). Silver Nanoparticles. InTech; Rijeka, Croatia.

Seydlova, G., Cabala, R., Svobodova, J. 2011. Surfactin—Novel solutions for global issues, biomedical engineering, trends, research and technologies. IntechOpen. doi: 10.5772/13015. Available from: https://www.intechopen.com/books/biomedical-engineering-trends-research-and-technologies/surfactin-novel-solutions-for-global-issues.

Shak, K., Pang, Y.L., Mah, S.K. 2018. Nanocellulose: Recent advances and its prospects in environmental remediation. Beilstein J. Nanotech. 9: 2479–2498. doi:10.3762/bjnano.9.232.

Shekhar, S., Sundaramanickam, A., Balasubramanian, T. 2015. Biosurfactant producing microbes and their potential applications: a review. Crit. Rev. Env. Sci. Tec. 45: 1522–1554.

Shikha, S., Haque, F., Bhattacharyya, M.S. 2017. Greener synthesis of gold nanoparticles (AuNPs-SL) using sophorolipid and its antimicrobial activity. J. Nanomed. Nanotech. doi: 10.4172/2157-7439-C1-065.

Shim, G.Y., Kim, S.H., Han, S.E., Kim, Y.B., Oh, Y.K. 2009. Cationic surfactin liposomes for enhanced cellular delivery of siRNA. Asian J. Pharm. Sci. 4: 207–214.

Siegrist, M., Cousin, M.E., Kastenholz, H., Wiek, A. 2007. Public acceptance of nanotechnology foods and food packaging: The influence of affect and trust. Appetite. 49: 459–466.

Singh, S., Patel, P., Jaiswal, S., Prabhune, A.A., Ramana, C.V., Prasad, B.L.V. 2009. A direct method for the preparation of glycolipid–metal nanoparticle conjugates: sophorolipids as reducing and capping agents for the synthesis of water re-dispersible silver nanoparticles and their antibacterial activity. New J. Chem. 33: 646–652.

Singh, A., Malviya, R., Sharma, P.K. 2011a. Novasome-a breakthrough in pharmaceutical technology a review article. Adv. Biol. Res. 5: 184–189.

Singh, B.R., Dwivedi, S., Al-Khedhairy, A.A., Musarrat, J. 2011b. Synthesis of stable cadmium sulfide nanoparticles using surfactin produced by *Bacillus amyloliquifaciens* strain KSU-109. Colloids Surf. B. 85: 207–213.

Singla, R.K., Dubey, H.D., Dubey, A.K. 2014. Therapeutic spectrum of bacterial metabolites. Indo Global J. Pharmaceut. Sci. 2: 52–64.

Socaciu, C. 2009. New techonolgies to synthesize. Extract and encapsulate natural food colorants. Bull. Univ. Agric. Sci. Vet. Med. Cluj Napoca. 64: 1–7. doi: http://dx.doi.org/10.15835/buasvmcn-asb:64:1-2:2185.

Spicer, C.D., Jumeaux, C., Gupta, B., Stevens, M.M. 2018. Peptide and protein nanoparticle conjugates: versatile platforms for biomedical applications. Chem. Soc. Rev. 47: 3574–3620.

Srinivas, K. 2016. The current role of nanomaterials in cosmetics. J. Chem. Pharm. Res. 8: 906–914.

Tapas, K.P., Oli, M. 2014. Prospect of nanotechnology in cosmetics: benefit and risk assessment. World. J. Pharm. Res. 3: 1909–1919.

Tartaj, P., Morales, M.P., González-Carreni, T., Veintemillas-Verdaguer, S., Serna, C.J. 2005. Advances in magnetic nanoparticles for biotechnology applications. J. Magn. Magn. Mater. 290: 28–34.

Thakur, R.S., Agrawal, R. 2015. Appliacation of nanotechnology in pharnaceutical formulation design and development. Curr. Drug. Ther. 15: 20–34.

Van Der Elst, M., Klein, C.P.A.T., Blieck-Hogervorst, J.M., Patka, P., Haarman, H.J. 1999. Bone tissue response to biodegradable polymers used for intra medullary fracture fixation: A long-term in vivo study in sheep femora. Biomaterials. 20: 121–128. https://doi.org/10.1016/S0142-9612(98)00117-3.

Vasileva-Tonkova, E., Gesheva, V. 2005. Glycolipids produced by *Antarctic Nocardioides* sp. during growth on n-paraffin. Process Biochem. 40: 2387–2391. doi: 10.1016/j.procbio.2004.09.018.

Vasudevan, S., Prabhune, A.A. 2018. Photophysical studies on curcumin-sophorolipid nanostructures: applications in quorum quenching and imaging. Royal Soc. Open Sci. 5. doi: 10.1098/rsos.170865.

Venturini, C.G., Nicolini, J., Machado, C., Machado, V.G. 2008. Propriedades e aplicações recentes das ciclodextrinas. Química Nova. 31: 360–368. https://dx.doi.org/10.1590/S0100-40422008000200032.

Viana, R.L.D.S. 2017. Síntese verde de nanopartículas contendo prata e xilana do sabugo de milho: caracterização físico-química e avaliação das atividades antioxidante e antimicrobiana frente a protozoário e a fungos (Master's thesis, Brasil).

Vu-Quang, H., Yoo, M.K., Jeong, H.J., Lee, H.J., Muthiah, M., Rhee, J.H., Lee, J.H., Cho, C.S., Jeong, Y.Y., Park, I.K. 2011. Targeted delivery of mannan-coated superparamagnetic iron oxide nanoparticles to antigenpresenting cells for magnetic resonance-based diagnosis of metastatic lymph nodes *in vivo*. Acta Biomater. 7: 3935–3945. https://doi.org/10.1016/j.actbio.2011.06.044.

Wang, C., Gao, X., Chen, Z., Chen, Y., Chen, H. 2017. Preparation, characterization and application of polysaccharide-based metallic nanoparticles: a review. Polymers. 9: 689. https://doi.org/10.3390/polym9120689.

Westwood, M., Roberts, D., Parker, R. 2011. Enzymatic degradation of poly-L-lysine-polygalacturonic acid multilayers. Carbohydr. Polym. 84: 960–969.

Willis, C. 2012. Biomass Assisted Synthesis of Antibacterial Gold Nanoparticles and Commentary on its Future Potential and Applications in Medicine. Thesis Project (USA).

Wu, Y.-S., Ngai, S.-C., Goh, B.-H., Chan, K.-G., Lee, L.-H., Chuah, L.-H. 2017. Anticancer activities of surfactin and potential application of nanotechnology assisted surfactin delivery. Front. Pharmacol. 8. doi:10.3389/fphar.2017.00761.

Xie, Y., Ye, R., Liu, H. 2006. Synthesis of silver nanoparticles in reverse micelles stabilized by natural biosurfactant, Colloids Surf. A Physicochem. Eng. Asp. 279: 175–178.

Yang, L.L., Mao, H., Wang, Y.A., Cao, Z.H., Peng, X.H., Wang, X.X., Duan, H.W., Ni, C.C., Yuan, Q.G., Adams, G., Smith, M.Q., Wood, W.C., Gao, X.H., Nie, S.M. 2009. Single chain epidermal growth factor receptor antibody conjugated nanoparticles for *in vivo* tumor targeting and imaging. Small 5: 235.

Yu, C.H., Tam, K., Tsang, E.S.C. 2008. Chemical methods for preparation of nanoparticles in solution. pp. 113–141. *In*: Blackman, J.A. (ed.). Handbook of Metal Physics. Vol. 5. Amsterdan.

Zare, D., Akbarzadeh, A., Bararpour, N. 2010. Synthesis and functionalization of gold nanoparticles by using of poly functional amino acids. Int. J. Nanosci. Nanotech. 6: 223–230.

Zhang, Y., Liu, C., Dong, B., Ma, X., Hou, L., Cao, X. 2015. Anti-inflammatory activity and mechanism of surfactin in lipopolysaccharide-activated macrophages. Inflammation. 38: 756–764. doi: 10.1007/s10753-014-9986-y.

Chapter 13

Synergistic Activity of Nanoparticles and Other Antimicrobials Against Pathogenic Bacteria

*Krystyna I. Wolska** and *Anna M. Grudniak*

Introduction

The alarming spread of microbial resistance to classic antimicrobial agents has greatly increased the search and study of novel, alternative agents such as plant compounds, antimicrobial peptides, genetically modified viruses, vaccines, therapeutic antibodies, quorum sensing quencher and nanoparticles, to mention only the most important and extensively studied tools (Kurek et al. 2010, Fernebro 2011, Chatterjee et al. 2016). Among the huge variety of nanoparticles, metal nanoparticles are of special interest, for example, silver, copper, gold, zinc oxide, titanium oxide, and magnesium oxide. The biological activity of NPs depends on their shape, size, surface-to-volume ratio and chemical and physical surroundings. The effectiveness of nanoparticles and nanosized carriers for antibiotic delivery led to the creation of the term "nanoantibiotics" as an example of new antibacterial drugs (Huh and Kwon 2011). The antimicrobial potential of NPs is still not entirely resolved in spite of a huge number of studies. Three main mechanisms are postulated: (1) direct interaction with cell envelopes leading to their damage, (2) production of reactive oxygen species causing oxidative stress, and (3) formation of complexes with intracellular compounds, mainly polymers, and their subsequent inactivation (Lemire et al. 2013). The antibiofilm activity of nanoparticles has also been reported, including biofilms formed by *Pseudomonas aeruginosa* and *Staphylococcus aureus*; this quality potentiates the importance of NPs as antimicrobial agents (Ramalingam et al. 2013). The ability of NPs to inhibit

Department of Bacterial Genetics, Institute of Microbiology, Faculty of Biology, University of Warsaw, 02-096 Warsaw, Miecznikowa St. 1, Poland.
* Corresponding author: izabelaw@biol.uw.edu.pl

the development of microbial biofilms is mainly based on their potential to hamper quorum sensing signaling pathways (Wolska et al. 2017). The direct interaction of NPs with extracellular polymeric substance interfering with biofilm integrity was also proved (Lundqvist et al. 2008). In spite of the well-documented antimicrobial activity of NPs, clinical trials to cure bacterial infections using nanotechnology are scarce (Caster et al. 2017). Some studies examine the application of nanoparticles not only in therapy but also in diagnosis, for example, pathogen detection in macroorganisms (Kim et al. 2017).

The pleiotropic effect of NPs on bacterial cell *a priori* has an additional desirable outcome due to the restricted ability of bacterial resistance development. However, a few reports describing the emergence of bacterial mutants resistant to AgNPs have been already published (Graves et al. 2015, Panaček et al. 2018). It should be mentioned that some microbial species intrinsically tolerate NPs and this trait is very useful in the biological synthesis of nanoparticles (Parikh et al. 2008).

Nanoparticles, especially gold nanoparticles, when conjugated with small antimicrobial molecules such as antibiotics, drugs, vaccines and antibodies, served as the vehicles for their delivery and also potentiates their antibacterial effect (Tao 2018). Their advantage as vectors is based on small size, ability to protect microbes against the destructive activity of antimicrobial cellular compounds, active delivery of antibiotics, and simultaneous interaction with several antimicrobial substances (Baptista et al. 2018).

Nanoparticles can be synthesized by chemical, physical, physicochemical and biological methods. Chemical methods are usually based on the reduction of silver nitrate in the presence of a stabilizer. Physicochemical methods use microwaves, radiation and ultrasonic approaches. The description of the variety of applied techniques can be found in the paper of Iravani and coworkers (2014). The environment-friendly, biological synthesis of NPs uses mainly plants, bacteria and fungi and leads to the production of well-defined particles characterized by broad activity against pathogenic bacteria, fungi and even viruses (Rai et al. 2009). The extracellular nitrate reductase is an enzyme crucial for the reduction of metal ions by various microorganisms (Khodashenas and Ghorbani 2016) but the intracellular, non-enzymatic synthesis of NPs was also described (Sukanya et al. 2013). Biological synthesis provides NPs naturally capped and stabilized by biological components and therefore the particles are not prompt to aggregate, which positively influences their activity (Kora and Rastogi 2013). The capping layers also constitute an active surface for interaction with the other biological compounds (Singh et al. 2018). In some cases it was experimentally proven that biologically synthesized NPs have higher antimicrobial potential than chemically synthesized particles (Singh et al. 2016).

In the development of NP synthetic pathways, the focus is on the enhancement of their antimicrobial activity and, if needed, on the reduction of their cytostatic activity. In spite of that, many preparations do not have sufficient antimicrobial potential to be valuable therapeutic agents. One way to resolve this problem is an extensive search for the potential synergy between NPs and other compounds of known antimicrobial effect (Fischbach 2011). This article describes the synergy between NPs, especially metal NPs, obtained by biological synthesis with several other antimicrobial agents.

Synergistic activity of nanoparticles and other antimicrobials

Three mechanisms of antimicrobial interactions can be distinguished that use inhibition of: (1) different synthetic pathways within the cell, (2) different targets in the same cell, and (3) the same target but in different ways. Four types of interaction of different compounds are known, characterized by their antimicrobial effect: synergistic, additive, neutral and antagonistic (Wagner and Ulrich-Merzenich 2009). Among them, synergy is the most important and promising effect from the therapeutic perspective. Several definitions of synergy are known; according to Berenbaum (1989), synergy is observed when the activity of the compound combinations exceeds the sum of individual compound activities (Berenbaum 1989). The main advantages of using antimicrobial compound combinations involve the reduction of their active dose, the reduction of side effects of individual compounds, and the control of the emergence of antibiotic-resistant organisms (Bollenbach 2015).

The interaction between two compounds is usually studied using the checkerboard method (Rand et al. 1993). This method is based on the examination of bacterial growth in the wells of titration plate consisting of columns containing one antimicrobial agent diluted along the x-axis and lines containing the second agent diluted along the y-axis. The determination of synergy between two compounds is based on calculation of fractional inhibitory concentration index; a value of ≤ 0.5 indicates synergistic interactions (EUCAST 2000). Synergy can also be determined by time-kill assay. The kinetics of microorganism survival in the presence of individual compounds is compared to survival in the presence of a combination. A difference (usually measured at 8 or 24 h incubation) equal to or higher than $2 \log_{10}$ indicates synergistic interaction (Matsumura et al. 1999).

The most frequent examples concern the synergistic activity of nanoparticles with antibiotics but synergy with other antimicrobial compounds has also been reported. Examples of synergy and in some cases the possible explanation of synergistic activity of nanoparticles and other compounds is described in this chapter. The majority of cited examples concern metal nanoparticles, especially AgNPs and AuNPs, and their biological origin is underlined. Because the ability of NPs to deliver other antimicrobials can be considered as a special case of synergy, the publications dealing with this subject are also quoted in this review.

Synergy of NPs with conventional antibiotics

The first reports on synergy between NPs, mainly AgNPs, and various types of antibiotics were published more than a decade ago. Li and coworkers (2005) demonstrated the synergy between AgNPs and β-lactam antibiotic amoxicillin against *Escherichia coli*. The authors hypothesized that AgNPs facilitated the penetration of antibiotic across the cell wall (Li et al. 2005). Shahverdi and coworkers (2007) described the synergy between bio-AgNPs synthesized by the reduction of Ag^+ ions by the culture supernatant of *Klebsiella pneumoniae* and various antibiotics against *S. aureus* and *E. coli*. The highest enhancing effect was observed for vancomycin, amoxicillin and penicillin G against *S. aureus* (Shahverdi et al. 2007). In contrast, Birla and coworkers (2009) found that the combined effect of AgNPs synthesized by *Phoma glomerata* and ampicillin, gentamicin, kanamycin or streptomycin was greater against *E. coli* and *P. aeruginosa* than against *S. aureus* (Birla et al. 2009). Subsequently, it was proved that AgNPs produced using the extracellular extracts of fungus *Trichoderma viridae* acted synergistically with kanamycin, erythromycin and chloramphenicol but the highest effect was observed for ampicillin (Fayaz et al. 2010). According to the authors, the observed synergy can be explained by the strong activity of NPs and ampicillin on bacterial cell wall leading to its destruction and cell lysis. Gosh and coworkers (2012) reported the synergy between AgNPs synthesized using *Discorea bulbifera* tuber extract and several classes of antibiotics against both Gram-positive and Gram-negative bacteria including multidrug-resistant *Acinetobacter baumanii.*

Synergy of fungal synthesized AgNPs with erythromycin, methicillin, chloramphenicol and ciprofloxacin was demonstrated; the combination of two compounds was more efficient in inhibiting the growth of test microorganisms including *Enterococcus faecalis* than the individual antimicrobials (Devi and Joshi 2012). The synergistic effect of several classes of antibiotics and AgNPs was also observed by Hwang and coworkers (2012). These authors reported the antibiofilm activity of the combination studied, although in varying degrees (Hwang et al. 2012). It was also demonstrated that AgNPs biosynthesized using cell-free extract of *Acinetobacter* spp. when combined with antibiotics displayed an enhanced antibacterial activity against multidrug-resistant strains: *Enterobacter aerogenes*, *A. baumanii* and *Streptococcus mutans* (Singh et al. 2013). In turn, AgNPs synthesized by the culture filtrate of *Aspergillus flavus* combined with five conventional antibiotics resulted in average 2.8-fold increase of their antibacterial activity against eight different multidrug-resistant bacteria (Naqvi et al. 2013). The enhanced activity of AgNPs synthesized by *Cryphonectria* sp. showed synergy with several standard antibiotics as determined by disk diffusion method (Dar et al. 2013). The ability of AgNPs synthesized extracellularly by soil bacterium *Bacillus flexus* in combination with several antibiotics to efficiently inhibit the growth of multidrug-resistant, biofilm-forming, coagulase-negative staphylococci from clinical samples and other pathogenic bacteria was proved; the highest synergistic effect was observed with chloramphenicol against *Salmonella typhi* (Thomas et al. 2014). The synergy of citrate-capped silver nanoparticles and aztreonam (monobactam) against *P. aeruginosa* PAO1 biofilms was also shown and this phenomenon was explained as a result of better penetration of small AgNPs into biofilm matrix, which enhanced the activity of antibiotic against *P. aeruginosa* cells (Habash et al. 2014). In contrast, the modulation of antibiotic susceptibility of free-living *P. aeruginosa* cells only, but not cells living in biofilms, by AgNPs was observed by Markowska et al. (2014). The possible reason for this observation was given based on the previously observed AgNP aggregation (Choi et al. 2010) leading to the increase of particle size and retardation of their diffusion into biofilm structure.

Out of several more recent papers on synergy between AgNPs and antibiotics only a few are mentioned here. The ability of AgNPs to enhance and even to restore bactericidal activity of inactive antibiotics (cefotaxime, cefiaridime, meropenem, ciprofloxacin and gentamicin) against multi-resistant β-lactamase- and carbapenemase-producing *Enterobacteriaceae* was described by Pánăcek and coworkers (2016). Smelakova and others (2016) observed the enhanced antibacterial action of antibiotics in combination with AgNPs against animal pathogens such as *Acinobacillus pleuropneumoniae* and *Pasteurella multocida* (Smelakova et al. 2016). It was also demonstrated that antifungal macrocyclic polyene amphotericin B conjugated with biogenic AgNPs displayed a high antifungal activity against *Candida albicans*, *Aspergillus niger* and *Fusarium culmorum* species (Ahmad et al. 2016, Tutaj et al. 2016). In a very interesting review, Gupta and coworkers (2017) described the role of metallic nanoparticles as efflux pump inhibitors that can help to restore the bactericidal effect of many antibiotics. The authors quoted the possible mechanism of direct binding to the efflux pump active site and/or the disruption of efflux kinetics through which the inhibition could be executed. Moreover, the inhibition of biofilm-forming capacity was also observed. The mechanism of this phenomenon was not resolved. The most probable justification is based on the role of efflux pumps to extrude critical components required in quorum sensing, which is the main regulator of biofilm development (Gupta et al. 2017).

A paper describing *in vitro* synergy of silver nanoparticles with ampicillin and amikacin against clinical isolates of multidrug-resistant Gram-negative and Gram-positive uropathogens was published (Lopez-Carrizales et al. 2018). It was shown that the activity of ampicillin–AgNP combination against 12 microorganisms showed one synergetic and seven partial synergetic effects in three cases and four additive effects; amikacin–AgNP combination showed three synergetic effects, eight partial synergetic effects and one additive effect. The authors concluded that compound combinations could be useful in treating multidrug infections of urinary tract (Lopez-Carrizales et al. 2018). The enhancement effect of gentamicin and chloramphenicol after addition of AgNPs biosynthesized by *E. coli* culture supernatant was demonstrated against biofilm formed by clinical isolates of *E. coli*, *K. pneumoniae*, *P. aeruginosa* and, though to lesser extent, against *S. aureus* (Neihaya and Zaman 2018). Katva and coworkers (2018) showed the synergism of AgNPs synthesized by cell-free culture supernatant of *K. pneumoniae* with gentamicin and chloramphenicol against *E. faecalis*. Silver nanoparticles combined with naphthoquinones showed synergistic activity against *S. aureus* reference strain and four clinical isolates. The direct interactions of both compounds leading to membrane damage was confirmed (Krychowiak et al. 2018).

Examples of synergistic interactions of biosynthesized AgNPs with antibiotics and other antimicrobials are collected in Table 13.1.

NPs other than AgNPs can also be applied as antimicrobial agents and their synergy with other compounds was described. Gold nanoparticles (AuNPs) are generally considered as deprived of biological activity unless they are functionalized by cationic and hydrophobic groups (Pissuwan et al. 2009, Li et al. 2014, Zong et al. 2017). In spite of that, AuNPs were used for diagnosis purposes using the color change phenomenon (Elghanian et al. 1997), improving the sensitivity of flow cytometry (Zharov et al. 2007) and targeted photothermal lysis of bacteria (Norman et al. 2008). AuNPs are particularly suitable, and therefore are commonly used, as a stable platform for attachment of various compounds, including antibiotics (Murakami and Tsuchida 2008). The resulting complexes have been shown to potentiate the antimicrobial effect of therapeutics due to their ability to bind and penetrate the cell wall and also allow delivery into a specific site (macrotargeting) (Grace and Pandian 2007). The enhancement of antimicrobial activity of antibiotics while they are conjugated with AuNPs could be considered as a synergistic relationship between these compounds. Gu and coworkers (2003) showed that vancomycin-capped AuNPs were active

Table 13.1. Selected examples of antimicrobial synergistic interactions of biosynthesized AgNPs with other compounds.

Antimicrobial	Example	Reference
Antibiotics	AMX, PEN G, VAN	Shahverdi et al. 2007
	AMP, GEN, KAN, STR	Birla et al. 2009
	AMP, CHL, ERY, KAN	Fayaz et al. 2010
	CHL, CIP, ERY, MET	Devi and Joshi 2012
	Many standard antibiotics	Dar et al. 2013
	CHL	Thomas et al. 2014
	CHL, GEN	Katva et al. 2018
	CHL, GEN	Neihaya and Zaman 2018
Plant compounds	Copaiba oil	Otaguiri et al. 2017
	Oregano essential oil	Scandorieiro et al. 2016
Chitosan		Sanupi et al. 2008, Potara et al. 2011
Short peptide		Reithofer et al. 2014

AMX – amoxicillin, AMP – ampicillin, CIP – ciprofloxicin, CHL – chloramphenicol, ERY – erythromycin, GEN – gentamicin, KAN – kanamycin, MET– methicillin, PEN G – penicillin G, STR – streptomycin, VAN – vancomycin.

against vancomycin-resistant enterococci and also *E. coli* strains (Gu et al. 2003). Brown and coworkers (2012) demonstrated that AuNP–ampicillin complexes were effective against a broad spectrum of Gram-positive bacteria such as methicillin-resistant *Staphylococcus aureus* (MRSA) and Gram-negative bacteria *E. coli*, *P. aeruginosa*, *Vibrio cholerae* and *E. aerogenes*. Payne and others (2016) described a single-step synthesis of kanamycin-capped AuNPs and showed their dose-dependent activity against Gram-positive (*Streptococcus bovis*, *Streptococcus epidermidis*) and Gram-negative (*E. aerogenes*, *P. aeruginosa*, *Yersinia pestis*) bacteria, including kanamycin-resistant strains. The increased efficacy of Kan-AuNPs could be the result of disruption of bacterial envelope (Payne et al. 2016). The paper presenting contradictory results should be mentioned. Burygin and coworkers (2009) failed to demonstrate the enhanced activity of the mixture of gentamicin and colloidal-gold against *E. coli* K12.

The synergy between zinc oxide nanoparticles (ZnONPs) and several antibiotics was also proved. Banoee and coworkers (2010) demonstrated that ZnONPs enhanced antibacterial activity of ciprofloxacin against *S. aureus* and *E. coli*. The observed effect was not very pronounced: 27% and 22% of increase in inhibition zone area was observed for *S. aureus* and *E. coli*, respectively, in a disk diffusion method (Banoee et al. 2010). Fathi and others (2016) described the synergy between ZnONPs and oxacillin against *Yersinia intermedia* determined by double-disk method. Padwal and coworkers (2014) proved the synergistic effect of polyacrylic acid–coated iron oxide (magnetic) nanoparticles (PAA-NPs) with rifampicin and other anti-TB drugs against mycobacteria. This combination caused 4-fold higher growth inhibition over rifampicin alone. The increase of antibiotic accumulation inside the cells was interpreted as the result of inhibition of rifampicin efflux (Padwal et al. 2014). The synergism between copper oxide nanoparticles (CuONPs) and cephalexin against *E. coli* was observed. The authors considered that the antibiotic concentrated on the surface of CuONPs interacted more strongly with *E. coli* cells to weaken the cell wall (Zhang et al. 2018).

The synergistic activity was demonstrated not only for metallic nanoparticles and antibiotics. For example, it was shown that the linkage of penicillin-G to the functionalized silica nanoparticles containing carboxyl groups located either on the surface or on polymer chains extending from it enhances the killing of *E. coli* and MRSA (Wang et al. 2014). Gounani and coworkers (2018) reported that the mesoporous silica nanoparticles, either bare or carboxyl-modified, carrying polymyxin B and vancomycin were effective against *S. aureus, E. coli, P. aeruginosa, Klebsiella oxytoca* and *A. baumanii* strains. Liposome-mediated antibiotic delivery was also reported; for example, vancomycin encapsulated in liposomes were very efficient in killing MRSA (Pumerantz et al. 2011). It was reported that chitosan nanoparticles loaded with the third generation fluoroquinolone levofloxacin not only increased antibacterial activity of the antibiotic but also reduced the emergence of bacterial resistance suppressing the occurrence of mutations in several quinolone resistance determining regions (Hadiya et al. 2018).

Examples of synergy between NPs, excluding AgNPs, and other antimicrobials are presented in Table 13.2.

Table 13.2.   Selected examples of synergy between NPs (excluding AgNPs) and other antimicrobials.

NPs	Partner	Reference
AuNPS	VAN	Gu et al. 2003
	AMP	Brown et al. 2012
	KAN	Payne et al. 2016
	Antimicrobial peptides	Rai et al. 2016
	Cinnamaldehyde	Ramasamy et al. 2017
ZnONPs	CIP	Banoee et al. 2010
	OXA	Fathi et al. 2016
FeONPs	RIF	Padwal et al. 2014
CuONPs	CEP	Zhang et al. 2018
sNPs	Phage endolysin	Solanki et al. 2013
	PEN G	Wang et al. 2014

sNPs – silica nanoparticles, AMP – ampicillin, CEP – cephalexin, CHL – chloramphenicol, CIP – ciprofloxacin, KAN – kanamycin, PEN G – penicillin G, RIF – rifampicin, VAN – vancomycin.

Synergy of NPs with compounds other than antibiotics

Combinations of NPs with plant compounds, mainly essential oils, are studied in the hope of finding a formulation of high antimicrobial potential. The most active plant compounds are phenolics and terpenoids (Wolska et al. 2012). It was proved that phenolics, especially flavones, cause disruption of bacterial membranes, inhibit energy metabolism and quorum sensing and neutralize bacterial toxins (Cushnie and Lamb 2011). Terpenoids influence biofilm formation preferentially in Gram-positive bacteria (Walencka et al. 2007). The literature on synergy between NPs and plant compounds is not abundant. It was shown that a combination of silver nanoparticles and an essential oil (cinnamaldehyde) acted synergistically against spore forming *Bacillus cereus* and *Clostridium perfringens*, which are the species releasing cytotoxins in contaminated food. The bactericidal action was very fast and extensive damage of the bacterial cell envelope was observed (Gosh et al. 2013). The spectacular enhancement of bactericidal activity of *Drosea binata* extract or its pure compound (3-chloroplumbagin) when combined with AgNPs was reported (Krychowiak et al. 2014). The authors suggested that this combination was a possible alternative for antibiotic treatment of burn wound infections caused by resistant *S. aureus*. It was also demonstrated that the combination of silver nanoparticles and curcumin nanoparticles is more potent in elimination of biofilms formed by *P. aeruginosa* and *S. aureus* compared to the individual compounds, and it was shown that none of these compounds was toxic to healthy human bronchial epithelial cells (Loo et al. 2016). The synergistic antibacterial effect of oleoresin from *Copaifera multijuga* (copaiba oil) and nanoparticles produced by *Fusarium oxysporum* against planctonic cells and biofilms of *Streptococcus agalactiae* (group B Streptococcus or GSB) was also demonstrated. This pathogen is a leading cause of neonatal infections as well as an infection in adults. The combination of two compounds resulted in the reduction of minimal inhibitory concentration values of both compounds and therefore can be considered as a new alternative strategy for controlling GBS infections (Otaguiri et al. 2017). Scandorieiro and coworkers (2018) described a synergistic and additive effect of oregano essential oils and biological silver nanoparticles in combination against multidrug-resistant bacterial strains such as MRSA (methicillin-resistant *S. aureus*) and *E. coli* and *A. baumanii* strains producing β-lactamase and carbapenemase (Scandorieiro et al. 2018). Oregano essential oil was extracted from leaves of *Oregano vulgare*; it was previously shown that its main antibacterial compounds are carvacrol and thymol (Nostro et al. 2004). In another study, bio-AgNPs were prepared according to the modified method of Duran et al. (2005) from *Fusarium oxysporum* strain 551. It was proved that the combination of these two compounds showed synergistic or additive effect and reduced MIC values and time of action compared to either compound used separately. Moreover, this composition efficiently disrupted *S. aureus* cells, as was shown by scanning electron microscopy.

The synergistic activity of antimicrobial compounds other than those of plant origin with NPs was also reported. One such poly-cationic biopolymer is chitosan prepared from the shells of marine crustaceans and cell wall of fungi (Rabea et al. 2003). Because of the presence of amine and hydroxyl groups, chitosan is able to bind and simultaneously reduce Ag^+ ions to AgNPs (Murugadoss and Chattopadhyay 2008). The resulting chitosan–AgNP composite has a potent antibacterial activity that is much higher than the activity of separate components at least against *E. coli*. Fluorescence confocal laser scanning and scanning electron microscopy images demonstrated destabilization of bacterial cell membrane upon treatment with this composite (Sanupi et al. 2008). The synergistic activity of chitosan–silver nanocomposite was also shown against two strains of *S. aureus*. The disruption of bacterial cell integrity was also demonstrated (Potara et al. 2011). Moreover, these authors showed that the preparation can be used to identify bacteria and to monitor their biochemical status. It was also demonstrated that carboxymethyl chitosan–AgNP hydrogels had high antibacterial activity against Gram-negative and Gram-positive bacterial species (Mohamed and Sabaa 2014). Shao and coworkers (2017) proved that silver nanoparticles incorporated into chitosan-based membranes displayed high and dose-dependent antibacterial potential against *Porphyromonas gingivalis* and *Fusobacterium*, which suggests their future application for tissue regeneration.

Of special interest are reports describing synergy between antibacterial peptides of different origin that are therapeutics designed for topical and surface application (Walkenhorst 2016) and various types of nanoparticles. In an early study, the synergy between AgNPs applied along with silver nitrate and antimicrobial peptides polymyxin B and gramicidin S was observed. It was known that both AgNPs and the peptides studied are able to permeabilize bacterial membranes, which helps to access internal target sites (Ruden et al. 2009). It was also shown that combination of nanoparticles and antmicrobial peptides displayed synergy against bacterial resistance (Allahverdyiev et al. 2011). Also, biomaterial composed of stable silver nanoparticles within ultrashort peptide hydrogels efficiently inhibited the growth of *E. coli*, *P. aeruginosa* and *S. aureus*. It was proposed that these hydrogels could have great potential for application in wound healing (Reithofer et al. 2014). Subsequent *in vivo* studies demonstrated synergistic activity of polymyxin B and AgNPs against *P. aeruginosa* isolates from the lungs of cystic fibrosis patients. Treatment of *P. aeruginosa* with this combination induced cytosolic release of green fluorescent protein, increase of cellular reactive oxygen species, and permeabilization of cellular membranes greater than that caused by each compound alone (Jasmin et al. 2017). As in the case of classical antibiotics, there are reports on antimicrobial activity of nanocomposites containing nanoparticles other than AgNPs and enzymes, antimicrobial peptides and plant compounds. Solanki and coworkers (2013) proved that *Listeria* bacteriophage endolysin Ply500 incorporated onto silica nanoparticles was very effective in killing *Listeria innocua*. Rai and coworkers (2016) demonstrated that one-step synthesized antimicrobial peptides conjugated with AuNPs had higher antimicrobial activity and stability in serum than soluble peptides, both *in vitro* and *in vivo*. At the same time it was shown that AuNPs conjugated with DNA aptamer efficiently delivered antimicrobial peptides into HeLa cells infected with *Salmonella enterica* serovar Typhimurium, leading to pathogen elimination, and also inhibited *S. typhimurium* colonization in mouse organs (Yeom et al. 2016). Antibiofilm activity was proved for cinnamaldehyde immobilized on AuNPs. The efficient damage of biofilms formed by *S. aureus* (methicillin-sensitive and methicillin-resistant), *E. coli* and *P. aeruginosa* strains was observed (Ramasamy et al. 2017). Also, solid lipid nanoparticles encapsulated with endogenous host defense peptide LL37 and elastase inhibitor Serpin A1 were proved to accelerate wound healing and potentiate antibacterial activity against *S. aureus* and *E. coli* in comparison to LL37 and A1 alone (Fumakia and Ho 2016). The very interesting results of Westmeier and coworkers (2018) should also be mentioned. This group showed that coating of the spores of *Aspergillus fumigatus*, known to cause severe respiratory diseases, by various nanoparticles impacted fungal pathology and reduced their sensitivity against defensions (Westmeier et al. 2018). Recently, lysozyme-immobilized magnetic nanoparticles were found to efficiently kill *Micrococcus lysodeikticus* cells within 5 min (Orhan et al. 2019).

Conclusions

The ability of various types of nanoparticles to act synergistically with other antimicrobial compounds has been clearly proved in many studies. This characteristic is of special interest because it allows us to potentiate the effect of individual compounds along with the restoration of the activity of otherwise inactive antibiotics, to diminish the dose applied, and to reduce the eventual cytotoxic effect. The majority of research papers describe the synergy between AgNPs and other antimicrobials, but the synergistic relationship of other metal NPs was also reported. Of these, AuNPs have potential to serve as a platform to other therapeutics as an example of synergistic interactions. The mechanism of synergistic effect is mainly based on the enhancement of cell wall permeability and efflux activity. In clinical practice, the effective application of antimicrobial compounds is only now being explored, and further study demands not only the characterization of their effectiveness but also the detailed exploration of their mode of action and biocompatibility.

References

Ahmad, A., Wei, Y., Syed, F., Tahir, K., Khan, A.U., Hameed, M.U., Yuan, Q. 2016. Amphotericin B-conjugated biogenic silver nanoparticles as an innovative strategy for fungal infections. Microb. Pathog. 99: 271–281.

Allahverdiyev, A.M., Kon, K.V., Abamor, E.S., Bagirova, M., Rafailovich, M. 2011. Coping with antibiotic resistance: combining nanoparticles with antibiotics and other animal agents. Expert Rev. Anti. Infect. Ther. 9: 1035–1052.

Banoee, M., Seif, S., Nazari, Z.E., Jafari-Fesharaki, P., Shahverdi, H.R., Moballegh, A., Moghaddam, K.M., Shahverdi, A.R. 2010. ZnO nanoparticles enhanced antibacterial activity of ciprofloxacin against *Staphylococcus aureus* and *Escherichia coli*. J. Biomed. Mater. Res. B Appl. Biomater. 93: 557–561.

Baptista, P.V., McCusker, M.P., Carvalho, A., Ferreira, D.A., Mohan, N.M., Martins, M., Fernandez A.R. 2018. Nano-strategies to fight multidrug resistant bacteria—"a battle of the titans". Front. Microbiol. 9: 1441.

Berenbaum, M.C. 1989. What is synergy? Pharmacol. Rev. 41: 93–141.

Birla, S.S., Tiwari, V.V., Gade, A.K., Ingle, A.P., Yadav, A.P., Rai, M.K. 2009. Fabrication of silver nanoparticles by *Phoma glomerata* and its combined effect against *Escherichia coli*, *Pseudomonas aeruginosa* and *Staphylococcus aureus*. Lett. Appl. Microbiol. 48: 173–179.

Bollenbach, T. 2015. Antimicrobial interactions mechanisms and implications for drug discovery and resistance evolution. Curr. Opin. Microbiol. 27: 1–9.

Brown, A.N., Smith, K., Samuels, T.A., Lu, J., Obare, S.O., Scott, M.E. 2012. Nanoparticles functionalized with ampicillin destroy multiple-antibiotic-resistant isolates of *Pseudomonas aeruginosa* and *Enterobacter aerogenes* and methicillin-resistant *Staphylococcus aureus*. Appl. Environ. Microbiol. 78: 2768–2774.

Burygin, G.L., Khlebstov, B.N., Shantrokha, A.N., Dykman, L.A., Bogatyrev, V.A., Khlebstov, N.G. 2009. On the enhanced activity of antibiotics mixed with gold nanoparticles. Nanosc. Res. Lett. 4: 794–801.

Caster, J.M., Patel, A.N., Zhang, T., Wang, A. 2017. Investigational nanomedicines in 2016: a review of nanotherapeutics currently undergoing clinical trials. Wiley Interdiscip. Nanomed. Nanobiotechnol. 9: e1416.

Chatterjee, M., Anju, C.P., Biswas, L., Kumar, V.A., Mohan, C.G., Biswas, R. 2016. Antibiotic resistance in *Pseudomonas aeruginosa* and alternative therapeutic options. Int. J. Med. Microbiol. 306: 48–58.

Choi, O., Yu, C.P., Esteban Fernández, G., Hu, Z. 2010. Interaction of nanosilver with *Escherichia coli* cells in planktonic and biofilm cultures. Water Res. 44: 6095–6103.

Cushnie, T.P., Lamb, A.J. 2011. Recent advances in understanding the antibacterial properties of flavonoids. Int. J. Antimicrob. Agents 38: 99–107.

Dar, M.A., Ingle, A., Rai, M. 2013. Enhanced antimicrobial activity of silver nanoparticles synthesized by *Cryphonectria* sp. evaluated singly and in combination with antibiotics. Nanomedicine 9: 105–110.

Devi, L.S., Joshi, S.R. 2012. Antimicrobial and synergistic effects of silver nanoparticles synthesized using soil fungi of high altitudes of eastern Himalaya. Mycobiology 40: 27–34.

Duran, N., Marcato, P.D., Alves, O.L., Souza, G.I.H., Eposito, E. 2005. Mechanistic aspects of biosynthesis of silver nanoparticles by several *Fusariun oxysporum* strains. J. Nanobiotechnol. 3: 8.

Elghanian, R., Storhoff, J.J., Mucic, R.C., Lestinger, R.L., Mirkin, C.A. 1997. Selective colorimetric detection of polynucleotides based on the distance-dependent optical properties of gold nanoparticles. Science 277: 1078–1081.

EUCAST. Terminology relating to methods for determination of susceptibility of bacteria to antimicrobial agents. 2000. http//www.escmid.org/Files/E.Def.1.2.03.2000.pdf.

Fathi, M., Najafi, M., Shakibapour, Z., Zaeifi, D. 2016. Kinetic activity of *Yersinia intermedia* against ZnO nanoparticles either synergism antibiotics by double-disc synergy test method. Iran J. Microbiol. 14: 39–44.

Fayaz, A.M., Balaji, K., Girilal, M., Yadav, R., Kalaichelvan, T., Venketesan, R. 2010. Biogenic synthesis of silver nanoparticles and their synergistic effect with antibiotics: a study against gram-positive and gram-negative bacteria. Nanomedicine 6: 103–109.

Fernebro, J. 2011. Fighting bacterial infections—future treatment options. Drug Res. Updates 14: 125–139.

Fischbach, M.A. 2011. Combinational therapies for combating antimicrobial resistance. Curr. Opin. Microbiol. 14: 519–523.

Fumakia, M., Ho, E.A. 2016. Nanoparticles encapsulated with LL37 and serpin A1 promotes wound healing and synergistically enhances antibacterial activity. Mol. Pharm. 13: 2318–2331.

Gosh, S., Patil, S., Ahire, M., Kitture, R., Kale, S., Pardesi, K., Cameotra, S.S., Bellare, J., Dhavale, D.D., Jabgunde, A., Chopade, B.A. 2012. Synthesis of silver nanoparticles using *Discorea bulbifera* extract and evaluation of its synergistic potential in combination with antimicrobial agents. Int. J. Nanomed. 7: 483–496.

Gosh, I.N., Patil, S.D., Sharma, T.K., Srivastava, S.K., Pathania, R., Navani, N.K. 2013. Synergistic action of cinnamaldehyde with silver nanoparticles against spore-forming bacteria: a case for judicious use of silver nanoparticles for antibacterial applications. Int. J. Nanomed. 8: 4721–4731.

Gounani, Z., Asadollahi, M.A., Pedersen, J.N., Lyngso, J., Pedersen, J.S., Arpanaei, A., Meyer, R.L. 2018. Mesoporous silica nanoparticles carrying multiple antibiotics provide enhanced synergistic effect and improved biocompatibility. Coll. Surf. B Biointerfaces 175: 498–508.

Grace, N.A., Pandian, K. 2007. Antibacterial efficacy of aminoglycosidic antibiotics protected gold nanoparticles – A brief study. Coll. Surf. A Physicochem. Eng. Asp. 297: 63–70.

Graves, J.L., Tajkarimi, M., Cunningham, Q., Campbell, A., Nouga, H., Harrison, S.H., Barrick, J.E. 2015. Rapid evolution of silver nanoparticle resistance in *Escherichia coli*. Front. Genet. 6: 42.

Gu, H., Ho, P.L., Tong, E., Wang, L., Xu, B. 2003. Presenting vancomycin on nanoparticles to enhance antimicrobial activities. Nano Lett. 3: 1261–1263.

Gupta, D., Singh, A., Khan, A.U. 2017. Nanoparticles as efflux pump and biofilm inhibitor to rejuvenate bactericidal effect of conventional antibiotics. Nanosc. Res. Lett. 12: 454.

Habash, M.B., Park, A.J., Vis, E.C., Harris, R.J., Khursigara, C.M. 2014. Synergy of silver nanoparticles and aztreonam against *Pseudomonas aeruginosa* PAO1 biofilms. Antimicrob. Agents Chemother. 58: 5818–5830.

Hadiya, S., Liu, X., Abd El-Hammed, W., Elsabahy, M., Aly, S.A. 2018. Levofloxacin-loaded nanoparticles decrease emergence of fluoroquinolone resistance in *Escherichia coli*. Microb. Drug. Resist. 24: 1098–1107.

Huh, A.J., Kwon, Y.J. 2011. "Nanoantibiotics": a new paradigm for treating infectious diseases using nanomaterials in the antibiotics resistant era. J. Contr. Release 156: 128–145.

Hwang, I., Hwang, J.H., Choi, H., Kim, K.-J., Lee, D.G. 2012. Synergistic effects between silver nanoparticles and antibiotics and the mechanisms involved. J. Med. Microbiol. 61: 1719–1726.

Iravani, S., Korbekendi, H., Mirmohanmmadi, S.V., Zolfaghari, B. 2014. Synthesis of silver nanoparticles chemical, physical and biological methods. Res. Pharm. Sci. 9: 385–406.

Jasmin, R., Schneider, E.K., Han, M., Azad, M.A.K., Hussein, M., Nowell, C., Baker, M.A., Wang, J., Li, J., Velkov, T. 2017. A fresh shine on cystic fibrosis inhalation therapy: antimicrobial synergy of polymyxin B in combination with silver nanoparticles. J. Biomed. Nanotechnol. 13: 447–457.

Katva, S., Das, S., Moti, H.S., Jyoti, A., Kaushik, S. 2018. Antibacterial synergy of silver nanoparticles with gentamicin and chloramphenicol against *Enterococcus faecalis*. Pharmacogn. Mag. 13: S828–S833.

Khodashenas, B., Ghorbani, H.R. 2016. Optimisation of nitrate reductase enzyme activity to synthesise silver nanoparticles. IET Nanotechnol. 10: 158–161.

Kim, J., Kim, M., Shinde, S., Sung, J., Ghodake, G. 2017. Cytotoxity and antibacterial assessment of gallic acid capped gold nanoparticles. Coll. Surf. B Biointerfaces 149: 162–167.

Kora, A.J., Rastogi, L. 2013. Enhancement of antibacterial activity of capped silver nanoparticles in combination with antibiotics, on model of gram-negative and gram-positive bacteria. Bioinorg. Chem. Appl. 871097.

Krychowiak, M., Grinholc, M., Banasiuk, R., Krauze-Baranowska, M., Głód, D., Kawiak, A., Królicka, A. 2014. Combination of silver nanoparticles and *Drosea banita* extract as a possible alternative for antibiotic treatment of burn wound infections caused by resistant *Staphylococcus aureus*. PLoS One 9: e115727.

Krychowiak, M., Kawiak, M., Narajczyk, M., Borowik, A., Królicka, A. 2018. Silver nanoparticles combined with naphthoquinones as an effective synergistic strategy against *Staphylococcus aureus*. Front. Pharmacol. 9: 816.

Kurek, A. Grudniak, A.M., Kraczkiewicz-Dowjat, A., Wolska, K.I. 2010. New antibacterial therapeutics and strategies. Pol. J. Microbiol. 60: 3–12.

Lemire, J.A., Harrison, J.J., Furner, R.J. 2013. Antimicrobial activity of metals: mechanisms, molecular targets and applications. Nat. Rev. Microbiol. 11: 371–374.

Li, P., Li, J., Wu, C., Wu, Q., Li, J. 2005. Synergistic antibacterial effects of β-lactam antibiotic combined with silver nanoparticles. Nanotechnology 16: 1912–1917.

Li, X., Robinson, S.M., Gupta, A., Saha, K., Jiang, Z., Moyano, D.F., Sahar, A., Riley, M.A., Rotello, V.M. 2014. Functional gold nanoparticles as potent antimicrobial agents against multidrug resistant bacteria. ACS Nano. 8: 10682–10686.

Lundqvist, M., Stigler, J., Elia, G., Lynch, I., Cedervall, T., Dawson, K.A. 2008. Nanoparticle size and surface properties determine the protein corona with possible implication for biological impact. Proc. Natl. Acad. Sci. USA 105: 14265–1470.

Loo, C.Y., Rohanizadeh, R., Young, P.M., Traini, D., Cavaliere, R., Whitchurch, C.B., Lee, W.H. 2016. Combination of silver nanoparticles and curcumin nanoparticles for enhanced anti-biofilm activities. J. Agric. Food Chem. 64: 2513–2522.

Lopez-Carrizales, M., Velasco, K.I., Castillo, C., Flores, A., Magaña, M., Martinez-Castanon, A., Martinez-Gutierrez, F. 2018. *In vitro* synergism of silver nanoparticles with antibiotics as an alternative treatment in multiresistant uropathogens. Antibiotics 7: 50.

Markowska, K., Grudniak, A.M., Krawczyk, K., Wróbel, I., Wolska, K.I. 2014. Modulation of antibiotic resistance and induction of a stress response in *Pseudomonas aeruginosa* by silver nanoparticles. J. Med. Microbiol. 63: 649–654.

Matsumura, S.O., Louie, L., Louie, M., Simor, A.E. 1999. Synergy testing of vancomycin-resistant *Enterococcus faecium* against quinupristin-dalfopristin in combination with other antimicrobial agents. Antimicrob. Agents Chemother. 43: 2776–2779.

Mohamed, R.R., Sabaa, M.W. 2014. Synthesis and characterization of antimicrobial crosslinked carboxymethyl chitosan nanoparticles loaded with silver. Int. J. Biol. Macromol. 69: 95–99.

Murakami, T., Tsuchida, K. 2008. Recent advances in inorganic nanoparticle-based drug delivery systems. Mini-Rev. Med. Chem. 8: 175–183.

Murugadoss, A., Chattopadhyay, A. 2008. A 'green' chitosan-silver nanoparticle composite as a heterogenous as well as micro-heterogenous catalyst. Nanotechnology 19: 015603/1-015603/9.

Naqvi, S.Z., Kiran, U., Ali, M.I., Jamal, A., Hameed, A., Ahmed, S., Ali, N. 2013. Combined efficacy of biologically synthesized silver nanoparticles and different antibiotics against multidrug-resistant bacteria. Int. J. Nanomed. 8: 3187–3195.

Neihaya, H.Z., Zaman, H.H. 2018. Investigating the effect of biosynthesized silver nanoparticles as antibiofilm on bacterial clinical isolates. Microb. Pathog. 116: 200–208.

Norman, R.S., Stone, J.W., Gole, A., Murphy, C.J., Sabo-Atlwood, T.L. 2008. Targeted photothermal lysis of the pathogenic bacteria, *Pseudomonas aeruginosa*, with gold nanorods. Nano Lett. 8: 302–306.

Nostro, A., Blanco, A.R., Canatelli, M.A., Enea, V., Flamini, G., Morelli, I., Sudano Roccardo, A., Alonzo, V. 2004. Susteptibility of methicillin-resistant staphylococci to oregano essential oil, carvacrol and thymol. FEMS Microbiol. Lett. 230: 191–195.

Orhan, H., Evil, S., Dabanca, M.B., Basbülbül, G., Uygun, M., Uygun, D.A. 2019. Bacteria killer enzyme attached magnetic nanoparticles. Mater. Sci. Eng. C Mater. Biol. Appl. 94: 558–564.

Otaguiri, E.S., Morguette, A.E.B., Biasi-Garbin, R.P., Morey, A.T., Lancheros, C.A.C., Kian, D., de Oliveira, A.G., Kerbauy, G., Perugini, M.R.E., Duran, N., Nakamura, C.V., da Veiga, V.F., Nakazato, G., Pinge-Filho, P., Yamauchi, L.M., Yamada-Ogatta, S.F. 2017. Antimicrobial combination of oleoresin from *Copaifera multijuga* Hayne and biogenic silver nanoparticles towards *Streptococcus agalactiae*. Curr. Pharm. Biotechnol. 18: 177–190.

Padwal, P., Bandyopadhyaya, R., Mehra, S. 2014. Polyacrylic acid-coated iron oxide nanoparticles for targeting drug resistance in mycobacteria. Langmuir 30: 15266–15276.

Panáček, A., Smékalová, R., Večeřová, R., Bogdanová, K., Röderová, M., Kolář, M., Kilianová, M., Hradilová, S., Froning, J.P., Havrdová, M., Prucek, R., Zbořil, R., Kvitek, L. 2016. Silver nanoparticles strongly enhance and restore bactericidal activity of inactive antibiotics against multiresistant *Enterobacteriaceae*. Coll. Surf. B: Interfaces 142: 392–399.

Panáček, A., Kvítek, L., Smékalová, R., Kolá, M., Röderova, M., Dyčka, F., Šebela, M., Prucek, R., Tomanec, O., Zbořil, R. 2018. Bacterial resistance to silver nanoparticles and how to overcome it. Nat. Nanotechnol. 13: 65–71.

Parikh, R.Y., Singh, S., Parasad, B.L.V., Patole, M.S., Sastry, M., Shouche, Y.S. 2008. Extracellular synthesis of crystalline silver nanoparticles and molecular evidence of silver resistance from *Morganella* sp.: Towards understanding biochemical synthesis mechanism. Chem. Bio. Chem. 9: 1415–1422.

Payne, J.N., Waghhwani, H.K., Connor, M.G., Hamilton, W., Tockstein, S., Moolani, H., Chavda, F., Badwaik, V., Lawrenz, M.B., Dakshinamurthy, R. 2016. Novel synthesis of kanamycin conjugated gold nanoparticles with potent antibacterial activity. Front. Microbiol. 7: 607.

Pissuwan, D., Cortie, C.H., Valenzuela, S.M., Cortie, M.B. 2009. Functionalised gold nanoparticles for controlling pathogenic bacteria. Trends Biotechnol. 8: 207–213.

Potara, M., Jakab, E., Damert, A., Popescu, O., Canpean, V., Astilean, S. 2011. Synergistic antibacterial activity of chitosan-silver nanocomposites on *Staphylococcus aureus*. Nanotechnology 22: 135101.

Pumerantz, A., Muppidi, K., Agnihotri, S., Guerra, C., Venketraman, V., Wang, J., Betageri, G. 2011. Preparatiom of liposomal vancomycin and intracellular killing of methicillin-resistant *Staphylococcus aureus* (MRSA). Int. Antimicrob. Agents 37: 140–144.

Rabea, E.I., Badawy, M.E.T, Stevens, C.V., Smagghe, G., Steurbaut, W. 2003. Chitosan as a antimicrobial agent: applications and a mode of action. Biomacromolecules 4: 1457–1465.

Rai, A., Pinto, S., Velho, T.R., Ferreira, A.F., Moita, C., Trivedi, U., Evangelista, M., Comune, M., Rumbaugh, K.P., Simões, P.N., Moita, L., Ferreira, L. 2016. One-step synthesis of high-density peptide-conjugated gold nanoparticles with antimicrobial efficacy in a systemic infection model. Biomaterials 85: 99–110.

Rai, M., Yadav, A., Gade, A. 2009. Silver nanoparticles as a new generation of antimicrobials. Biotechnol. Adv. 27: 76–83.

Ramalingam, V., Rajaram, R., PremKumar, C., Santhanam, P., Dhinesh, P., Vinothkumar, S., Kaleshkumar, K. 2013. Biosynthesis of silver nanoparticles from deep sea bacterium *Pseudomonas aeruginosa* JQ 989348 for antimicrobial, antibiofilm and cytotoxic activity. J. Basic Microbiol. 53: 1–9.

Ramasamy, M., Lee, J.H., Lee, J. 2017. Direct one-pot synthesis of cinnamaldehyde immobilized on gold nanoparticles and their antibiofilm properties. Coll. Sur. B Bionterfaces 160: 639–648.

Rand, K.H., Houck, H.J., Brown, P., Bennett, D. 1993. Reproducibility of the microdilution checkerboard method for antibiotic synergy. Antimicrob. Agents Chemother. 37: 613–615.

Reithofer, M.R., Lakshmanan, A., Ping, A.T., Chin, J.M., Hauser, C.A. 2014. *In situ* synthesis of size-controlled, stable silver nanoparticles within ultrashort peptide hydrogels and their anti-bacterial properties. Biomaterials 35: 7535–7542.

Ruden, S., Hilpert, K., Berditsch, M., Wadhwani, P., Ulrich, A.S. 2009. Synergistic interaction between silver nanoparticles and membrane-permeabilizing antimicrobial peptides. Antimicrob. Agents Chemother. 53: 3538–3540.

Sanupi, P., Murugadoss, A., Durga Prasad, P.V., Ghosh, S.S., Chattopadadhyay, A. 2008. The antibacterial properties of a novel chitosan–Ag-nanocomposite. Int. J. Food Microbiol. 124: 142–146.

Scandorieiro, S., de Camargo, L.C., Lancheros, C.A.C., Yamada-Ogatta, S.F., Nakamura, C.V., de Oliveira, A.G., Andrade, C.G.T.J., Duran, N., Nakazato, G., Kobayashi, R.K.T. 2016. Synergistic and additive effect of oregano essential oil and biological silver nanoparticles against multidrug-resistant bacterial strains. Front. Microbiol. 7: 760.

Shahverdi, A.R., Fakhimi, A., Shakerverdi, H.R., Minaiam, S. 2007. Synthesis and effect of silver nanoparticles on the antibacterial activity of different antibiotics against *Staphylococcus aureus* and *Escherichia coli*. Nanomedicine 3: 168–171.

Shao, J., Yu, N., Kolwijck, E., Wang, B., Tan, K.W., Jansen, J.A., Walboomers, X.F., Yang, F. 2017. Biological evaluation of silver nanoparticles incorporated into chitosan-based membranes. Nanomedicine 12: 2771–2785.

Singh, P., Garg, A., Pandit, S., Mokkapati, V.R.S.S., Mijakovic, I. 2018. Antimicrobial effect of biogenic nanoparticles. Nanomaterials p8: 3390.

Singh, P., Kim, Y.J., Zhang, D., Yang, D.C. 2016. Biological synthesis of nanoparticles from plants and microorganisms. Trends Biotechnol. 34: 588–816.

Singh, R., Wagh, P., Wadhwani, S., Gaidhani, S., Kumbhar, A., Bellare, J., Chopade, B.A. 2013. Synthesis, optimization, and characterization of silver nanoparticles from *Acinetobacter calcoaceticus* and their enhanced antibacterial activity when combined with antibiotics. Int. J. Nanomed. 8: 4277–4290.

Smelakova, M., Aragon, V., Panacek, A., Prucek, R., Zboril, R., Kvitek, L. 2016. Enhanced antibacterial effect of antibiotics in combination with silver nanoparticles against animal pathogens. Vet. J. 209: 174–179.

Solanki, K., Grover, N., Downs, P., Paskaleva, E.E., Mehta, K.K., Lee, L., Schadler, L.S., Kane, R.S., Dordick, J.S. 2013. Enzyme-based listericidal nanocomposites. Sci. Rep. 3: 1584.

Sukanya, M.K., Saju, K.A., Praseetha, P.K., Sakthivel, G. 2013. Therapeutic potential of biologically reduced silver nanoparticles from actinomycete cultures. J. Nanosci. 2013: 1–8.

Tao, C. 2018. Antimicrobial activity and toxicity of gold nanoparticles: research progress, challenges and prospect. Lett. Appl. Microbiol. 67: 537–543.

Thomas, R., Nair, A.P., Mathew, S.K.J., Ek, R. 2014. Antibacterial activity and synergistic effect of biosynthesized AgNPs with antibiotics against multidrug-resistant biofilm-forming coagulase-negative staphylococci isolated from clinical samples. Appl. Biochem. Biotechnol. 173: 449–460.

Tutaj, K., Szlazak, R., Szalapata, K., Starzyk, J., Luchowski, R., Grudziński, W., Osińska-Jaroszuk, M., Jarosz-Wilkolazka, A., Szuster-Ciesielska, A., Gruszecki, W.I. 2016. Amphotericin B-silver hybrid nanoparticles: synthesis, properties and antifungal activity. Nanomedicine 12: 1095–1103.

Wagner, H., Ulrich-Merzenich, G. 2009. Synergy research: approaching a new generation of phytopharmaceuticals. Phytomedicine 16: 97–111.

Walkenhorst, W.F. 2016. Using adjuvants and environmental factors to modulate the activity of antimicrobial peptides. Biochim. Biophys. Acta 1858: 926–935.

Walencka, E., Różalska, S., Wysokińska, H., Różalski, M., Kuźma, L., Różalska, B. 2007. Salvipisone and aethopinone from *Salvia sclarea* hairy roots modulate staphylococcal antibiotic resistance and express anti-biofilm activity. Planta Med. 73: 545–551.

Wang, L., Chen, Y.P., Miller, K.P., Cash, B.M., Jones, S., Glenn, S., Benicewicz, B.C., Decho, A.W. 2014. Functionalized nanoparticles complexed with antibiotic efficiently kill MRSA and other bacteria. Chem Commun. (Camb). 50: 12030–12033.

Westmeier, D., Solouk-Saran, D., Vallet, C., Siemer, S., Docter, D., Gotz, H., Mann, L., Hasenberg, A., Hahlbrock, A., Erler, K., Reinhardt, C., Schilling, O., Becker, S., Gunzer, M., Hasenberg, M., Knauer, S.K., Stauber, R.H. 2018. Nanoparticle decoration impacts airborne fungal pathobiology. Proc. Natl. Acad. Sci. USA 115: 7087–7092.

Wolska, K.I., Grudniak, A.M., Markowska, K. 2017. Inhibition of bacterial quorum sensing system by metal nanoparticles. pp. 123–138. *In*: Rai, M., Shegokar, R. (eds.). Metal Nanoparticles in Pharma. Springer Nature, Switzerland.

Wolska, K.I., Grześ, K., Kurek, A. 2012. Synergy between novel antimicrobials and conventional antibiotics and bacteriocins. Pol. J. Microbiol. 61: 95–104.

Yeom, J.H., Lee, B., Kim, D., Lee, J.K., Kim, S., Bae, J., Park, Y., Lee, K. 2016. Gold-nanoparticle-DNA aptamer conjugate-assisted delivery of antimicrobial peptide effectively eliminate intracellular *Salmonella enterica* serovar Typhimurium. Biomaterials 104: 43–51.

Zharov, V.P., Galanzha, E.I., Tuchin, V.P. 2007. Photothermal flow cytometry *in vitro* for detection and imaging of individual moving cells. Cytometry A. 71A: 191–206.

Zhang, Y., Wang, L., Xu, X., Li, F., Wu, Q. 2018. Combined systems of different antibiotics with nano-CuO against *Escherichia coli* and the mechanism involved. Nanomedicine 13: 339–351.

Zong, J., Cobb, S.L., Cameron, N.R. 2017. Peptide-functionalized gold nanoparticles: versalite biomaterials for diagnosis and therapeutic applications. Biomater. Sci. 5: 872–886.

Chapter 14

Synthesis of Copper Nanomaterials by Microbes and Their Use in Sustainable Agriculture

Sudhir Shende,[1,*] *Vishnu Rajput,*[2] *Avinash P. Ingle,*[3]
Aniket Gade,[1] *Tatiana Minkina*[2] and *Mahendra Rai*[1]

Introduction

Nanoscience and nanotechnology offer an exciting prospect to study a state of matter that is intermediate between bulk and isolated atoms or ions, as well as the spatial confinement effect on electron behavior. It also provides an opportunity to investigate the problems associated with surface or interface because of their interfacial nature (Rai and Duran 2011). Nanotechnology is a multidisciplinary branch that includes research and technological evolution in areas of chemistry, physics and the biological sciences (McNeil 2005, Uskokovic 2008). Nanoparticles are metal or metal oxide particles smaller than 100 nm that can be synthesized in various shapes (e.g., spherical, triangular, hexagons, rods) from different metal ions. The advantages of nanotechnology encourage us to apply the concepts in a variety of fields and hence nanoparticles (NPs) are used in pharmacology, medicine, electronics, agriculture, environmental monitoring and many other fields (Liu et al. 2006, Thuesombat et al. 2014, Dimpka et al. 2015, Shende et al. 2017, Paralikar and Rai 2018).

[1] Nanobiotechnology Laboratory, Department of Biotechnology, Sant Gadge Baba Amravati University, Amravati, Maharashtra, India 444602.
[2] Academy of Biology and Biotechnology, Southern Federal University, Rostov-on-Don 344090, Russia.
[3] Department of Biotechnology, Engineering School of Lorena, University of Sao Paulo, Lorena, SP, Brazil.
* Corresponding author: sudhirsshende13884@gmail.com

Globally, nanotechnology has been extensively used in food, biomedical and agricultural sectors (FAO/WHO 2010, Safiuddin et al. 2014, Paralikar and Rai 2018, Shende et al. 2017). Although NPs are naturally present in the surroundings, the application of nanotechnology has resulted in a considerable increase in the fabrication of engineered nanoparticles. It is anticipated that over 800 products are currently available in the market (Safiuddin et al. 2014, Al-Halafi 2014, Zhou et al. 2015, Rajapaksha et al. 2015). Engineered nanoparticles have a typical dimension of ≤ 100 nm and diverse properties from bulk particles with a similar chemical composition (Auffan et al. 2009). The NPs could penetrate into soils through the application of paints, pesticides, and direct discharge from industries (Grieger et al. 2009, Zhang and Fang 2010, Liu and Lal 2015). Because of the stagnant nature of NPs and their interaction with soil and air, plants are exposed to NPs released into the environment. In the past few years, both the advantages and limitations of NPs on crop growth have been reported. These effects were reliant on the type, source and size of the NPs, the plant species, and the exposure time (An et al. 2008, Roghayyeh et al. 2010, Rico et al. 2014, Bandyopadhyay et al. 2015, Lalau et al. 2015). For instance, silver (Ag) NPs increased ascorbate and chlorophyll content in leaves of asparagus (*Asparagus officinalis* L.) and iron (Fe) NPs increased the biomass of soybean (*Glycine max* L.) (An et al. 2008, Roghayyeh et al. 2010). Likewise, application of silica NPs enhanced the seed germination, root and shoot length, photosynthesis and dry weight of maize seedlings grown in the field (Suriyaprabha et al. 2012). Titanium dioxide (TiO$_2$) NPs decreased the hydrogen peroxide (H$_2$O$_2$), electrolyte and malondialdehyde leakage in chickpea (*Cicer arietinum* L.) over the control plants (Mohammadi et al. 2014). Alternatively, numerous reports indicated that metal NPs adversely affect the growth and physiology of wheat (*Triticum aestivum* L.), canola (*Brassica napus*), rice (*Oryza sativa* L.), maize (*Zea mays* L.), barley (*Hordeum vulgare* L.), soybean (*Glycine max* L.) and spring barley (*Hordeum sativum* distichum) (Dimkpa et al. 2012, Feizi et al. 2013b, Mahmoodzadeh et al. 2013a, Mahmoodzadeh et al. 2013b, Nair and Chung 2014a, Shaw et al. 2014, Antisari et al. 2015, Rajput et al. 2018).

In addition, metal and metal oxide NPs were more noxious to crops than bulk metal particles (Wang et al. 2011). It has been reported that crops exposed to NPs might uptake and translocate NPs to different plant parts (Du et al. 2011, Le et al. 2014, Rico et al. 2015). The NPs reduced seed germination of various crop plants (Moon et al. 2014, Thiruvengadam et al. 2015, Xiang et al. 2015). The toxic effects of NPs in several plant species can be observed in reduced plant growth, biomass, fruit or grain yield, and anatomical and ultrastructural changes (Servin et al. 2012, Antisari et al. 2015, Rico et al. 2015, Minkina et al. 2019, Rajput et al. 2019a,b). The excess of NPs caused physiological disorders in crops, i.e., reduction in photosynthesis and gas exchange attributes, oxidative stress and reduction in the activities of antioxidant enzymes leading to cytotoxicity and genotoxicity (Foltête et al. 2011, Atha et al. 2012, Vannini et al. 2013, Cui et al. 2014, Shaw et al. 2014). A number of studies reported that NPs decreased mitotic index and impaired stages of cell division in root tips of various plants and altered the expression of genes associated with root growth (Kumari et al. 2009, Kumari et al. 2011, Vannini et al. 2014, Nagaonkar et al. 2015, Wang et al. 2015). NPs not only directly affect the plants, but also indirectly affect the growth medium, shifting soil bacterial communities, damaging the roots and causing uptake of co-contaminants by plants (Ma et al. 2011, Ge et al. 2014, Dimkpa et al. 2015, Rajput et al. 2017). Therefore, NPs might enhance or reduce crop growth and yield and could easily enter the food chain with unidentified consequences to humans and animals (Foltête et al. 2011, Atha et al. 2012, Vannini et al. 2013, Cui et al. 2014, Moon et al. 2014, Shaw et al. 2014, Vannini et al. 2014, Thiruvengadam et al. 2015, Xiang et al. 2015, Bramhanwade et al. 2016, Shende et al. 2017, Rajput et al. 2019b).

The positive and toxic effects of metal as well as metal oxide NPs on the growth, yield, and physiology of agricultural crops have comprehensively been reported (Moon et al. 2014, Thuesombat et al. 2014, Thiruvengadam et al. 2015, Dimkpa et al. 2015, Zhang et al. 2015b). In this chapter, we focus on the synthesis of copper nanomaterials and their application in agriculture for sustainable development.

Methods for the synthesis of copper nanomaterials

One of the most vital aspects of nanotechnology is the synthesis of nanoparticles, which forms the core component of the nanomaterials. Because NPs have more surface atoms than microparticles, their practical capabilities are enhanced. The biosynthesis of metal NPs is gaining importance as an environment-friendly tool in material science, due to their size-dependent chemical and physical characteristics. These nanomaterials have the desired features that can overcome the limitations of the bulk materials. Nanomaterials are useful in applications associated with energy (Venugopal et al. 2010), electronics (Maruccio et al. 2004), pharmacology, medicine (Barreto et al. 2011), environment (Englert 2007) and agricultural science (Bhattacharyya et al. 2010, Durán and Marcato 2013, Shende et al. 2015, Bramhanwade et al. 2016).

The basis of nanomaterials is metal NPs. At present, the fabrication of NPs is a fascinating area of research because of their exceptional optoelectronic (Chandrasekharan and Kamat 2000), chemical and physical (Kumar et al. 2003), catalytic (Chen et al. 2011), photonics and biosensing (Garcia-Parajo 2012) properties. The chemical methods are capital-intensive and noxious; hence, the significance of biological synthesis has been globally acknowledged. The need for clean, biocompatible, cost-effective and eco-friendly synthesis of metal NPs encouraged researchers to exploit biological systems as nano-factories (Rai et al. 2009, Seshadri et al. 2011, Wei et al. 2012, Christopher et al. 2015, El-Sheekh and El-Kassas 2016, Patil et al. 2016, Plaza et al. 2016, Owaid and Ibraheem 2017).

Numerous methods have been proposed for the synthesis of copper nanomaterials. Khodashenas and Ghorbani (2014) reviewed various chemical, physical, and biological methods. Some of these methods are described in detail in later sections.

Chemical methods

The chemical synthetic methods for copper nanomaterials are again classified into five different procedures reported so far. It has also been reported that the growth and morphology of NPs could be optimized by variable factors such as temperature and concentration of the surfactant precursor (Aslam et al. 2002). Brief accounts of chemical methods for the synthesis of copper nanomaterials are given in Table 14.1.

Table 14.1. Various chemical methods reported for the synthesis of copper nanomaterials.

Method of synthesis	Size of NPs (in nm)	Reference
Chemical reduction	45	Athawale et al. 2005
Chemical reduction	2–10	Lisiecki et al. 1996
Chemical reduction	5	Wu and Chen 2004
Chemical reduction	100	Wang et al. 2006
Chemical reduction	3	Goia and Matijevic 1998
Chemical reduction	70	Yang et al. 2007
Chemical reduction	3	Kanninen et al. 2007
Chemical reduction	DLS (20–100) HR-TEM, STEM, XRD (5–50)	Grouchko et al. 2009
Chemical reduction	> 10 nm	Dang et al. 2011
Chemical reduction	30	Rahimi et al. 2010
Chemical reduction	2	Xiong et al. 2011
Electrochemical method	40–60	Raja et al. 2008
Electrochemical method	24	Theivasanthi and Alagar 2011
Microemulsion/colloidal method	3–13	Salzemann et al. 2004
Microemulsion/colloidal method	70–80	Kaminskiene et al. 2013
Microwave method	10	Zhu et al. 2004b
Photochemical method	15	Yeh et al. 1999

Chemical reduction method

In the chemical reduction method, various copper (Cu (II)) salts may be reduced by different reducing agents, for example, ascorbic acid (Wu et al. 2006), hydrazine (Su et al. 2007), hypophosphite (Zhu et al. 2004a), sodium borohydride (Song et al. 2004) and polyol (Park et al. 2007). These reducing agents are used to produce copper nanoparticles (CuNPs) with controlled size as well as morphology. Gotoh and coworkers (2000) reported the production of NPs of Cu/PAA composite films using long-chain poly(acrylic) acid (MW 150,000), which were deposited on glass plates. Ostaeva et al. (2008) reported the synthesis of stable CuNPs with a particle size < 10 nm by reduction of Cu^{2+} in poly(acrylic acid)–pluronic blends solution. Prucek and coworkers (2009) demonstrated copper nanocrystals with a diameter of 14 nm synthesized using Cu reduction by sodium borohydride ions. Chatterjee et al. (2012) used $CuCl_2$ resurgence in the existence of gelatin as stabilizer, devoid of any special procedures (for example, nitrogen purging), to produce stable CuNPs with a size of 50–60 nm.

Microemulsion/colloidal method

In the synthesis of NPs, microemulsion is a technique in which two immiscible liquids (oil in water, water in oil, or water in supercritical carbon dioxide) become a thermodynamically stable dispersion with the help of a surfactant (Jackelen et al. 1999, Kapoor et al. 2002, Chen et al. 2006). The reverse micelles method, which essentially consists of two inverted emulsions of oil and water, is also placed in this class (Dadgostar et al. 2010). Salzemann et al. (2004) reported the synthesis of CuNPs by reverse micelles method, which produced CuNPs of size 3–13 nm. Kaminskiene et al. (2013) synthesized Cu colloidal solution with a size of 70–80 nm using the same technique.

Photochemical method

Nanoparticles can be synthesized using microwaves, which is a form of electromagnetic energy that ranges from the frequency MHz 300 to GHz 300. Zhu et al. (2004b) deliberated a rapid method for the synthesis of CuNPs using copper sulfate salt as a precursor and sodium hypophosphite as a reducing agent in ethylene glycol under microwave irradiation, which produced NPs of 10 nm size (Zhu et al. 2004b). Kapoor and Mukherjee (2003) achieved a mean size of 15 nm of CuNPs synthesized using photochemical method with poly(N-vinyl pyrrolidone) as a stabilizer. Giuffrida et al. (2008) used photochemical method in the synthesis of CuNPs and found that the technique is expedient and cost-effective. These authors reported that light intensity, the concentration of surfactant polyvinyl pyrrolidone, sensitizer nature, and other factors may affect the size of NPs. CuNP recovery rate was affected by light intensity, as it is a key factor and is directly associated with the reduction rate.

Electrochemical method

Electricity is used as a controlling force in electrochemical synthesis of CuNPs. Electrochemical synthesis of NPs occurs when an electric current is passed between two electrodes separated by an electrolyte. The synthesis takes place at the electrode/electrolyte interface. Generally, an electrolytic solution of copper salt along with sulfuric acid is used for the production of CuNPs. Raja et al. (2008) reported the method for CuNP synthesis using copper sulfate and sulfuric acid as an electrolytic solution and supplied current 4V, 5A for 30 min, which successfully produced CuNPs of size 40–60 nm (Raja et al. 2008).

Thermal decomposition

In the thermal decomposition method, the chemical reactions take place in a pressure- and temperature-controlled container such as an autoclave, in which the solvent reaches a temperature above its boiling point. If water is used as a solvent in this process, then this method is termed a hydrothermal process (Yu 2001, Rajamathi and Seshadri 2002). Chen and coworkers (2010) found the hydrothermal method useful in the synthesis of CuNPs of different sizes. Authors used sodium dodecyl benzene sulfonate (SDBS) as a surfactant for stabilization as well as to control the shape and size of the NPs. In this method, they found that the reaction temperature and SDBS play a significant role in shaping the final CuNPs.

Physical methods

Pulse laser ablation/deposition

The laser ablation method is characteristically designed for the synthesis of colloidal silver NPs in different types of solvents. NPs are provided in the colloidal form to prevent oxidation (Amendola et al. 2006, Amendola and Meneghetti 2009). A chamber under vacuum and in the presence of some inert gases was used to carry out the process. The final product is influenced by the type of laser, the duration of pulsing and the type of solvent (Raja et al. 2008). The various physical methods for the synthesis of copper nanomaterials that are reported are given in Table 14.2.

Table 14.2. Various physical methods reported for the synthesis of copper nanomaterials.

Method of Synthesis	Size of NPs (in nm)	Reference
Pulse laser ablation/deposition	2–20	Saito and Yasukawa 2008
Pulse laser ablation/deposition	5–15	Song and Yamaguchi 2007
Pulse laser ablation/deposition	10–50	Kim and Jang 2007
Pulse laser ablation/deposition	3–9 (small particles) 10–30 (larger particles)	Muniz-Miranda et al. 2011
Pulse laser ablation/deposition	5–30	Lee et al. 2006
Exploding wire method	62.2	Sen et al. 2003
Pulsed wire discharge method	27–72	Raffi et al. 2010
Exploding wire method	55	Das et al. 2012

Mechanical/ball milling method

The milling process makes some alterations in micron-sized material. In this method, different kinds of milling machines are used for the synthesis of NPs. This process has some restrictions as it takes a long time and it is very difficult to produce ultrafine particles. On the other hand, the advantages include ease of operation, potential for large-scale production and lower production cost (Suryanarayana 2001). It has also been noticed that variable factors determine the quality of the final output, i.e., the kind of mill, milling speed, time, temperature, atmosphere, and container (Benjamin 1989).

Pulsed wire discharge method

Another physical method used to synthesize NPs is the pulsed wire discharge method (Lisiecki et al. 2000, Barrabes et al. 2006). This method is normally not used in industries because of its high cost and probable inefficiency for some metals. For metals with high electrical conductivity, this method is found to be appropriate, as they can be easily produced into thin wires (Jiang and Yatsui 1998). Muraia et al. (2007) demonstrated that CuNPs coated with organic materials can be effectively synthesized through evaporation of copper wire in oleic acid vapor or mist, and a coating layer with a thickness of a few nanometers, of size 10–25 nm. Dash et al. (2011) used pulse wire discharge method by applying 22 kV power with 3 µF capacitance under 0.1 MPa pressure and synthesized CuNPs with a size of 16.11–43.06 nm. The mechano-chemical method for synthesis of NPs may also be mentioned (Gaffet et al. 1993, Ding et al. 1996, Shen and Koch 1996), along with metal-organic chemical vapor condensation (Cioffi et al. 2011). The disadvantages of physical methods are that they usually require costly vacuum systems or equipment.

Biological methods

It has been reported that living organisms, for example, bacteria, fungi, and plants, have an enormous potential for the synthesis of metal nanomaterials. Microbes can serve as a nano-biofactory and can also be employed for the fabrication of metal nanomaterials. Additionally, many researchers have a preference for biological synthesis, since it makes it easier to control the size distribution of synthesized NPs than do other methods (Bansal et al. 2005, Ingle et al. 2008, Varshney et al. 2012, Ingle et al. 2013, Shende et al. 2015). Biological methods have not shown a toxic impact on the environment (Bansal et al. 2005, Varshney et al. 2012). The advancement of trustworthy green protocols for the fabrication of NPs over an array of chemical composition, size, and elevated mono-dispersity is a challenging problem (Mandal et al. 2006). Chemical and physical methods of copper nanomaterial synthesis are quite expensive, include noxious chemicals, and show low bioactivity compared to biogenic synthesis. Thus, there is a need for cost-effective, clean, biocompatible, and eco-friendly synthesis of NPs, and researchers have explored the use of biological resources as nano-factories during the synthesis process.

It was found that microbes such as fungus and bacteria as well as plant system score over other biological systems. With regard to myco- and phytosynthesis of NPs, it was observed that NPs with good dispersity and dimensions could be synthesized. As fungi are found to produce a high quantity of protein they might result in the considerable mass productivity of NPs. The proteins of fungi are competent to hydrolyze the metal ions (Rai et al. 2010), and fungi are very easy to isolate and culture on a medium. Furthermore, the downstream processing and management of fungal biomass are less complicated than with synthetic methods. Fungi are able to accumulate metal ions by physicochemical and biological mechanisms, for example, binding to specific polypeptides, metabolism-dependent accumulation and extracellular binding by metabolites and polymers (Volesky and Holan 1995). Moreover, the extracellular secretion of enzymes has an added advantage in the downstream processing and biomass handling (Gade et al. 2008) as compared to bacterial fermentation, which includes the application of sophisticated instruments to get clear filtrate from the colloidal broth. Similarly, fungi are excellent protein secretors, compared to bacteria and actinomycetes, resulting in higher yield of NPs (Sastry et al. 2003). Regarding plant systems, they are a good source of secondary metabolites and are also rich in carbohydrates, proteins, flavonoids, and phenolic compounds, which act as capping as well as reducing agents during the synthesis of NPs.

There are a number of publications on the biosynthesis of metal NPs using prokaryotic microbes such as bacteria (Deshmukh et al. 2012), myxobacteria (Dahm et al. 2015, Bhople et al. 2016) and eukaryotic cell systems including actinomycetes (Deepa et al. 2013, Sunitha et al. 2013, Golińska et al. 2015, Anasane et al. 2016, Golińska et al. 2016, Wypij et al. 2017a, Wypij et al. 2017b), algae (Narayanan and Sakthivel 2010, Rajeshkumar et al. 2014), yeasts (Namasivayam et al. 2011, Apte et al. 2013), fungi (Birla et al. 2013, Gade et al. 2014) and plants (Song et al. 2010, Mittal et al. 2013, Shende et al. 2015, Ahmed et al. 2016, Chung et al. 2016, Dauthal and Mukhopadhyay 2016, Agarwal et al. 2017, Elemike et al. 2017, Rajendran and Sengodan 2017, Saha et al. 2017, Santhoshkumar et al. 2017).

The literature on the biological synthesis of copper nanomaterials is very limited when compared with the available literature for AgNP synthesis. The biological synthesis of copper nanomaterials is a quite challenging task as copper salts can easily get oxidized and lose stability when they come into contact with environmental oxygen. So the selection of suitable antioxidants, which could act as stabilizing and capping agents, is a very important task during copper nanomaterial synthesis. Apart from this, several researchers have reported the synthesis of CuNPs using biological resources.

Using bacteria

Some researchers have published data on their findings with respect to biological synthesis of copper nanomaterials. Varshney et al. (2010) used a rapid biological method to synthesize CuNPs by a non-pathogenic bacterium of *Pseudomonas stutzeri*. The NPs produced by this method were spherical and 8–15 nm in size as examined by high-resolution transmission electron microscopy. Varshney et al. (2011) again used *P. stutzeri* isolated from electroplating industry wastewaters to synthesize CuNPs, obtaining a cubic shape and particle size of 50–150 nm. *Salmonella typhimurium* supernatant, when added to aqueous copper nitrate solution (1 mM), forms CuNPs extracellularly, with an average diameter of 49 nm (Ghorbani et al. 2015).

Using fungi

A large number of fungi are used for the synthesis of metal NPs, such as *Fusarium oxysporum* (Durán et al. 2007, Gaikwad et al. 2013), *Fusarium acuminatum* (Ingle et al. 2008), *Fusarium graminearum, Fusarium scirpi* (Gaikwad et al. 2013), *Phoma glomerata* (Birla et al. 2009, Gade et al. 2014), *Fusarium culmorum* (Bawaskar et al. 2010), and *Alternaria alternata* (Gajbhiye et al. 2009, Sarkar et al. 2011). The higher fungi have also been studied for the *in vitro* synthesis of metal NPs, and in such a study Sanghi and Verma (2009) reported the role of white-rot fungus *Trametes versicolor* in synthesis of AgNPs.

Earlier studies established that the NADH- and NADPH-dependent enzymes are essential factors in the biosynthesis of metal NPs (Bansal et al. 2005). The use of fungal species for the myco-synthesis of metallic NPs has the advantages of being cost-effective, involving a simple synthesis process, and offering the possibility of glazing large surfaces because of the fungal growth. Pavani et al. (2013) used *Aspergillus* species to synthesize CuNPs. Salvadori et al. (2014) developed a biological method using dead biomass of the yeast *Rhodotorula mucilaginosa* for removal of copper from wastewater to synthesize CuNPs.

Using plants

The exploitation of a variety of plants for the synthesis of metal NPs has been studied by several researchers because it is inexpensive and involves non-toxic and easily available materials (Mallikarjuna et al. 2011, Awwad et al. 2012, Khalil et al. 2013, Yallappa et al. 2013). Positive results have been achieved. The extracts of biological resources such as plants often include flavonoids, terpenoids, proteins, polyphenols, and other biomolecules, which may act as reducing agents as well as capping agents to control morphology, minimize particle accumulation and protect and stabilize the synthesized NPs. Bali and coworkers (2006) demonstrated the synthesis of varied types of NPs including CuNPs in *Medicago sativa* (alfalfa), *Brassica juncea* (Indian mustard) and *Helianthus annuus* (sunflower). Lee et al. (2013) employed the leaf extract of magnolia as a reducing agent for the conversion of $Cu^{+2} \rightarrow Cu^0$ to synthesize stable CuNPs of size 40–100 nm using $CuSO_4 \cdot 5H_2O$ in aqueous solution and leaf extract as a precursor. Valodkar et al. (2011) used stem latex of medicinal plant *Euphorbia nivulia* for the synthesis of CuNPs. Subhankari and Nayak (2013a) used *Zingiber officinale* (ginger) extract to reduce copper sulfate and fabricate CuNPs as a method for synthesis of CuNPs, and they also analyzed the antibacterial effect of the NPs produced. In another study, Subhankari and Nayak (2013b) used *Syzygium aromaticum* (cloves) in an aqueous extract for synthesis of spherical CuNPs of 5–40 nm size. The study identified that existing biomolecules in bio-copper not only reduce metal ions, but also prevent oxidation through stabilization of the CuNPs produced.

Naika and coworkers (2015) reported the green synthesis of copper oxide nanoparticles (CuONPs) from *Gloriosa superba* L. Prabhu et al. (2017) synthesized CuNPs using *Garcinia mangostana* leaf extract as reducing agent with copper nitrate. They found cubic phase CuNPs of size 26.51 nm by X-ray diffraction analysis. Mittal et al. (2014) published a review on the plants that can be used as nano-factories for the synthesis of CuNPs and other metal NPs. Shende et al. (2015) reported the synthesis of CuNPs using *Citrus medica* L. (citron) juice extract. In other studies, *Magnolia kobus* leaf extract (Lee et al. 2013), *Ocimum sanctum* leaf extract (Kulkarni and Kulkarni 2013), *Nerium oleander* (Gopinath et al. 2014), *Artabotrys odoratissimus* (Kathad and Gajera 2014), *Capparis zeylanica* (Saranyaadevi et al. 2014), *Vitis vinifera* (Angrasan and Subbaiya 2014), aqueous *Phyllanthus embilica* (gooseberry) extract (Caroling et al. 2015a), and *Psidium guajava* (guava) extract (Caroling et al. 2015b) were reported for CuNP synthesis.

CuNPs can be an alternative to Ag and Au NPs and have potential applications in biofungicides and biofertilizers. Many artificial routes have been documented for the preparation of CuNPs, but few are eco-friendly and scalable. There have been several activities in this direction, resulting in different synthesis methods that make use of natural products involving bacteria, fungi and plants (Shende et al. 2015). Moreover, among different metal NPs such as iron and their oxides, silver, gold, and zinc oxide, CuNPs are most attractive because of their environmental, energy, biomedical, and agricultural applications (Chinnamuthu and Boopathi 2009, Kailasam et al. 2011, Ingle et al. 2013, Nguyen 2014, Bramhanwade et al. 2016), their unique physical and chemical properties, and the low cost of their preparation. So far, many biological systems have been exploited for the synthesis of copper nanomaterials.

Traditional uses of copper nanomaterials

In ancient periods, copper and silver vessels were used to store water and make it potable (Richard et al. 2002, Castellano et al. 2007). Copper and copper compounds such as copper carbonate, copper silicate, copper oxide, copper chloride, and copper sulfate were used in the treatment of diseases in the ancient Egyptian, Greek, Roman and Aztec civilizations. Many civilizations in completely different geographical locations and mostly independent from one another discovered the capacity of copper to improve or resolve skin and other tissue maladies (Dollwet and Sorenson 2001, Milanino 2006). Copper has potent biocidal properties (Borkow and Gabbay 2005, Borkow 2012) and has been used as a biocide for centuries (Dollwet and Sorenson 2001, Milanino 2006). Both Gram-positive and Gram-negative bacteria, including antibiotic-resistant bacteria and bacterial spores, fungi and viruses, are killed when exposed to copper (Borkow and Gabbay 2005, Borkow 2012). In several cases, they are destroyed within minutes of exposure to copper or copper compounds (Borkow et al. 2007, Espirito et al. 2008). Accordingly, copper biocides have become obligatory and several thousands of tons are used per annum worldwide, in antifouling paints (Cooney 1995), wood preservation (Schultz et al. 2007), and agriculture (La Torre et al. 2008).

In Ayurveda, copper nanopowder ("Tamra Bhasma") was used for the preparation of traditional medicines (Pal et al. 2014). The ancient Chinese (~ 3000 B.C. to 1100 A.D.) used copper sulfate or copper sulfide for topical treatment of skin and eye diseases and also treated systemic infections by oral administration of copper. Mayan, Aztec, and Inca societies (~ 600 B.C. to 150 A.D.) used gauzes soaked in a copper sulfate solution for the disinfection of surgical wounds suffered during the common practice of drilling a hole in the skull as a physical, mental, or spiritual treatment, with estimated survival rates above 50%. The ancient Greeks (1300–323 B.C.) used copper preparations for the purification of drinking water and also in the treatment of various cutaneous and eye diseases, and vaginal, pulmonary and gastrointestinal disorders. Copper bracelets were used to treat arthritis. The early Phoenicians (1550 B.C. to 300 B.C.) nailed copper strips to the bottom of their ships to restrain fouling to boost speed and maneuverability. The ancient Romans (~ 600 B.C. to 476 A.D.) used various copper compounds to treat inflammation of the tonsils, eye and skin diseases, hemorrhoids and wounds. For centuries, Hindus have stored "holy water" from the Ganges in copper utensils to keep the water clean. Finally, copper sulfate is widely used by many inhabitants of the African continent for healing sores and skin diseases (Borkow 2014).

Metallic copper has been used for many years in dental fillings (Ferracane 2001) and in copper intrauterine devices for reversible contraception (Bilian 2002, O'Brien et al. 2008). Copper compounds are extensively used in anthroposophical medicine (Gorter et al. 2004), via subcutaneous injections, oral or topical applications, consecutively to stimulate the body to heal itself. Cu-containing ointments release copper ions, which are absorbed through the skin (Gorter et al. 2004, Hostynek et al. 2006) and are used in the treatment of cramps, disturbances of renal function, rheumatic disease and swelling associated with trauma and peripheral, venous hypostatic circulatory disturbances (Heilmittel 1994). There are also commercially available cosmetic facial creams that contain copper as their active ingredient (e.g., Neutrogena Visibly Firm® Face Lotion SPF 20).

Copper and its compounds were widely used by ancient farmers to control crop diseases because they showed bactericidal and fungicidal activity. Copper sulfate was used to treat cereal seed in 1761 (Copper Development Association 2003). However, it was not until the 1880s that copper sulfate as a fungicide was developed from the "accidental" discovery of the Bordeaux mixture. At this time, farmers in the region of Bordeaux, France, were applying a paste of copper sulfate and lime mixture to grapes infected with downy mildew. The French botany professor Pierre-Marie-Alexis

Millardet from the University of Bordeaux observed grapes free of disease (Copper Development Association 2019). By 1885, Prof. Millardet completed experiments that confirmed the applicability of this mixture against downy mildew disease at a lower cost. Bordeaux mixture was subsequently used globally as a fungicide on large scale (Schneiderhan 1933). Copper-based fungicides have been used to efficiently manage a broad range of plant pathogens on a variety of crops (Zwieten et al. 2007). Several bacteria and fungi deal with excess copper via intra- as well as extracellular sequestration through the cell envelopes and membrane efflux pumps. Moreover, tolerance in addition to adaptation occurs by upregulation of essential genes in the presence of Cu and by the precipitation of copper by secreted metabolites (Borkow and Gabbay 2005). Since microorganisms cannot tolerate the copper burden, when they are exposed to high concentrations of Cu they are irreversibly damaged and killed (Borkow 2014). Copper and copper compounds are now widely used in medical and agriculture applications.

Antifungal activities of copper nanomaterials

Studies have demonstrated the role of NPs as nano-fungicides, nano-fertilizers and nano-pesticides (Khan and Rizvi 2014, Singh et al. 2015, Manjunatha et al. 2016, Chhipa 2017, Duhan et al. 2017, Iavicoli et al. 2017). CuNPs showed efficient antifungal activity against fungi that are resistant to antibiotics. Less work on antifungal effects of CuNPs has been reported than on antibacterial effects. Consequently, there are very few fungicidal agents available in the market containing CuNPs. Moreover, the regular use of similar antimycotic agents may result in the development of resistance against them. Antimycotic agents functionalized with CuNPs have the potential to address emerging challenges, predominantly reducing infections caused by fungi in animals as well as plants.

Kim et al. (2012) demonstrated the potential antifungal activity of AgNPs against *Fusarium, Monosporascus, Botrytis, Corynespora, Didymella, Cladosporium, Pythium,* and other fungi. Likewise, several other NPs have been successfully evaluated against plant pathogenic fungi: *Ag*/chitosan NPs against seed-borne fungi *Aspergillus flavus, Alternaria alternata* and *Rhizoctonia solani* of chickpea seeds (Kaur et al. 2012); zinc oxide NPs against *A. flavus, A. nidulans, Trichoderma harzianum* and *Rhizopus stolonifer* (Gunalan et al. 2012); copper-based NPs (CuO, Cu_2O) against *Phytophthora infestans* (Giannousi et al. 2013); and CuNPs against *F. culmorum, F. oxysporum* and *F. equiseti* (Bramhanwade et al. 2016).

Copper and its compounds were introduced into textiles and solid surfaces for odor and microbial control (Borkow and Gabbay 2004, Borkow and Gabbay 2009, Borkow et al. 2009, Borkow et al. 2010), including for reduction of microbial bio-burden in medical institutions (Borkow and Gabbay 2006, Borkow 2012, Borkow and Monk 2012, Borkow 2014, Lazary et al. 2014, Monk et al. 2014, Schmidt et al. 2014). Cu exerts its toxicity against microorganisms through several parallel mechanisms. These include direct contact killing and damage caused by exposure to released copper ions (Borkow and Gabbay 2005, Grass et al. 2011, Borkow 2012, Santo et al. 2012, Bleichert et al. 2014). The damage is nonspecific and includes damage to the envelope phospholipids or intracellular proteins as well as nucleic acids of microorganism (Kim et al. 2000, Rifkind et al. 2001, Nan et al. 2008, Santo et al. 2012, Bleichert et al.

2014). Cu can exist in several oxidation states, i.e., metallic copper (Cu^0), monovalent copper (cuprous, Cu^+) or divalent copper (cupric, Cu^{2+}) ions. Cu^+ ions may exhibit more cytotoxic effects with respect to bacteria as compared to Cu^{2+} ions (Abicht et al. 2013). It has been reported that cuprous ions demonstrated greater cytotoxicity with respect to fungi. Significantly, redox cycling between Cu^{2+} and Cu^+ can catalyze the production of short-lived hydroxyl radicals, which in turn probably will contribute to superior biocidal activity and thus the combined activity of both cuprous and cupric ions provide superior cytocidal activity.

Apart from these, NPs may be effectively used as pesticides. Liu et al. (2006) developed efficient water-soluble pesticide using porous hollow silica NPs loaded with validamycin (pesticide) for its controlled release. The controlled release behavior of nano-pesticides makes them promising carriers in agriculture. According to Wang et al. (2007), oil in water (nano-emulsions) can be efficiently used as pesticides against various insect pests in agriculture. In addition, metal NPs can also be used as nano-fertilizers for targeted delivery or slow/controlled and conditional release. Zeolites and nano clays are a set of naturally occurring minerals with a honeycomb-like layered crystal structure that can be used as nano-fertilizers (Chinnamuthu and Boopathi 2009). In one study of assimilation of chemical fertilizers in NPs, the authors incorporated NPK fertilizers into chitosan NPs for the deliberate and constant release of NPK fertilizers in adequate amount (Corradini et al. 2010). Rui et al. (2016) demonstrated effective use of iron oxide NPs as iron fertilizer for peanut (*Arachis hypogaea*). Similarly, Liu and Lal (2015) reported that engineered nanomaterials can be used as potential nano-fertilizer and effective alternatives to chemical fertilizers. There is a general recognition that nanotechnology has the potential to move science forward in addition to improving quality of life and producing extensive financial gains, but a number of reports have also suggested that potential toxicity should be considered for safe use of NPs (Bawa et al. 2016). The current knowledge of the toxic effects of NPs is relatively incomplete. A defensive approach is therefore required for evaluation of potential risks to health and environment associated with the use of nanomaterials (Khan and Shanker 2015).

Effects of copper nanomaterials on plants

Because of the extensive use of metal and metal oxide NPs and their entry into the food chain through plants, the growth as well as the yield of important crops gets compromised. Therefore, there is an urgent need to study the effect of NPs on crop plants. In spite of several studies reporting the advantageous effects of NPs on agricultural crops, toxic effects on crops are also reported, which may limit the use of NP applications in agricultural crops. Toxic effects of NPs in plants include decrease in seed germination, growth inhibition, and reduction in yield and quality (Fig. 14.1) (Rizwan et al. 2017).

Effect of nanoparticles on seed germination

Seed germination is the foremost step in plant development and determines the success of crop growth in soils contaminated with metal and metal oxide NPs. Toxic effects of metal NPs on seed germination of food crops have been extensively reported

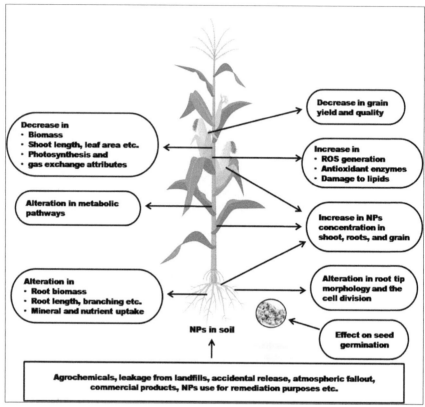

Figure 14.1. Possible sources of nanoparticles (NPs) in the soil and their effects on the growth and physiology of plants. ROS = reactive oxygen species (Source: Rizwan et al. 2017).

(Lin and Xing 2007, Mandeh et al. 2012, Moon et al. 2014, Thiruvengadam et al. 2015). Generally, toxicity of NPs reduced the seed germination rate and dispersed the germination events of many crop species (Wu et al. 2014, Thiruvengadam et al. 2015, Xiang et al. 2015). There are reports on the effects of AgNPs on seed germination of rice, barley (*Hordeum vulgare* L.), turnip (*Brassica rapa* L.) and faba bean (*Vicia faba* L.) (Ma et al. 2011, El-Temsah and Joner 2012, Thiruvengadam et al. 2015). Results demonstrated that AgNPs decreased seed germination of turnip and faba bean in a dose-dependent manner. AgNPs decreased seed germination in barley, by 10–20% over the control. Rice seed germination was decreased by increasing dose and size of AgNPs. Metal oxide NPs such as CuO, NiO, TiO_2, iron oxide (Fe_2O_3) and Co_3O_4 reduced the seed germination percentage in lettuce, radish (*Raphanus sativus* L.) and cucumber by adsorbing and releasing free metal ions on surfaces of seed and near the seeds respectively (Wu et al. 2014). The authors reported that the toxic effects of NPs on seed germination vary with crops (i.e., lettuce > cucumber > radish) and seed size.

Conversely, Yasur and Rani (2013) demonstrated that AgNPs (up to 4000 mg L^{-1}) did not have an effect on the seed germination of castor bean (*Ricinus communis* L.). Only a few reports are available related to seed germination of crop plants in soil (El-Temsah and Joner 2012). For instance, inhibition of seed germination was less manifested in rye grass, flax (*Linum usitatissimum* L.) and barley exposed to zerovalent iron NPs (0–5000 mg L^{-1}) or AgNPs (0–100 mg L^{-1}) in soil compared to water. Furthermore, seed germination varied in different soils, i.e., it was less pronounced in clayey soil than in sandy soil (El-Temsah and Joner 2012). Sorghum (*Sorghum bicolor* L.) seed germination was improved by the reduction of surface area attributes, which led to better soil aggregation as well as sorption of dissolved silver ion with AgNPs (Lee et al. 2012). Likewise, Gruyer et al. (2013) reported that germination of radish and lettuce seeds was less inhibited by AgNPs when applied in soil, as compared to water, in a dose-dependent manner. CuONPs significantly reduced seed germination in cucumber and lowered the germination rate up to 23.3% at 600 mg L^{-1} (Moon et al. 2014). In addition, seed germination of rice was notably reduced under CuONP stress in a dose-dependent approach (Shaw and Hossain 2013).

López-Moreno et al. (2010) demonstrated the effect of cerium oxide (CeO$_2$) NPs at a concentration of 0–4000 mg L^{-1} on seeds of alfalfa (*Medicago sativa* L.), cucumber, maize and tomato. Results showed that 2000 mg L^{-1} CeO$_2$ NPs reduced the seed germination of maize, tomato and cucumber by 30%, 30% and 20% as compared to the control respectively, whereas there was no significant effect on seed germination in alfalfa. In tomato (*Lycopersicum esculentum* Mill), application of SiO$_2$NPs (8 g L^{-1}, size 12 nm) increased seed germination, seed germination index, and mean germination time along with seed vigor index (Siddiqui and Al-Whaibi 2014). Numerous studies reported that NP-mediated inhibition in seed germination might be due to the partial dissolution of NPs and the discharge of toxic metal ions or surface modifications in exposure solution or plant tissues (Ma et al. 2011, Nair and Chung 2014b).

Metal and metal oxide NPs both do affect the seed germination of several food crops (Feizi et al. 2013a, Feizi et al. 2013b, Haghighi and DaSilva 2014). For instance, reports on the effect of TiO$_2$ NPs on seed germination of fennel (*Foeniculum vulgare* L.), maize, onion (*Allium cepa* L.), parsley (*Petroselinum crispum* L.), radish, rapeseed, tomato and wheat are documented (Feizi et al. 2012, Feizi et al. 2013a, Song et al. 2013a, Song et al. 2013b, Haghighi and DaSilva 2014, Dehkourdi and Mosavi 2013, Kim et al. 2015). Haghighi and DaSilva (2014) reported 100% seed germination in tomato and onion at 100 mg L^{-1} and in radish at 400 mg L^{-1} concentration of TiO$_2$ NPs.

Zhang et al. (2015) have reported that ZnONPs did not affect the seed germination of maize and cucumber. Boonyanitipong et al. (2011) observed that ZnO and TiO$_2$ NPs (100–1000 ppm and seed soaking for 1–3 days) did not affect seed germination in rice. Alternatively, the frequency of seed germination decreased in Chinese cabbage (*Brassica pekinensis* L.), maize and garden cress (*Lepidium sativum* L.) with increasing ZnONP concentration (Lin and Xing 2007, Jósko and Oleszczuk 2013, Xiang et al. 2015).

In a nutshell, metal and metal oxide NPs reduced or enhanced seed germination in numerous crop plants; plant responses varied considerably among NPs and were partially interrelated to dose and the size of the NPs. Although NPs positively or negatively affected seed germination of tested plants, the mechanism behind the germination is still poorly understood.

Effect of nanoparticles on plant growth and morphology

The morphological parameters of plant, i.e., leaf area, shoot and root length, as well as shoot and root weight are the indicators of plant health. The effects of metal-based NPs such as ZnO, Fe_2O_3, aluminum dioxide (Al_2O_3), and CuO on shoot or root growth and elongation have been reported for barley, maize, tomato, rice and wheat. The NP toxicity might be due to the enhanced release of metal ions from NPs (Dimkpa et al. 2012, Mahmoodzadeh et al. 2013a, Nair and Chung 2014a). It is worth mentioning that AgNPs inhibited the root development and biomass of several crops such as wheat, rice, sorghum and tomato (Mazumdar and Ahmed 2011, Dimkpa et al. 2012, Lee et al. 2012, Song et al. 2013b, Vannini et al. 2014, Nair and Chung 2015). The root cell wall and vacuoles of rice were damaged when plants were exposed for 12 days to AgNPs of size 25 nm with concentration of 1000 ppm, which might be due to the penetration of large NPs through the small pores of the cell wall (Mazumdar and Ahmed 2011). Comparable results in rice seedlings exposed to AgNPs were reported (Mirzajani et al. 2013). Rice seedling exposed to AgNPs (0, 0.2, 0.5, and 1 mg L^{-1} for one week) showed noteworthy reductions in root elongation as well as shoot and root fresh weights (Nair and Chung 2014c). In wheat, AgNPs reduced the growth in a dose-dependent manner in distinct cultures (Dimkpa et al. 2012, Vannini et al. 2014). AgNP-mediated dose- and size-dependent decreases were observed in shoot and root, and fresh and dry weight of rice. It has been reported that lower AgNP treatments, up to 30 mg L^{-1}, accelerated the rice root growth, though higher concentration, at or above 60 mg L^{-1}, reduced the root growth and caused root loss. Shoot growth was more vulnerable to NP stress (Mirzajani et al. 2013, Thuesombat et al. 2014).

Shoot height as well as root biomass of cotton seedlings linearly declined with increasing concentration of CeO_2 and SiO_2 NPs in growth media (Le et al. 2014, Li et al. 2014). Earlier studies reported both the positive and negative effects of CeO_2 NPs on crop growth and development (Wang et al. 2012, Wang et al. 2013, Zhang et al. 2015). In tomatoes, the application of CeO_2NPs (0.1–10 mg L^{-1} concentration) slightly enhanced the plant height along with biomass (Wang et al. 2012). In addition, CeO_2NPs (10 mg L^{-1}) affected the second-generation tomato seedlings, which were smaller and weaker with lower biomass plus water transpiration as compared to control (Wang et al. 2013). Alternatively, the application of SiO_2 NPs at 5–20 kg ha^{-1} concentration in sandy loam soil improved shoot as well as root length, and stem height along with leaf area of 20-day-old seedlings of maize (Suriyaprabha et al. 2012). Optimistic effects of silicon sustenance on plant growth have been extensively reported, particularly under stressful circumstances in assorted growth medium (Rizwan et al. 2012, Adrees et al. 2015, Keller et al. 2015).

The toxic effect of NPs on plant growth depends on the size, shape, concentration and duration of NP exposure in the growth medium (Lee et al. 2008, Pokhrel and Dubey 2013, Song et al. 2013a, Thuesombat et al. 2014). For example, Antisari et al. (2015) reported that TiO_2 NPs (size ~ 20 nm) at lower concentrations, i.e., 10 and 100 ppm, considerably increased the fresh weight of shoots and roots in wheat, while higher TiO_2 NP concentration (> 100 ppm) decreased fresh weight in a dose-dependent manner. In another instance, TiO_2 NPs (1000–5000 mg L^{-1}) did not show any significant change in the biomass of tomato (Song et al. 2013a).

Several studies revealed the toxic effects of ZnONPs on the growth of various crops grown in diverse ecological conditions (Mahajan et al. 2011, Kim et al. 2012, Pavani et al. 2014, Mousavi Kouhi et al. 2015). In alfalfa, the treatment of ZnONPs at 500 and 750 mg kg^{-1} of soil under pot experiment reduced the root and shoot biomass of seedlings by 80% and 25%, respectively (Bandyopadhyay et al. 2015, Mousavi Kouhi et al. 2015). The anatomical and ultrastructural modifications in the roots and leaves of rapeseed treated with ZnONPs at 100 mg L^{-1} concentration for 2 months resulted in the decreased diameter of root tip along with the size of epidermal and pericycle cells. These NPs also diminished the size of mitochondria and plastoglobuli and the number of chloroplasts in the leaves, while increasing the size of starch grains and the number of plastoglobuli. All these modifications were recorded compared to control, which showed reduced crop growth and biomass. Pavani et al. (2014) reported that ZnONPs increased the shoot and root length and fresh as well as dry weight of chickpea. Kim et al. (2012) observed that the treatment of CuO and ZnO NPs at 1000 ppm concentration for 5 days on cucumber seedlings resulted in biomass decrease by 75% and 35% respectively. ZnONPs at a concentration of 50 or 500 mg kg^{-1} of soil reduced the root and shoot length, root surface area and volume of soybean over the control (Yoon et al. 2014). Dhoke et al. (2013) studied the effect of ZnO (20 ppm) and FeO (50 ppm) NPs by foliar application, which enhanced the growth of mung bean (*Vigna radiata* L.). Additionally, ZnONPs at concentrations 100–1000 mg L^{-1} reduced the root length and number of roots of rice plants. TiO$_2$ NPs at similar concentrations had no effect on root length (Boonyanitipong et al. 2011).

Information related to the toxicity of NPs in crop plants under field conditions is very rare (Du et al. 2011, Ngo et al. 2014). It has been reported that TiO$_2$ and ZnO NPs shrank the growth of wheat (Du et al. 2011), whereas application of zero-valent Cu, Co and Fe augmented the growth of soybean under field conditions (Ngo et al. 2014). Even though NPs have positive as well as negative effects on plant growth and morphology, the responses were assorted with applied dose, plant species, duration of exposure and experimental conditions. Moreover, most of the studies were performed under controlled conditions with short-term exposure to NPs and long-term studies are still lacking in this regard.

Effect of nanoparticles on grain or fruit yield and quality

Toxicity of NPs was mainly studied with respect to the grain or fruit yield, as well as the nutritional quality of many food crops. The effect of CeO$_2$NPs on yield and quality of crops varies with plants and cultivars (An et al. 2008, Peralta-Videa et al. 2014, Rico et al. 2015). For instance, application of CeO$_2$ NPs, 500 ppm in soil, increased yield, spike length, the number of spikelets per spike as well as the number of grains per spike of wheat. An et al. (2008) demonstrated amplification in grain yield wheat by 36.6% when the 500 ppm CeO$_2$ NPs were directly applied in the soil. In another study, in barley, authors observed contrary results with the application of the same NPs under similar conditions. When barley seedlings were exposed to 500 ppm CeO$_2$ NPs, seed setting was affected (Rico et al. 2015). In rice grains, CeO$_2$ NP treatment decreases Fe, sulfur (S), lauric and valeric acids, prolamin, and glutelin along with the starch content (Rico et al. 2013). CeO$_2$ NPs (800 mg kg^{-1}) reduced cucumber yield by 31.6% (Zhao et al. 2013). Additionally, Zhao et al. (2014) reported that CeO$_2$ and ZnO NPs

(400 and 800 mg kg^{-1} of soil) distorted the quality of proteins, carbohydrates and mineral nutrients in cucumber. Moreover, Zhao et al. (2015) reported that yield of maize was reduced by 38% and 49% with the application of CeO$_2$ and ZnO NPs, respectively, and these NPs also changed the quality of corn by varying the mineral elements in cobs as well as kernels. Comparable results were observed in soybean plants exposed to CeO$_2$ and ZnO NPs in soil (Peralta-Videa et al. 2014). CeO$_2$ NPs reduced the molybdenum (Mo) concentration in cucumber fruits, changed non-reducing sugars, and also affected fractionation of proteins and phenolic content (Zhao et al. 2014). TiO$_2$ NPs reduced potassium (K) and phosphorus (P) content in cucumber fruits (Servin et al. 2012). In tomato fruits, AgNPs increased the content of K, while reducing that of magnesium (Mg), P and S (Antisari et al. 2015).

In brief, grain or fruit yield, as well as quality, was considerably affected when plants were exposed to different types of NPs in alternative time- and dose-dependent manner. Nevertheless, the data overall regarding the toxic effects of NPs on grain yield, quality and development are insufficient; more exhaustive study is required concerning the effects of NPs on grain or fruit development and quality of several crop plants. In these studies, precise endpoints need to be incorporated and associated with the grain yield parameters, number, and mass of fruits or grains and nutritional contents, for example, proteins, carbohydrates, amino acids, and metals. Long-term experimental conditions might effectively assess the grain or fruit quality and nutritional status.

Conclusions and future perspectives

It was stated that copper nanomaterials have drawn tremendous attraction because of their valuable properties such as optical, electrical, medical, sensor and catalytic applications and also because they are more economical than silver and gold. The use of microorganisms in the synthesis of copper nanomaterials is eco-friendly, cheaper, and easier to use than chemical or physical synthesis; they offer a superior range of potential methods of novel routes to synthesize the CuNPs. With both direct and indirect input into agricultural ecosystems, copper nanomaterials can enter soil and plant systems and remain active. Because of the unique physical and chemical properties of CuNPs, a full understanding and prediction of their environmental behavior remain a challenge to the scientific community. The presence of soil organic matter, plants, soil colloids, and microorganisms further complicates analysis of the fate and transport of these nanomaterials in the soil matrix. The critical factors that determine the mobility of metal NPs comprise the variety of core metals of NPs and their speciation, surface coating, size, surface charge, and chemical and physical condition of the environmental medium (e.g., cation condition, pH, and organic acid). Metal NPs are able to accumulate in plant roots and are further translocated to different plant parts. The characteristics of metal NPs, plants and their interactions all play significant functions in NP translocation.

Metal NPs such as copper nanomaterials may have positive as well as negative impacts on the plant system. Since toxicity occurs at all physiological levels, to biochemical and molecular, the advantages include accelerating growth, photosynthetic yield, pest control, and improved fruit yields. Nevertheless, the mechanism of CuNP phytotoxicity and translocation pathways inside plants are still chiefly unidentified.

The interaction of metal NPs with soil, plant root and soil microbiome in root rhizosphere should also be emphasized in future studies to truly reflect the effect of metal-based NPs in natural environmental conditions. The fate and transfer of metal NPs are established to be extremely sensitive towards elements of the environment, for example, solution pH, redox state, ionic strength, and organic matter present. Moreover, these NPs will probable undergo various chemical, physical, or biological transformations concurrently (e.g., aggregation and oxidation). Most of the laboratory experiments are too simplified to represent agricultural environments. Thus, research on metal-based NPs in environmentally significant deliberations and conditions is urgently needed.

Field-based life studies to better understand the effect of metal NPs in agriculture system are a burning need. Nanobiotechnology has been considered as a key constituent of sustainable development, though the promise of nanobiotechnology can be attained only if exposure and toxicity can be completely evaluated and appropriately administered. Understanding the fate, transport, and toxicity of metal NPs is critical for environmental risk assessment. Additional information on the interaction of NPs with soil, plant and soil microbiome in the agriculture system is needed for better analysis and prediction of ecotoxicity of metal NPs such as copper nanomaterials.

References

Abicht, H.K., Gonskikh, Y., Gerber, S.D., Solioz, M. 2013. Nonenzymic copper reduction by menaquinone enhances copper toxicity in *Lactococcus lactis* IL1403. Microbiol. 159: 1190–97.

Adrees, M., Ali, S., Rizwan, M., Rehman, M.Z., Ibrahim, M., Abbas, F., Farid, M., Qayyum, M.F., Irshad, M.K. 2015. Mechanisms of silicon-mediated alleviation of heavy metal toxicity in plants: a review. Ecotoxicol. Environ. Saf. 119: 186–97.

Agarwal, H., Venkat Kumar, S., Rajeshkumar, S. 2017. A review on green synthesis of zinc oxide nanoparticles—An eco-friendly approach. Res. Eff. Technol. 3(4): 406–13.

Ahmed, S., Ahmad, M., Swami, B.L., Ikram, S. 2016. A review on plants extract mediated synthesis of silver nanoparticles for antimicrobial applications: A green expertise. J. Adv. Res. 7: 17–28.

Al-Halafi, A.M. 2014. Nanocarriers of nanotechnology in retinal diseases. Saudi J. Ophthalmol. 28: 204–309.

Amendola, V., Polizzi, S., Meneghetti, M. 2006. Laser ablation synthesis of gold nanoparticles in organic solvents. J. Phys. Chem. B. 110: 7232–7237.

Amendola, V., Meneghetti, M. 2009. Laser ablation synthesis in solution and size manipulation of noble metal nanoparticles. Phys. Chem. Chem. Phys. 11(20): 3805–3821. Doi: 10.1039/b900654k.

An, J., Zhang, M., Wang, S., Tang, J. 2008. Physical, chemical and microbiological changes in stored green asparagus spears as affected by coating of silver nanoparticles-PVP. LWT-Food Sci. Technol. 41: 1100–07.

Anasane, N., Golińska, P., Wypij, M., Rathod, D., Dahm, H., Rai, M. 2016. Acidophilic actinobacteria synthesised silver nanoparticles showed remarkable activity against fungi-causing superficial mycoses in humans. Mycoses. 59(3): 157–66.

Angrasan, J.K.V.M, Subbaiya, R. 2014. Biosynthesis of copper nanoparticles by *Vitis vinifera* leaf aqueous extract and its antibacterial activity. Int. J. Curr. Microbiol. Appl. Sci. 3(9): 768–74.

Antisari, L.V., Carbone, S., Gatti, A., Vianello, G., Nannipieri, P. 2015. Uptake and translocation of metals and nutrients in tomato grown in soil polluted with metal oxide (CeO_2, Fe_3O_4, SnO_2, TiO_2) or metallic (Ag, Co Ni) engineered nanoparticles. Environ. Sci. Pollut. Res. 22(3): 1841–53.

Apte, M., Girme, G., Bankar, A., Ravikumar, A., Zinjarde, S. 2013. 3, 4-dihydroxy-L-phenylalanine-derived melanin from Yarrowialipolytica mediates the synthesis of silver and gold nanostructures. J. Nanobiotechnol. 11: 2.

Aslam, M., Gopakumar, G., Shoba, T.L., Mulla, I.S., Vijayamohanan, K., Kulkarni, S.K., Urban, J., Vogel, W. 2002. Formation of Cu and Cu$_2$O nanoparticles by variation of the surface ligand: preparation, structure, and insulating-to-metallic transition. Colloid Interface Sci. 205(1): 79.

Atha, D.H., Wang, H., Petersen, E.J., Cleveland, D., Holbrook, R.D., Jaruga, P., Dizdaroglu, M., Xing, B., Nelson, B.C. 2012. Copper oxide nanoparticle mediated DNA damage in terrestrial plant models. Environ. Sci. Technol. 46: 1819–27.

Athawale, A.A., Katre, P.P., Kumar, M., Majumdar M.B. 2005. Synthesis of CTAB–IPA reduced copper nanoparticles. Mater. Chem. Phys. 91(2-3): 507–12.

Auffan, M., Rose, J., Bottero, J.Y., Lowry, G.V., Jolivet, J.P., Wiesner, M.R. 2009. Towards a definition of inorganic nanoparticles from an environmental, health and safety perspective. Nat. Nanotechnol. 4: 634–41.

Awwad, A.M., Salem, N.M., Abdeen, A.O. 2012. Biosynthesis of silver nanoparticles using *Olea europaea* leaves extract and its antibacterial activity. Nanosci. Nanotechnol. 2(6): 164–70.

Bali, R., Razak, N., Lumb, A., Harris, A.T. 2006. The synthesis of metallic nanoparticles inside live plants. IEEE Xplore (2006), DOI:10.1109/ICONN.2006.340592.

Bandyopadhyay, S., Plascencia-Villa, G., Mukherjee, A., Rico, C.M., José-Yacamán, M., Peralta-Videa, J.R., Gardea-Torresdey, J.L. 2015. Comparative phytotoxicity of ZnO NPs, bulk ZnO, and ionic zinc onto the alfalfa plants symbiotically associated with *Sinorhizobium meliloti* in soil. Sci. Total Environ. 515-516: 60–69.

Bansal, V., Rautaray, D., Bharde, A., Ahire, K., Sanyal, A., Ahmad, A., Sastry M. 2005. Fungus-mediated biosynthesis of silica and titania particles. J. Mater. Chem. 15: 2583–89.

Barrabes, N., Just, J., Dafinov, A., Medina, F., Fierro, J.L.G., Sueiras, J.E., Salagre, P., Cesteros, Y. 2006. Catalytic reduction of nitrate on Pt-Cu and Pd-Cu on active carbon using continuous reactor: The effect of copper nanoparticles. Appl. Catal. B: Environ. 62(1-2): 77–85.

Barreto, J., O'Malley, W., Kubeil, M., Graham, B., Stephan, H., Spiccia, L. 2011. Nanomaterials: applications in cancer imaging and therapy. Adv. Mater. 23: H18–40.

Bawa, R., Bawa, S.R., Mehra, R.N. 2016. The translational challenge in medicine at the nanoscale. pp. 1291–1346. *In*: Bawa, R., Audette, G.F., Reese, B.E. (eds.). Handbook of Clinical Nanomedicine: Law, Business, Regulation, Safety, and Risk. 978-981-4669-23-8 (eBook).

Bawaskar, M., Gaikwad, S., Ingle, A., Rathod, D., Gade, A., Duran, N., Marcato, D., Rai, M. 2010. A new report on mycosynthesis of silver nanoparticles by *Fusarium culmorum*. Curr. Nanosci. 6: 376–80.

Benjamin, J.S. 1989. In new materials by mechanical alloying techniques. pp. 1–18. *In*: Arzt, E., Schultz, L. (eds.). DGM Information Gesellschaft, Oberursel, Germany.

Bhattacharyya, A., Bhaumik, A., Rani, P.U., Mandal, S., Epidi, T.T. 2010. Nanoparticles—A recent approach to insect pest control. African J. Biotechnol. 9: 3489–93.

Bhople, S., Gaikwad, S., Deshmukh, S., Bonde, S., Gade, A., Sen, S., Brezinska, A., Dahm, H., Rai, M. 2016. Myxobacteria-mediated synthesis of silver nanoparticles and their impregnation in wrapping paper used for enhancing shelf life of apples. IET Nanobiotechnol. 10(6): 389–94.

Bilian, X. 2002. Intrauterine devices. Best. Pract. Res. Clin. Obstet. Gynaecol. 16: 155–68.

Birla, S.S., Gaikwad, S.C., Gade, A.K., Rai, M.K. 2013. Rapid synthesis of silver nanoparticles from *Fusarium oxysporum* by optimizing physicocultural conditions. Sci. World J. 2013: 796018.

Birla, S.S., Tiwari, V.V, Gade, A.K., Ingle, A.P., Yadav, A.P., Rai, M.K. 2009. Fabrication of silver nanoparticles by *Phoma glomerata* and its combined effect against *Escherichia coli*, *Pseudomonas aeruginosa* and *Staphylococcus aureus*. Lett. Appl. Microbiol. 48: 173–79.

Bleichert, P., Espirito, S.C., Hanczaruk, M., Meyer, H., Grass, G. 2014. Inactivation of bacterial and viral biothreat agents on metallic copper surfaces. Biometals. 27: 1179–89.

Boonyanitipong, P., Kositsup, B., Kumar, P., Baruah, S., Dutta, J. 2011. Toxicity of ZnO and TiO$_2$ nanoparticles on germinating rice seed. Int. J. Biosci. Biochem. Bioinform. 1: 282–85.

Borkow, G., Gabbay, J. 2005. Copper as a biocidal tool. Curr. Med. Chem. 12: 2163–75.

Borkow, G., Gabbay, J. 2006. Endowing textiles with permanent potent biocidal properties by impregnating them with copper oxide. JTATM 2006: 5.

Borkow, G., Sidwell, R.W., Smee, D.F., Barnard, D.L., Morrey, J.D., Lara-Villegas, H.H., Shemer-Avni, Y., Gabbay, J. 2007. Neutralizing viruses in suspensions by copper oxide based filters. Antimicrob. Agents Chemother. 51: 2605–07.

Borkow, G., Gabbay, J. 2009. An ancient remedy returning to fight microbial, fungal and viral infections. Curr. Chem. Biol. 3: 272–78.

Borkow, G., Zatcoff, R.C., Gabbay, J. 2009. Reducing the risk of skin pathologies in diabetics by using copper impregnated socks. Med. Hypotheses. 73: 883–86.

Borkow, G., Zhou, S.S., Page, T., Gabbay, J. 2010. A novel anti-influenza copper oxide containing respiratory face mask. PLoS One. 5: e11295.

Borkow, G. 2012. Using copper to fight microorganisms. Curr. Chem. Biol. 6: 93–103.

Borkow, G., Monk, A.B. 2012. Fighting nosocomial infections with biocidal non-intrusive hard and soft surfaces. World J. Clin. Infect. Dis. 12: 77–90.

Borkow, G. 2014. Biocidal hard and soft surfaces containing copper oxide particles for the reduction of health-care acquired pathogens. In book: Use of biocidal surfaces for reduction of healthcare acquired infections (Ed. Borkow, G.). Springer International Publishing. Switzerland 85–102.

Bramhanwade, K., Shende, S., Bonde, S., Gade, A., Rai, M. 2016. Fungicidal activity of Cu nanoparticles against *Fusarium* causing crop diseases. Environ. Chem. Lett. 14: 229–35.

Caroling, G., Vinodhini, E., Ranjitham, A.M., Shanthi, P. 2015a. Biosynthesis of copper nanoparticles using aqueous *Phyllanthus embilica* (Gooseberry) extract-characterization and study of antimicrobial effects. Int. J. Nano. Chem. 1(2): 53–63.

Caroling, G., Priyadharshini, M.N., Vinodhini, E., Ranjitham, A.M., Shanthi, P. 2015b. Biosynthesis of copper nanoparticles using aqueous Guava extract–characterisation and study of antibacterial effects. Int. J. Pharm. Bio. Sci. 5(2): 25–43.

Castellano, J., Shafii, S., Ko, F., Donate, G., Wright, T., Mannari, R., Payne, W., Smith, D.J., Al., E. 2007. Comparative evaluation of silver containing antimicrobial dressings and drugs. Int. Wound. 4: 14–22.

Chandrasekharan, N., Kamat, P.V. 2000. Improving the photoelectrochemical performance of nanostructured TiO_2 films by adsorption of gold nanoparticles. J. Phys. Chem. B. 104: 10851–57.

Chatterjee, A.K., Sarkar, R.K., Chattopadhyay, A.P., Aich, P., Chakraborty, R., Basu, T. 2012. A simple robust method for synthesis of metallic copper nanoparticles of high antibacterial potency against *E. coli*. Nanotechnol. 23(8): 085103.

Chen, H., Lee, J.H., Kim, Y.H., Shin, D.W., Park, S.C., Meng, X., Yoo, J.B. 2010. Metallic copper nanostructures synthesized by a facile hydrothermal method. J. Nanosci. Nanotechnol. 10(1): 629–36.

Chen, L., Zhang, D., Zhou, H., Wan, H. 2006. The use of CTAB to control the size of copper nanoparticles and the concentration of alkylthiols on their surfaces. Mater. Sci. Eng.: A 415(1-2): 156–61.

Chen, X., Wu, G., Chen, J., Xie, Z., Wang, X. 2011. Synthesis of "clean" and well-dispersive Pd nanoparticles with excellent electrocatalytic property on graphene oxide. J. American Chem. Soc. 133: 3693–95.

Chhipa, H. 2017. Nanofertilizers and nanopesticides for agriculture. Environ. Chem. Lett. 15: 15–22.

Chinnamuthu, C.R., Boopathi, P.M. 2009. Nanotechnology and agroecosystem. Madras Agric. J. 96(1-6): 17–31.

Christopher, J.S.G., Banerjee, S., Panneerselvam, E. 2015. Optimization of parameters for biosynthesis of silver nanoparticles using leaf extract of *Aegle marmelos*. Braz. Arch. Biol. Technol. 58(5): 702–10.

Chung, I.-M., Park, I., Seung-Hyun, K., Thiruvengadam, M., Rajakumar, G. 2016. Plant-mediated synthesis of silver nanoparticles: their characteristic properties and therapeutic applications. Nanoscale Res. Lett. 11: 40.

Cioffi, N., Colaianni, L., Ieva, E., Pilolli, R., Ditaranto, N., Angione, M.D., Cotrone, S., Buchholt, K., Spetz, A.L., Sabbatini, L., Torsi, L. 2011. Electrosynthesis and characterization of gold nanoparticles for electronic capacitance sensing of pollutants. Electrochimica Acta. (56)10: 3713–20. http://dx.doi.org/ 10.1016/j.electacta. 2010.12.105.

Cooney, T.E. 1995. Bactericidal activity of copper and noncopper paints. Infect. Control Hosp. Epidemiol. 16: 444–50.

Copper Development Association 2003. Copper Development Association Inc. https://www.copper. org/ (Accessed on 20.11.2019).

Copper Development Association 2019. Copper Development Association Inc. https://www.copper.org/ (Accessed on 20.11.2019).

Corradini, E., Moura, M.R., Mattoso, L.H.C. 2010. A preliminary study of the incorporation of NPK fertilizer into chitosan nanoparticles express. Polymer Lett. 4: 509–15.

Cui, D., Zhang, P., Ma, Y., He, X., Li, Y., Zhang, J., Zhao, Y., Zhang, Z. 2014. Effect of cerium oxide nanoparticles on *Asparagus lettuce* cultured in an agar medium. Environ. Sci. Nano. 1: 459–65.

Dadgostar, N., Ferdous, S., Henneke, D.E. 2010. Colloidal synthesis of copper nanoparticles in a two-phase liquid-liquid system. Mater. Lett. 64(1): 45–8.

Dahm, H., Brzezińska, A.J., Wrótniak-Drzewiecka, W., Golińska, P., Różycki, H., Rai, M. 2015. Myxobacteria as a potential biocontrol agent effective against pathogenic fungi of economically important forest trees. Dendrobiol. 74: 13–4.

Dang, T.D., Le, T.T.T., Fribourg-Blanc, E., Dang, M.C. 2011. Synthesis and optical properties of copper nanoparticles prepared by a chemical reduction method. Adv. Nat. Sci. Nanosci. Nanotechnol. 2(015009): 1–6.

Das, R., Das, B.K., Shyam, A. 2012. Synthesis and characterization of copper nanoparticles by using the exploding wire method. J. Korean Phys. Soc. 61(5): 710–12.

Dash, P.K., Balto, Y.J. 2011. Generation of Nano-copper particles through wire explosion method and its characterization. Res. J. Nanosci. Nanotechnol. 1: 25–33.

Dauthal, P., Mukhopadhyay, M. 2016. Noble metal nanoparticles: plant-mediated synthesis, mechanistic aspects of synthesis, and applications. Ind. Eng. Chem. Res. 55: 9557–77.

Deepa, S., Kanimozhi, K., Panneerselvam, A. 2013. Antimicrobial activity of extracellularly synthesized silver nanoparticles from marine derived actinomycetes. Int. J. Curr. Microbiol. App. Sci. 2: 223–30.

Dehkourdi, E.H., Mosavi, M. 2013. Effect of anatase nanoparticles (TiO_2) on parsley seed germination (*Petroselinum crispum*) *in vitro*. Biol. Trace. Elem. Res. 155: 283–86.

Deshmukh, S.D., Deshmukh, S.D., Gade, A.K., Rai, M. 2012. *Pseudomonas aeruginosa* mediated synthesis of silver nanoparticles having significant antimycotic potential against plant pathogenic fungi. J. Bionanosci. 6: 90–4.

Dhoke, S.K., Mahajan, P., Kamble, R., Khanna, A. 2013. Effect of nanoparticles suspension on the growth of mung (*Vigna radiata*) seedlings by foliar spray method. Nanotechnol. Dev. 3(3): e1.

Dimkpa, C.O., Hansen, T., Stewart, J., McLean, J.E., Britt, D.W., Anderson, A.J. 2015. ZnO nanoparticles and root colonization by a beneficial pseudomonad influence essential metal responses in bean (*Phaseolus vulgaris*). Nanotoxicol. 9(3): 271–78.

Dimkpa, C.O., McLean, J.E., Latta, D.E., Manangon, E., Britt, D.W., Johnson, W.P., Boyanov, M.I., Anderson, A.J. 2012. CuO and ZnO nanoparticles: phytotoxicity, metal speciation, and induction of oxidative stress in sand-grown wheat. J. Nanopart. Res. 14: 1125.

Ding, J., Tsuzuki, T., McCormik, P.G., Street, R. 1996. Ultrafine Cu particles prepared by mechanochemical process. J. Alloys Compd. 234(2): L1–L3.

Dollwet, H.H.A., Sorenson, J.R.J. 2001. Historic uses of copper compounds in medicine. Trace Element. Med. 2: 80–7.

Du, W., Sun, Y., Ji, R., Zhu, J., Wu, J., Guo, H. 2011. TiO_2 and ZnO nanoparticles negatively affect wheat growth and soil enzyme activities in agricultural soil. J. Environ. Monit. 13: 822–28.

Duhan, J.S., Kumar, R., Kumar, N., Kaur, P., Nehra, K., Duhan, S. 2017. Nanotechnology: The new perspective in precision agriculture. Biotechnol. Rep. 15: 11–23.

Durán, N., Marcato, P.D. 2013. Nanobiotechnology perspectives. Role of nanotechnology in the food industry: A review. Int. J. Food Sci. Technol. 48: 1127–34.

Durán, N., Marcato, P.D., De Souza, G.I.H., Alves, O.L., Esposito, E. 2007. Antibacterial effect of silver nanoparticles produced by fungal process on textile fabrics and their effluent treatment. J. Biomed. Nanotechnol. 3: 203–08.

Elemike, E.E., Fayemi, O.E., Ekennia, A.C., Onwudiwe, D.C., Ebenso, E.E. 2017. Silver nanoparticles mediated by *Costusafer* leaf extract: synthesis, antibacterial, antioxidant and electrochemical properties. Molecules. 22: 701.

El-Sheekh, M.M., El-Kassas, H.Y. 2016. Algal production of nano-silver and gold: Their antimicrobial and cytotoxic activities: A review. J. Gene. Engin. Biotechnol. 14: 299–310.

El-Temsah, Y.S., Joner, E.J. 2012. Impact of Fe and Ag nanoparticles on seed germination and differences in bioavailability during exposure in aqueous suspension and soil. Environ. Toxicol. 27: 42–49.

Englert, B.C. 2007. Nanomaterials and the environment: uses, methods synthesized and measurement. J. Environ. Monit. 9: 1154–61.

Espirito, S.C., Taudte, N., Nies, D.H., Grass, G. 2008. Contribution of copper ion resistance to survival of *Escherichia coli* on metallic copper surfaces. Appl. Environ. Microbiol. 74: 977–86.

FAO/WHO. 2010. [Food and Agriculture Organization of the United Nations/World Health Organization]: FAO/WHO Expert meeting on the application of nanotechnologies in the food and agriculture sectors: potential food safety implications. Rome: Meeting report. (2010). (Accessed on 21/10/2019).

Feizi, H., Moghaddam, P.R., Shahtahmassebi, N., Fotovat, A. 2012. Impact of bulk and nanosized titanium dioxide (TiO_2) on wheat seed germination and seedling growth. Biol. Trace Elem. Res. 146: 101–6.

Feizi, H., Kamali, M., Jafari, L., Moghaddam, P.R. 2013a. Phytotoxicity and stimulatory impacts of nanosized and bulk titanium dioxide on fennel (*Foeniculum vulgare* Mill). Chemosphere. 91: 506–11.

Feizi, H., Moghaddam, P.R., Shahtahmassebi, N., Fotovat, A. 2013b. Assessment of concentrations of nano and bulk iron oxide particles on early growth of wheat (*Triticum aestivum* L.). Ann. Rev. Res. Biol. 3: 752–61.

Ferracane, J.L. 2001. Materials in Dentistry: Principles and Applications. Lippincott Williams & Wilkins.

Foltête, A.S., Masfaraud, J.F., Bigorgne, E., Nahmani, J., Chaurand, P., Botta, C., Labille, J., Rose, J., Férard, J.F., Cotelle, S. 2011. Environmental impact of sunscreen nanomaterials: ecotoxicity and genotoxicity of altered TiO_2 nanocomposites on *Vicia faba*. Environ. Pollut. 159: 2515–22.

Gade, A., Bonde, P., Ingle, A., Marcato, P., Duran, N., Rai, M. 2008. Exploitation of *Aspergillus niger* for synthesis of silver nanoparticles. J. Biobased Mater. Bioenergy. 2: 1–5.

Gade, A., Gaikwad, S., Duran, N., Rai, M. 2014. Green synthesis of silver nanoparticles by *Phoma glomerata*. Micron. 59: 52–9.

Gaffet, E., Harmelin, M., Faudot, F. 1993. Far-from-equilibrium phase transition induced by mechanical alloying in the Cu-Fe system. J. Alloys Compd. 194(1): 23–30.

Gaikwad, S.C., Birla, S.S., Ingle, A.P., Gade, A.K., Marcato, P.D., Rai, M., Duran, N. 2013. Screening of different *Fusarium* species to select potential species for the synthesis of silver nanoparticles. J. Brazilian Chem. Society. 24(12): 1974–82. Doi: 10.5935/0103-5053.20130247.

Gajbhiye, M., Kesharwani, J., Ingle, A., Gade, A., Rai, M. 2009. Fungus-mediated synthesis of silver nanoparticles and their activity against pathogenic fungi in combination with fluconazole. Nanomed: Nanotechnol. Biol. Med. 5: 382–86.

Garcia-Parajo, M.F. 2012. The role of nanophotonics in regenerative medicine. Methods Mol. Biol. 811: 267–84.

Ge, Y., Priester, J.H., Van De Werfhorst, L.C., Walker, S.L., Nisbet, R.M., An, Y.-J., Schimel J.P., Gardea-Torresdey, J.L., Holden, P.A. 2014. Soybean plants modify metal oxide nanoparticle effects on soil bacterial communities. Environ. Sci. Technol. 48: 13489–96.

Ghorbani, H.R., Mehr, F.P., Poor, A.K. 2015. Extracellular synthesis of copper nanoparticles using culture supernatants of *Salmonella typhimurium*. Orient J. Chem. 31(1): 527–29.

Giannousi, K., Avramidis, I., Dendrinou-Samara, C. 2013. Synthesis, characterization and evaluation of copper based nanoparticles as agrochemicals against *Phytophthora infestans*. RSC Adv. 3: 21743–52.

Giuffrida, S., Costanzo, L.L., Ventimiglia, G., Bongiorno, C. 2008. Photochemical synthesis of copper nanoparticles incorporated in poly(vinyl pyrrolidone). J. Nanopart. Res. 10(7): 1183–92.

Goia, D.V., Matijevic, E. 1998. Preparation of monodispersed metal particles. New J. Chem. 22: 1203–15.

Golińska, P., Wypij, M., Agarkar, G., Rathod, D., Dahm, H., Rai, M. 2015. Endophytic actinobacteria of medicinal plants: diversity and bioactivity. Antonie van Leeu. 108(2): 267–89.

Golińska, P., Wypij, M., Rathod, D., Tikar, S., Dahm, H., Rai, M. 2016. Synthesis of silver nanoparticles from two acidophilic strains of Pilimeliacolumellifera subsp. pallida and their antibacterial activities. J. Basic Microbiol. 56(5): 541–56.

Gopinath, M., Subbaiya, S., Selvam, M.M., Suresh, D. 2014. Synthesis of copper nanoparticles from *Nerium oleander* leaf aqueous extract and its antibacterial activity. Int. J. Curr. Microbiol. Appl. Sci. 3(9): 814–18.

Gorter, R.W., Butorac, M., Cobian, E.P. 2004. Examination of the cutaneous absorption of copper after the use of copper-containing ointments. Am. J. Ther. 11: 453–58.

Gotoh, Y., Igarashi, R., Ohkoshi, Y., Nagura, M., Akamatsu, K., Deki, S. 2000. Preparation and structure of copper nanoparticle/poly(acrylic acid) composite films. J. Mater. Chem. 10(11): 2548–52.

Grass, G., Rensing, C., Solioz, M. 2011. Metallic copper as an antimicrobial surface. Appl. Environ. Microbiol. 77: 1541–47.

Grieger, K.D., Hansen, S.F., Baun, A. 2009. The known unknowns of nanomaterials: describing and characterizing uncertainty within environmental health and safety risks. Nanotoxicol. 3: 222–33.

Grouchko, M., Kamyshny, A., Ben-Ami, K., Magdassi, S. 2009. Synthesis of copper nanoparticles catalyzed by pre-formed silver nanoparticles. J. Nanopart. Res. 11(3): 713–16.

Gruyer, N., Dorais, M., Bastien, C., Dassylva, N., Triffault-Bouchet, G. 2013. Interaction between silver nanoparticles and plant growth. *In*: International Symposium on New Technologies for Environment Control, Energy-Saving and Crop Production in Greenhouse and Plant. 1037: 795–800.

Gunalan, S., Sivaraj, R., Rajendran, V. 2012. Green synthesized ZnO nanoparticles against bacterial and fungal pathogens. Prog. Nat. Sci.: Mater. Inter. 22(6): 693–700.

Haghighi, M., DaSilva, J.A.T. 2014. The effect of N-TiO$_2$ on tomato onion, and radish seed germination. J. Crop Sci. Biotechnol. 17: 221–27.

Heilmittel, W.W. 1994. Therapeutic preparations (Handbook). In book: Heilmittel, W., Ed. Weleda Pharmacy Medicines, Handbook for Physicians. New York: Weleda, Congers, 80.

Hostynek, J.J., Dreher, F., Maibach, H.I. 2006. Human stratum corneum penetration by copper: *in vivo* study after occlusive and semi-occlusive application of the metal as powder. Food Chem. Toxicol. 44: 1539–43.

Iavicoli, I., Leso, V., Beezhold, D.H., Shvedova, A.A. 2017. Nanotechnology in agriculture: Opportunities, toxicological implications, and occupational risks. Toxicol. Appl. Pharmacol. 329: 96–111.

Ingle, A., Gade, A., Pierrat, S., Sonnichsen, C., Rai, M. 2008. Mycosynthesis of silver nanoparticles using the fungus *Fusarium acuminatum* and its activity against some human pathogenic bacteria. Curr. Nanosci. 4: 141–44.

Ingle, A.P., Duran, N., Rai, M. 2013. Bioactivity, mechanism of action, and cytotoxicity of copper-based nanoparticles: A review. Appl. Microbiol. Biotechnol. 98: 1001–09.

Jackelen, A.-M.L., Jungbaur, M., Glavee, G.N. 1999. Nanoscale materials synthesis.1. Solvent effects on hydridoborate reduction of copper ions. Langmuir. 15(7): 2322–26.

Jiang, W., Yatsui, K. 1998. Pulsed wire discharge for nanosize powder synthesis. IEEE Trans. Plasma Sci. 26(5): 1498–1501.

Jośko, I., Oleszczuk, P. 2013. Influence of soil type and environmental conditions on ZnO: TiO$_2$ and Ni nanoparticles phytotoxicity. Chemosphere 92: 91–99.

Kailasam, K., Epping, J.D., Thomas, A., Losse, S., Junge, H. 2011. Mesoporous carbon nitride-silica composites by a combined sol-gel/thermal condensation approach and their application as photocatalysts. Energy Environ. Sci. 4: 4668.

Kaminskiene, Z., Prosycevas, I., Stonkute, J., Guobiene, A. 2013. Evaluation of optical properties of Ag, Cu, and Co nanoparticles synthesized in organic medium. Acta Physica. Polonica. A. 123: 111–14.

Kanninen, P., Johans, C., Merta, J., Kontturi, K. 2007. Influence of ligand structure on the stability and oxidation of copper nanoparticles. J. Colloid Interface Sci. 318(1): 88–95.

Kapoor, S., Joshi, R., Mukherjee, T. 2002. Influence of I⁻ anions on the formation and stabilization of copper nanoparticles. Chem. Phys. Lett. 354(5-6): 443–48.

Kapoor, S., Mukherjee, T. 2003. Photochemical formation of copper nanoparticles in poly(N-vinylpyrrolidone). Chem. Phys. Lett. 370: 83–87.

Kathad, U., Gajera, H.P. 2014. Synthesis of copper nanoparticles by two different methods and size comparison. Int. J. Pharm. Bio. Sci. 5(3): 533–40.

Kaur, P., Thakur, R., Choudhary, A. 2012. An *in vitro* study of the antifungal activity of silver/chitosan nanoformulations against important seed borne pathogens. Inter. J. Sci. Technol. Res. 1(6): 83–6.

Keller, C., Rizwan, M., Davidian, J.C., Pokrovsky, O.S., Bovet, N., Chaurand, P., Meunier, J.D. 2015. Effect of silicon on wheat seedlings (*Triticum turgidum* L.) grown in hydroponics under Cu stress. Planta. 241: 847–60.

Khalil, M.M.H., Ismail, E.H., El-BaghdadyKh., Z., Mohamed, D. 2013. Green synthesis of silver nanoparticles using olive leaf extract and its antibacterial activity. Arab. J. Chem. 7(6): 1131–39.

Khan, H.A., Shanker, R. 2015. Toxicity of nanomaterials. BioMed. Res. Inter. Article ID 521014: 1–3.

Khan, M.R., Rizvi, T.F. 2014. Nanotechnology: Scope and Applications in plant disease management. Plant Pathol. J. 13(3): 214–31.

Khodashenas, B., Ghorbani, H.R. 2014. Synthesis of copper nanoparticles: An overview of the various methods. Korean J. Chem. Eng. 31: 1105–09.

Kim, D., Jang, D. 2007. Synthesis of nanoparticles and suspensions by pulsed laser ablation of microparticles in liquid. Appl. Surf. Sci. 253(19): 8045–49.

Kim, J.H., Cho, H., Ryu, S.E., Choi, M.U. 2000. Effects of metal ions on the activity of protein tyrosine phosphatise VHR: highly potent and reversible oxidative inactivation by Cu^{2+} ion. Arch. Biochem. Biophys. 382: 72–80.

Kim, J.H., Oh, Y., Yoon, H., Hwang, I., Chang, Y.S. 2015. Iron nanoparticle-induced activation of plasma membrane H^{+}-ATPase promotes stomatal opening in *Arabidopsis thaliana*. Environ. Sci. Technol. 49(2): 1113–19.

Kim, S.W., Jung, J.H., Lamsal, K., Kim, Y.S., Min, J.S., Lee, Y.S. 2012. Antifungal effects of silver nanoparticles (AgNPs) against various plant pathogenic fungi. Mycobiol. 40(1): 53–8.

Kulkarni, V.D., Kulkarni, P.S. 2013. Green synthesis of copper nanoparticles using *Ocimum sanctum* leaf extract. Int. J. Chem. Stud.1(3): 1–4.

Kumar, A., Mandal, S., Selvakannan, P.R., Pasricha, R., Mandale, A.B., Sastry, M. 2003. Investigation into the interaction between surface-bound alkylamines and gold nanoparticles. Langmuir. 19: 6277–82.

Kumari, M., Mukherjee, A., Chandrasekaran, N. 2009. Genotoxicity of silver nanoparticles in *Allium cepa*. Sci. Total Environ. 407: 5243–46.

Kumari, M., Khan, S.S., Pakrashi, S., Mukherjee, A., Chandrasekaran, N. 2011. Cytogenetic and genotoxic effects of zinc oxide nanoparticles on root cells of *Allium cepa*. J. Hazard. Mater. 190: 613–21.

La Torre, A., Talocci, S., Spera, G., Valori, R. 2008. Control of downy mildew on grapes in organic viticulture. Commun. Agric. Appl. Biol. Sci. 73: 169–78.

Lalau, C.M., Mohedano, R.A., Schmidt, E.C., Bouzon, Z.L., Ouriques, L.C., dos Santos, R.W., da Costa, C.H., Vicentini, D.S., Matias, W.G. 2015. Toxicological effects of copper oxide nanoparticles on the growth rate, photosynthetic pigment content, and cell morphology of the duckweed *Landoltia punctate*. Protoplasma 252: 221–29.

Lazary, A., Weinberg, I., Vatine, J.J., Jefidoff, A., Bardenstein, R., Borkow, G., Ohana, N. 2014. Reduction of healthcare-associated infections in a long-term care brain injury ward by replacing regular linens with biocidal copper oxide impregnated linens. Int. J. Infect. Dis. 24: 23–9.

Le, V.N., Rui, Y., Gui, X., Li, X., Liu, S., Han, Y. 2014. Uptake, transport, distribution and Bio-effects of SiO_2 nanoparticles in Bt-transgenic cotton. J. Nanobiotechnol. 12: 50.

Lee J., Kim D.K., Kang W. 2006. Preparation of Cu nanoparticles from Cu powder dispersed in 2-propanol by laser ablation. Bull. Korean Chem. Soc. 27(11): 1869–72.

Lee, H.J., Song, J.Y., Kim, B.S. 2013. Biological synthesis of copper nanoparticles using *Magnolia kobus* leaf extract and their antibacterial activity. J. Chem. Technol. Biotechnol. 88: 1971–77.

Lee, W.M., An, Y.J., Yoon, H., Kweon, H.S. 2008. Toxicity and bioavailability of copper nanoparticles to the terrestrial plants mung bean (*Phaseolus radiatus*) and wheat (*Triticum aestivum*): plant agar test for water-insoluble nanoparticles. Environ. Toxicol. Chem. 27: 1915–21.

Lee, W.M., Kwak, J.I., An, Y.J. 2012. Effect of silver nanoparticles in crop plants *Phaseolus radiatus* and *Sorghum bicolor*: media effect on phytotoxicity. Chemosphere. 86: 491–99.

Li, X., Gui, X., Rui, Y., Ji, W., Yu, Z., Peng, S. 2014. Bt-transgenic cotton is more sensitive to CeO_2 nanoparticles than its parental non-transgenic cotton. J. Hazard. Mater. 274: 173–80.

Lin, D., Xing, B. 2007. Phytotoxicity of nanoparticles: inhibition of seed germination and root growth. Environ. Pollut. 150: 243–50.

Lisiecki, L., Billoudet, F., Pileni, P. 1996. Control of the shape and the size of copper metallic particles. J. Phys Chem. 100(10): 4160–66.

Lisiecki, I., Filankembo, A., Sack-Kongehl, H., Weiss, K., Pileni, M.P., Urban, J. 2000. Structural investigations of copper nanorods by high-resolution TEM. Phys. Rev. B. Condens. Matter. 61(7): 4968–4974.

Liu, F., Wen, L.X., Li, Z.Z., Yu, W., Sun, H.Y., Chen, J.F. 2006. Porous hollow silica nanoparticles as controlled delivery system for water soluble pesticide. Mat. Res. Bull. 41: 2268–75.

Liu, R., Lal, R. 2015. Potentials of engineered nanoparticles as fertilizers for increasing agronomic productions. Sci. Total Environ. 514: 131–39.

López-Moreno, M.L., de la Rosa, G., Hernández-Viezcas, J.A., Peralta-Videa, J.R., Gardea-Torresdey, J.L. 2010. XAS corroboration of the uptake and storage of CeO_2 nanoparticles and assessment of their differential toxicity in four edible plant species. J. Agric. Food Chem. 58: 3689–93.

Ma, Y., He, X., Zhang, P., Zhang, Z. 2011. Phytotoxicity and biotransformation of La_2O_3 nanoparticles in a terrestrial plant cucumber (*Cucumis sativus*). Nanotoxicol. 5: 743–53.

Mahajan, P., Dhoke, S.K., Khanna, A.S. 2011. Effect of nano-ZnO particle suspension on growth of mung (*Vigna radiata*) and gram (*Cicer arietinum*) seedlings using plant agar method. J. Nanotechnol. Article ID 696535: 1–7. Doi: 10.1155/2011/696535.

Mahmoodzadeh, H., Aghili, R., Nabavi, M. 2013a. Physiological effects of TiO_2 nanoparticles on wheat (*Triticum aestivum*). Tech. J. Eng. Appl. Sci. 3: 1365–70.

Mahmoodzadeh, H., Nabavi, M., Kashefi, H. 2013b. Effect of nanoscale titanium dioxide particles on the germination and growth of canola (*Brassica napus*). J. Ornament. Horticult. Plants. 3: 25–32.

Mallikarjuna, K., Narasimha, G., Dillip, G.R., Praveen, B., Shreedhar, B., SreeLakshami, C., Reddy, B.V.S., Deva Prasad Raju, B. 2011. Green synthesis of silver nanoparticles using *Ocimum* leaf extract and their characterization. Digest J. Nanomater. Biostruct. 6(1): 181–86.

Mandal, D., Bolander, M., Mukhopadhyay, D., Sarkar, G., Mukherjee, P. 2006. The use of microorganisms for the formation of metal nanoparticles and their application. Appl. Microbiol. Biotechnol. 69: 485–92.

Mandeh, M., Omidi, M., Rahaie, M. 2012. *In vitro* influences of TiO_2 nanoparticles on barley (*Hordeum vulgare* L.) tissue culture. Biol. Trace Elem. Res. 150: 376–80.

Manjunatha, S.B., Biradar, D.P., Aladakatti, Y.R. 2016. Nanotechnology and its applications in agriculture: A review. J. Farm Sci. 29(1): 1–13.

Maruccio, G., Cingolani, R., Rinaldi, R. 2004. Projecting the nanoworld: Concepts, results and perspectives of molecular electronics. J. Mater. Chem. 14: 542.

Mazumdar, H., Ahmed, G.U. 2011. Phytotoxicity effect of silver nanoparticles on *Oryza sativa*. Int. J. Chem. Technol. Res. 3: 1494–1500.

McNeil, S.E. 2005. Nanotechnology for the biologist. J. Leukoc. Biol. 78: 585–93.

Milanino, R. 2006. Copper in medicine and personal care: a historical overview. pp. 149–60. *In*: Hostynek, J.J., Maibach, H.I. (eds.). Copper and the Skin. Informa Healthcare USA, Inc., New York, NY.

Minkina, T., Rajput, V., Fedorenko, G., Fedorenko, A., Mandzhieva, S., Sushkova, S., Morin, T., Yao, J. 2019. Anatomical and ultrastructural responses of *Hordeum sativum* to the soil spiked by copper. Environ. Geochem. Health. Doi:10.1007/s10653-019-00269-8.

Mirzajani, F., Askari, H., Hamzelou, S., Farzaneh, M., Ghassempour, A. 2013. Effect of silver nanoparticles on *Oryza sativa* L. and its rhizosphere bacteria. Ecotoxicol. Environ. Saf. 88: 48–54.

Mittal, A.K., Chisti, Y., Banerjee, U.C. 2013. Synthesis of metallic nanoparticles using plant extracts. Biotechnol. Adv. 31: 346–56.

Mittal, J., Batra, A., Singh, A., Sharma, M.M. 2014. Phytofabrication of nanoparticles through plant as nanofactories. Adv. Nat. Sci. Nanosci. Nanotechnol. 5: 043002.

Mohammadi, R., Maali-Amiri, R., Mantri, N.L. 2014. Effect of TiO_2 nanoparticles on oxidative damage and antioxidant defense systems in chickpea seedlings during cold stress. Rus. J. Plant Physiol. 61: 768–75.

Monk, A.B., Kanmukhla, V., Trinder, K., Borkow, G. 2014. Potent bactericidal efficacy of copper oxide impregnated non-porous solid surfaces. BMC Microbiol. 14: 57.

Moon, Y.S., Park, E.S., Kim, T.O., Lee, H.S., Lee, S.E. 2014. SELDI-TOF MS-based discovery of a biomarker in *Cucumis sativus* seeds exposed to CuO nanoparticles. Environ. Toxicol. Pharmacol. 38: 922–31.

Mousavi Kouhi, S.M.M., Lahouti, M., Ganjeali, A., Entezari, M.H. 2015. Long-term exposure of rapeseed (*Brassica napus* L.) to ZnO nanoparticles: anatomical and ultrastructural responses. Environ. Sci. Pollut. Res. 22(14): 10733–43.

Muniz-Miranda, M., Gellini, C., Giorgetti, E. 2011. Surface-enhanced raman scattering from copper nanoparticles obtained by laser ablation. J. Phys. Chem. C. 115: 5021–27. DOI:10.1021/jp1086027.

Nagaonkar, D., Shende, S., Rai, M. 2015. Biosynthesis of copper nanoparticles and its effect on actively dividing cells of mitosis in *Allium cepa*. Biotechnol. Prog. 31(2): 557–65.

Naika, H.R., Lingaraju, K., Manjunath, K., Kumar, D., Nagaraju, G., Suresh, D., Nagabhushana, H. 2015. Green synthesis of CuO nanoparticles using *Gloriosa superba* L. extract and their antibacterial activity. J. Taibah Univ. Sci. 9: 7–12.

Nair, P.M.G., Chung, I.M. 2014a. A mechanistic study on the toxic effect of copper oxide nanoparticles in soybean (*Glycine max* L.) root development and lignification of root cells. Biol. Trace Elem. Res. 162: 342–52.

Nair, P.M.G., Chung, I.M. 2014b. Impact of copper oxide nanoparticles exposure on *Arabidopsis thaliana* growth, root system development, root lignificaion, and molecular level changes. Environ. Sci. Pollut. Res. 21: 12709–22.

Nair, P.M.G., Chung, I.M. 2014c. Physiological and molecular level effects of silver nanoparticles exposure in rice (*Oryza sativa* L.) seedlings. Chemosphere. 112: 105–13.

Nair, P.M.G., Chung, I.M. 2015. Changes in the growth, redox status and expression of oxidative stress related genes in chickpea (*Cicer arietinum* L.) in response to copper oxide nanoparticle exposure. J. Plant Growth Regul. 34(2): 350–61.

Namasivayam, S.K.R., Ganesh, S., Avimanyu, S. 2011. Evaluation of anti-bacterial activity of silver nanoparticles synthesized from *Candida glabrata* and *Fusarium oxysporum*. Int. J. Med. Res. 1: 130–36.

Nan, L., Liu, Y., Lu, M., Yang, K. 2008. Study on antibacterial mechanism of copper-bearing austenitic antibacterial stainless steel by atomic force microscopy. J. Mater. Sci. Mater. Med. 19: 3057–62.

Narayanan, K.B., Sakthivel, N. 2010. Biological synthesis of metal nanoparticles by microbes. Adv. Colloid. Interface. Sci. 156: 1–13.

Ngo, Q.B., Dao, T.H., Nguyen, H.C., Tran, X.T., Van Nguyen, T., Khuu, T.D., Huynh, T.H. 2014. Effects of nanocrystalline powders (Fe Co and Cu) on the germination, growth, crop yield and product quality of soybean (Vietnamese species DT-51). Adv. Nat. Sci.: Nanosci. Nanotechnol. 5: 015–016.

Nguyen, V.H., Nguyen, H.V., Dollfus, P. 2014. Improved performance of graphene transistors by strain engineering. Nanotechnol. 25(16): 165201.

O'Brien, P.A., Kulier, R., Helmerhorst, F.M., Usher-Patel, M., d'Arcangues, C. 2008. Copper-containing, framed intrauterine devices for contraception: a systematic review of randomized controlled trials. Contraception 77: 318–327.

Ostaeva, G.Y., Selishcheva, E.D., Pautov, V.D., Papisov I.M. 2008. Pseudotemplate synthesis of copper nanoparticles in solutions of poly(acrylic acid)-pluronic blends. Polym. Sci., Ser. B. 50(5-6): 147–49.

Owaid, M.N., Ibraheem, I.J. 2017. Mycosynthesis of nanoparticles using edible and medicinal mushrooms. Eur. J. Nanomed. 9(1): 5–23.

Pal, D.K., Sahu, C.K., Haldar, A. 2014. Bhasma: The ancient Indian nanomedicine. J. Adv. Pharma. Technol. Res. 5(1): 4–12.

Paralikar, P., Rai, M. 2018. Bio-inspired synthesis of sulphur nanoparticles using leaf extract of four medicinal plants with special reference to their antibacterial activity. IET Nanobiotechnol. 12(1): 25–31.

Park, B.K., Jeong, S., Kim, D., Moon, J., Lim, S., Kim, J.S. 2007. Synthesis and size control of monodisperse copper nanoparticles by polyol method. J. Colloid Interface Sci. 311(2): 417–24.

Patil, H.B.V., Nithin, K.S., Siddaramaiah, S., Chandrashekara, K.T., Sathish Kumar, B.Y. 2016. Mycofabrication of bioactive silver nanoparticle: Photo catalysed synthesis and characterization to attest its augmented bio-efficacy. Arabian J. Chem. (Article in press). Doi: 10.1016/j.arabjc.2016.07.009.

Pavani, K.V., Srujana, N., Preethi, G., Swati, T. 2013. Synthesis of copper nanoparticles by *Aspergillus* species. Lett. Appl. NanoBioSci. 2(2): 110–13.

Pavani, K.V., Divya, V., Veena, I., Aditya, M., Devakinandan, G.V.S. 2014. Influence of bioengineered zinc nanoparticles and zinc metal on *Cicer arietinum* seedlings growth. Asian J. Agric. Biol. 2: 216–23.

Peralta-Videa, J.R., Hernandez-Viezcas, J.A., Zhao, L., Diaz, B.C., Ge, Y., Priester, J.H., Holden, P.A., Gardea-Torresdey, J.L. 2014. Cerium dioxide and zinc oxide nanoparticles alter the nutritional value of soil cultivated soybean plants. Plant Physiol. Biochem. 80: 128–35.

Plaza, D.O., Gallardo, C., Straub, Y.D., Bravo, D., Pérez-Donoso, J.M. 2016. Biological synthesis of fluorescent nanoparticles by cadmium and tellurite resistant Antarctic bacteria: exploring novel natural nanofactories. Microb. Cell Fact. 15: 76.

Pokhrel, L.R., Dubey, B. 2013. Evaluation of developmental responses of two crop plants exposed to silver and zinc oxide nanoparticles. Sci. Total Environ. 452-453: 321–32.

Prabhu, Y.T., Venkateswara Rao, K., Sesha Sai, V., Pavani, T. 2017. A facile biosynthesis of copper nanoparticles: A micro-structural and antibacterial activity investigation. J. Saudi Chem. Soc. 21: 180–85.

Prucek, R., Kvitek, L., Panacek, A., Vancurova, L., Soukupova, J., Jancik, D., Zboril, R. 2009. Polyacrylate-assisted synthesis of stable copper nanoparticles and copper (I) oxide nanocubes with high catalytic efficiency. J. Mater. Chem. 19: 8463–8469.

Raffi, M., Mehrwan, S., Bhatti, T.M., Akhter, J.I., Hameed, A., Yawar, W., Hasan, M.M. 2010. Investigations into the antibacterial behavior of copper nanoparticles against *Escherichia coli*. Ann. Microbiol. 60(1): 75–80.

Rahimi, P., Hashemipour, H., Ehtesham Zadeh, M., Pourakbari, R. 2010. Experimental investigation on the synthesis and size control of copper nanoparticle via chemical reduction method. Int. J. Phys. Sci. 6(18): 4331.

Rai, M., Duran, N. 2011. Metal nanoparticles in Microbiology. Springer: Verlag Berlin Heidelberg, Germany.

Rai, M., Yadav, A., Bridge, P., Gade, A. 2009. Myconanotechnology: A new and emerging science. pp. 258–267. *In*: Rai, M., Bridge, P.D. (eds.). Applied Mycology. CAB International, Oxfordshire UK.

Rai, M., Yadav, A., Gade, A. 2010. Mycofabrication, mechanistic aspect and multifunctionality of metal nanoparticles-Where are we? And where should we go? pp. 1343–54. *In*: Mendez-Vilas, A. (ed.). Current Research, Technology and Education Topics in Applied Microbiology and Microbial Biotechnology. FORMATEX, Spain.

Raja, M., Subha, J., Binti Ali, F., Ryu, S.H. 2008. Synthesis of copper nanoparticles by electroreduction process. Mater. Manuf. Process. 23(8): 782–85.

Rajamathi, M., Seshadri, R. 2002. Oxide and chalcogenide nanoparticles from hydrothermal/solvothermal reactions. Curr. Opin. Solid State Mater. Sci. 6(4): 337–45.

Rajapaksha, A.U., Ahmad, M., Vithanage, M., Kim, K.R., Chang, J.Y., Lee, S.S., Ok, Y.S. 2015. The role of biochar, natural iron oxides, and nanomaterials as soil amendments for immobilizing metals in shooting range soil. Environ. Geochem. Health. 37(6): 931–42.

Rajendran, S.P., Sengodan, K. 2017. Synthesis and characterization of zinc oxide and iron oxide nanoparticles using *Sesbania grandiflora* leaf extract as reducing agent. J. Nanosci. Article ID 8348507: 1–7.

Rajeshkumar, S., Malarkodi, C., Paulkumar, K., Vanaja, M., Gnanajobitha, G., Annadurai, G. 2014. Algae mediated green fabrication of silver nanoparticles and examination of its antifungal activity against clinical pathogens. Int. J. Met. 2014: 1–8.

Rajput, V.D., Minkina, T., Suskova, S., Mandzhieva, S., Tsitsuashvili, V., Chapligin, V., Fedorenko, A.G. 2017. Effects of copper nanoparticles (CuO NPs) on crop plants: a mini review. BioNanoSci. 8: 36–42.

Rajput, V., Minkina, T., Fedorenko, A., Sushkova, S., Mandzhieva, S., Lysenko, V., Duplii, N., Fedorenko, A.G., Dvadnenko, K., Ghazaryan, K. 2018. Toxicity of copper oxide nanoparticles on spring barley (*Hordeum sativum* distichum). Sci. Total Environ. 645: 1103–13.

Rajput, V., Minkina, T., Ahmed, B., Sushkova, S., Singh, R., Soldatov, M., Laratte, B., Fedorenko, A., Mandzhieva, S., Blicharska, E., Musarrat, J., Saquib, Q., Flieger, J., Gorovtsov, A. 2019a. Interaction of copper-based nanoparticles to soil, terrestrial, and aquatic systems: critical review of the state of the science and future perspectives. Rev. Environ. Contamin. Toxicol. 1–46. Doi: 10.1007/398_2019_34.

Rajput, V. D., Minkina, T., Sushkova, S., Mandzhieva, S., Fedorenko, A., Lysenko, V., Bederska, M., Olchowik, J., Tsitsuashvili, V., Chapligin, V.A. 2019b. Structural and ultrastructural changes in nanoparticle exposed plants. Springer, Cham. Doi: 10.1007/978-3-319-97852-9_13.

Richard, J., Spencer, B., McCoy, L. 2002. Acticoa versus Silver lon1: the truth. J. Burn. Surg. Wound. 1(1): 11–19.

Rico, C.M., Morales, M.I., Barrios, A.C., McCreary, R., Hong, J., Lee, W.Y., Nunez, J., Peralta-Videa, J.R., Gardea-Torresdey, J.L. 2013. Effect of cerium oxide nanoparticles on the quality of rice (*Oryza sativa* L.) grains. J. Agric. Food Chem. 67: 11278–85.

Rico, C.M., Lee, S.C., Rubenecia, R., Mukherjee, A., Hong, J., Peralta-Videa, J.R., Gardea-Torresdey, J.L. 2014. Cerium oxide nanoparticles impact yield and modify nutritional parameters in wheat (*Triticum aestivum* L.). J. Agric. Food Chem. 62: 9669–75.

Rico, C.M., Barrios, A.C., Tan, W., Rubenecia, R., Lee, S.C., Varela-Ramirez, A., Peralta-Videa, J.R., Gardea-Torresdey, J.L. 2015. Physiological and biochemical response of soil-grown barley (*Hordeum vulgare* L.) to cerium oxide nanoparticles. Environ. Sci. Pollut. Res. Int. 22(14): 10551–58.

Rifkind, J.M., Shin, Y.A., Hiem, J.M., Eichorn, G.L. 2001. Cooperative disordering of single stranded polynucleotides through copper crosslinking. Biopolymers. 15: 1879.

Rizwan, M., Meunier, J.D., Hélène, M., Keller, C. 2012. Effect of silicon on reducing cadmium toxicity in durum wheat (*Triticum turgidum* L. cv. Claudio W.) grown in a soil with aged contamination. J. Hazard. Mater. 209-210: 326–34.

Rizwan, M., Ali, S., Qayyum, M.F., Ok, Y.S., Adrees, M., Ibrahim, M., Zia-ur-Rehman, M., Farid, M., Abbas, F. 2017. Effect of metal and metal oxide nanoparticles on growth and physiology of globally important food crops: A critical review. J. Hazard. Mater. 15: 322(Pt A): 2–16. Doi:10.1016/j.jhazmat. 2016.05.061.

Roghayyeh, S.M.S., Mehdi, T.S., Rauf, S.S. 2010. Effects of nano-iron oxide particles on agronomic traits of soybean. Notulae Sci. Biol. 2: 112–13.

Rui, M., Ma, C., Hao, Yi., Guo, J., Rui, Y., Tang, X., Zhao, Q., Fan, X., Zhang, Z., Hou, T., Zhu, S. 2016. Iron oxide nanoparticles as a potential iron fertilizer for Peanut (*Arachis hypogaea*). Front. Plant Sci. 7: 815. 1–10.

Safiuddin, M., Gonzalez, M., Cao, J.W., Tighe, S.L. 2014. State of-the-art report on use of nano materials in concrete. Int. J. Pavement Eng. 15: 940–49.

Saha, J., Begum, A., Mukherjee, A., Kumar, S. 2017. A novel green synthesis of silver nanoparticles and their catalytic action in reduction of methylene blue dye. Sust. Environ. Res. 27(5): 245–50.

Saito, M., Yasukawa, K. 2008. Copper nanoparticles fabricated by laser ablation in polysiloxane. Opt. Mater. 30(7): 1201–04.

Salvadori, M.R., Ando, R.A., Nascimento, C.A.O., Correa, B. 2014. Intracellular biosynthesis and removal of copper nanoparticles by dead biomass of yeast isolated from the wastewater of a mine in the Brazilian Amazonia. PLoS ONE 9(1): e87968.

Salzemann, C., Lisiecki, I., Brioude, A., Urban, J., Pileni, M.P. 2004. Collections of copper nanocrystals characterized by different sizes and shapes: optical response of these nanoobjects. J. Phys. Chem. B. 108: 13242–48.

Sanghi, R., Verma, P. 2009. Biomimetic synthesis and characterization of protein capped silver nanoparticles. Bioresour. Technol. 100: 501–04.

Santhoshkumar, J., Venkat Kumar, S., Rajeshkumar, S. 2017. Synthesis of zinc oxide nanoparticles using plant leaf extract against urinary tract infection pathogen. Res. Eff. Technol. 3(4): 459–65.

Santo, C.E., Quaranta, D., Grass, G. 2012. Antimicrobial metallic copper surfaces kill *Staphylococcus haemolyticus* via membrane damage. Microbiologyopen. 1: 46–52.

Saranyaadevi, K., Subha, V., Ravindran, R.S.E., Renganathan, S. 2014. Synthesis and characterization of copper nanoparticles using *Capparis zeylanica* leaf extract. Int. J. ChemTech. Res. 6(10): 4533–41.

Sarkar, J., Chattopadhyay, D., Patra, S., Deo, S.S., Sinha, S., Ghosh, M., Mukherjee, A., Acharya, K. 2011. *Alternaria alternata* mediated synthesis of protein capped silver nanoparticles and their genotoxic activity. Dig. J. Nanomater. Biostruct. 6(2): 563–73.

Sastry, M., Ahmad, A., Khan, M.I., Kumar, R. 2003. Biosynthesis of metal nanoparticles using fungi and actinomycete. Curr. Nanosci. 85: 162–70.

Schmidt, M.G., Banks, A.L., Salgado, C.D. 2014. Role of the microbial burden in the acquisition and control of healthcare associated infections: the utility of solid copper surfaces. pp. 59–83. *In*: Borkow, G. (ed.). Use of Biocidal Surfaces for Reduction of Healthcare Acquired Infections. Springer International Publishing, Switzerland.

Schneiderhan, F.J. 1933. The discovery of bordeaux mixture: three papers: I. Treatment of mildew and rot. II. Treatment of mildew with copper sulphate and lime mixture. III. Concerning the history of the treatment of mildew with copper sulphate, by Pierre Marie Alexis Millardet, 1885; A translation from the French by Felix John Schneiderhan. American Pathological Society. Phytopathol. Class. 3.

Schultz, T.P., Nicholas, D.D., Preston, A.F. 2007. A brief review of the past, present and future of wood preservation. Pest. Manag. Sci. 63: 784–88.

Sen, P., Ghosh, J., Abdullah, A., Kumar, P., Vandana. 2003. Preparation of Cu, Ag, Fe and Al nanoparticles by the exploding wire technique. J. Chem. Sci. 115(5-6): 499–508. http://sci-hub.tw/10.1007/BF02708241.

Servin, A.D., Castillo-Michel, H., Hernandez-Viezcas, J.A., Diaz, B.C., Peralta-Videa, J.R., Gardea-Torresdey, J.L. 2012. Synchrotron micro-XRF and micro-XANES confirmation of the uptake and translocation of TiO_2 nanoparticles in cucumber (*Cucumis sativus*) plants. Environ. Sci. Technol. 46: 7637–43.

Seshadri, S., Saranya, K., Kowshik, M. 2011. Green synthesis of lead sulfide nanoparticles by the lead resistant marine yeast, *Rhodosporidium diobovatum*. Biotechnol. Prog. 27: 1464–69.

Shaw, A.K., Ghosh, S., Kalaji, H.M., Bosa, K., Brestic, M., Zivcak, M., Hossain, Z. 2014. Nano-CuO stress induced modulation of antioxidative defense and photosynthetic performance of Syrian barley (*Hordeum vulgare* L.). Environ. Exp. Bot. 102: 37–47.

Shaw, A.K., Hossain, Z. 2013. Impact of nano-CuO stress on rice (*Oryza sativa* L.) seedlings. Chemosphere. 93: 906–15.

Shen, T.D., Koch, C.C. 1996. Formation, solid solution hardening and softening of nanocrystalline solid solutions prepared by mechanical attrition. Acta Mater. 44(2): 753–61.

Shende, S., Ingle, A.P., Gade, A., Rai, M. 2015. Green synthesis of copper nanoparticles by *Citrus medica* Linn. (Idilimbu) juice and its antimicrobial activity. World J. Microbiol. Biotechnol. 31(6): 865–73.

Shende, S., Rathod, D.P., Gade, A., Rai, M. 2017. Biogenic copper nanoparticles promote the growth of pigeon pea (*Cajanus cajan* L.). IET Nanobiotechnol. 11(7): 773–81. Doi: 10.1049/iet-nbt.2016.0179.

Siddiqui, M.H., Al-Whaibi, M.H. 2014. Role of nano-SiO_2 in germination of tomato (*Lycopersicum esculentum* seeds Mill). Saudi J. Biol. Sci. 21: 13–17.

Singh, A., Singh, N.B., Hussain, I., Singh, H., Singh, S.C. 2015. Plant-nanoparticle interaction: An approach to improve agricultural practices and plant productivity. Inter. J. Pharma. Sci. Inven. 4(8): 25–40.

Song, R.G., Yamaguchi, M. 2007. Investigation of metal nanoparticles produced by laser ablation and their catalytic activity. Appl. Surf. Sci. 253(6): 3093–97.

Song, X.Y., Sun, S.X., Zhang, W.M., Yin, Z.L. 2004. A method for the synthesis of spherical copper nanoparticles in the organic phase. J. Colloid Interface Sci. 273(2): 463–69.

Song, J.Y., Kwon, E.-Y., Kim, B.S. 2010. Biological synthesis of platinum nanoparticles using *Diopyros kaki* leaf extract. Bioprocess Biosyst. Eng. 33: 159–64.

Song, U., Jun, H., Waldman, B., Roh, J., Kim, Y., Yi, J., Lee, E.J. 2013a. Functional analyses of nanoparticle toxicity: a comparative study of the effects of TiO_2 and Ag on tomatoes (*Lycopersicon esculentum*). Ecotoxicol. Environ. Saf. 93: 60–67.

Song, U., Shin, M., Lee, G., Roh, J., Kim, Y., Lee, E.J. 2013b. Functional analysis of TiO_2 nanoparticle toxicity in three plant species. Biol. Trace Elem. Res. 155: 93–103.

Su, X.D., Zhao, J.Z., Bala, H., Zhu, Y.C., Gao, Y., Ma, S.S., Wang, Z.C. 2007. Fast synthesis of stable cubic copper nanocages in the aqueous phase. J. Phys. Chem. C. 111: 14689–93.

Subhankari, I., Nayak, P.L. 2013a. Antimicrobial activity of copper nanoparticles synthesised by ginger (*Zingiber officinale*) extract. World J. NanoSci. Technol. 2(1): 10–13. DOI:10.5829/idosi. wjnst.2013.2.1.21133.

Subhankari, I., Nayak, P.L. 2013b. Synthesis of copper nanoparticles using *Syzygium aromaticum* (Cloves) aqueous extract by using green chemistry. World J. NanoSci. Technol. 2(1): 14–17.

Sunitha, A., Rimal, I.R.., Sweetly, G., Sornalekshmi, S., Arsula, R., Praseetha, P. 2013. Evaluation of antimicrobial activity of biosynthesized iron and silver nanoparticles using the fungi *Fusarium* and *Actinomycets* sp. on human pathogen. Nano Biomed. Eng. 5: 39–45.

Suriyaprabha, R., Karunakaran, G., Yuvakkumar, R., Prabu, P., Rajendran, V., Kannan, N. 2012. Growth and physiological responses of maize (*Zea mays* L.) to porous silica nanoparticles in soil. J. Nanopart. Res. 14: 1294–96.

Theivasanthi, T., Alagar, M. 2011. Nano-sized copper particles by electrolytic synthesis and characterizations. Int. J. Phys. Sci. 6(15): 3662.

Thiruvengadam, M., Gurunathan, S., Chung, I.M. 2015. Physiological, metabolic, and transcriptional effects of biologically-synthesized silver nanoparticles in turnip (*Brassica rapa* sp. *rapa* L.). Protoplasma. 252(4): 1031–46.

Thuesombat, P., Hannongbua, S., Akasit, S., Chadchawan, S. 2014. Effect of silver nanoparticles on rice (*Oryza sativa* L cv. KDML 105) seed germination and seedling growth. Ecotoxicol. Environ. Saf. 104: 302–09.

Uskokovic, V. 2008. Nanomaterials and nanotechnologies: approaching the crest of this big wave. Curr. Nanosci. 4: 119–29.

Valodkar, M., Jadeja, R.N., Thounaojam, M.C., Devkar, R.V., Thakore, S. 2011. Biocompatible synthesis of peptide capped copper nanoparticles and their biological effect on tumor cells. Mater. Chem. Phys. 128(1-2): 83–89.

Vannini, C., Domingo, G., Onelli, E., Prinsi, B., Marsoni, M., Espen, L., Bracale, M. 2013. Morphological and proteomic responses of *Eruca sativa* exposed to silver nanoparticles or silver nitrate. PLoS One 8: e68752.

Vannini, C., Domingo, G., Onelli, E., De Mattia, F., Bruni, I., Marsoni, M., Bracale, M. 2014. Phytotoxic and genotoxic effects of silver nanoparticles exposure on germinating wheat seedlings. J. Plant Physiol. 171: 1142–48.

Varshney, R., Bhadauria, S., Gaur, M.S., Pasricha, R. 2010. Characterization of copper nanoparticles synthesized by a novel microbiological method. J. Metals. 62(12): 102–04.

Varshney, R., Bhadauria, S., Gaur, M.S., Pasricha, R. 2011. Copper nanoparticles synthesis from electroplating industry effluent. Nano Biomed. Eng. 3(2): 115–19.

Varshney, R., Bhadauria, S., Gaur, M.S. 2012. A Review: biological synthesis of silver and copper nanoparticles. Nano Biomed. Eng. 4(2): 99–106. DOI: 10.5101/nbe.v4i2.

Venugopal, G., Hunt, A., Alamgir, F. 2010. Nanomaterials for energy storage in Lithium-ion battery applications. Mater. Matters. 5: 42–45.

Volesky, B., Holan, Z.R. 1995. Biosorption of heavy metals. Biotechnol. Prog. 11: 235–50.

Wang, Y., Chen, P., Liu, M. 2006. Synthesis of well-defined copper nanocubes by a one-pot solution process. Nanotechnol. 17: 6000–06.

Wang, L., Li, X., Zhang, G. Dong, J., Eastoe, J. 2007. Oil-in-water nanoemulsions for pesticide formulations. J. Colloid Interface Sci. 314: 230–35.

Wang, H., Kou, X., Pei, Z., Xiao, J.Q., Shan, X., Xing, B. 2011. Physiological effects of magnetite (Fe$_3$O$_4$) nanoparticles on perennial ryegrass (*Lolium perenne* L.) and pumpkin (*Cucurbita mixta*) plants. Nanotoxicol. 5: 30–42.

Wang, Q., Ma, X., Zhang, W., Pei, H., Chen, Y. 2012. The impact of cerium oxide nanoparticles on tomato (*Solanum lycopersicum* L.) and its implications for food safety. Metallom. 4: 1105–12.

Wang, Q., Ebbs, S.D., Chen, Y., Ma, X. 2013. Trans-generational impact of cerium oxide nanoparticles on tomato plants. Metallom. 5: 753–59.

Wang, S., Liu, H., Zhang, Y., Xin, H. 2015. The effect of CuO NPs on reactive oxygen species and cell cycle gene expression in roots of rice. Environ. Toxicol. Chem. 34(3): 554–61.

Wei, X., Luo, M., Li, W., Yang, L., Liang, X., Xu, L., Kong, P. 2012. Synthesis of silver nanoparticles by solar irradiation of cell-free *Bacillus amyloliquefaciens* extracts and AgNO$_3$. Bioresour. Technol. 103: 273–78.

Wu, C.W., Mosher B.P., Zeng T.F. 2006. One-step green route to narrowly dispersed copper nanocrystals. J. Nanopart. Res. 8(6): 965–69.

Wu, S.H., Chen D.H. 2004. Synthesis of high-concentration Cu nanoparticles in aqueous CTAB solutions. J. Colloid Interface Sci. 273(1): 165–69.

Wu, S.G., Huang, L., Head, J., Ball, M., Tang, Y.J., Chen, D.R. 2014. Electrospray facilitates the germination of plant seeds. Aerosol Air Qual. Res. 14: 632–41.

Wypij, M., Czarnecka, J., Dahm, H., Rai, M., Golinska, P. 2017a. Silver nanoparticles from *Pilimeliacolumellifera* subsp. *pallida* SL19 strain demonstrated antifungal activity against fungi causing superficial mycoses. J. Basic Microbiol. 57(9): 793–800. Doi: 10.1002/jobm.201700121.

Wypij, M., Golinska, P., Dahm, H., Rai, M. 2017b. Actinobacterial-mediated synthesis of silver nanoparticles and their activity against pathogenic bacteria. IET Nanobiotechnol. 11(3): 336–42.

Xiang, L., Zhao, H.M., Li, Y.W., Huang, X.P., Wu, X.L., Zhai, T., Yuan, Y., Cai, Q.-Y., Mo, C.-H. 2015. Effects of the size and morphology of zinc oxide nanoparticles on the germination of Chinese cabbage seeds. Environ. Sci. Pollut. Res. 22 (14): 10452–62.

Xiong, J., Wang, Y., Xue, Q., Wu, X. 2011. Synthesis of highly stable dispersions of nanosized copper particles using l-ascorbic acid. Green Chem. 13: 900.

Yallappa, S., Manjanna, J., Sindhe, M.A., Satyanarayan, N.D., Pramod, S.N., Nagaraja, K. 2013. Microwave assisted rapid synthesis and biological evaluation of stable copper nanoparticles using *T. arjuna* bark extract. Spectrochim. Acta A: Mole. Biomole. Spectro. 110: 108–15.

Yang, J.G., Yang, S.H., Tang, C.B., He, J., Tang, M.T. 2007. Synthesis of ultrafine copper particles by complex reduction-extraction method. Trans. Non-ferr. Metals Soc. China. 17: s1181–s1185.

Yasur, J., Rani, P.U. 2013. Environmental effects of nanosilver: impact on castor seed germination seedling growth, and plant physiology. Environ. Sci. Pollut. Res. 20: 8636–48.

Yeh, M. S., Yang, Y.S., Lee, H.F., Yeh, Y.H., Yeh C.S. 1999. Formation and characteristics of Cu colloids from CuO powder by laser irradiation in 2-propanol. J. Phys. Chem. B. 103(33): 6851–57.

Yoon, S.J., Kwak, J.I., Lee, W.M., Holden, P.A., An, Y.J. 2014. Zinc oxide nanoparticles delay soybean development: a standard soil microcosm study. Ecotoxicol. Environ. Saf. 100: 131–37.

Yu, S.H. 2001. Hydrothermal/Solvothermal processing of advanced ceramic materials. J. Ceram. Soc. Japan 109(5): S65–S75.

Zhang, L., Fang, M. 2010. Nanomaterials in pollution trace detection and environmental improvement. Nano Today. 5: 128–42.

Zhang, R., Zhang, H., Tu, C., Hu, X., Li, L., Luo, Y., Christie, P. 2015. Phytotoxicity of ZnO nanoparticles and the released Zn (II) ion to corn (*Zea mays* L.) and cucumber (*Cucumis sativus* L.) during germination. Environ. Sci. Pollut. Res. 22(14): 11109–17.

Zhang, W., Ebbs, S.D., Musante, C., White, J.C., Gao, C., Ma, X. 2015. Uptake and accumulation of bulk and nano-sized cerium oxide particles and ionic cerium by radish (*Raphanus sativus* L.). J. Agric. Food Chem. 63: 382–90.

Zhao, L., Hernandez-Viezcas, J.A., Peralta-Videa, J.R., Bandyopadhyay, S., Peng, B., Munoz, B., Keller, A.A., Gardea-Torresdey, J.L. 2013. ZnO nanoparticle fate in soil and zinc bioaccumulation in corn plants (*Zea mays*) influenced by alginate. Environ. Sci. Process. Impact. 15: 260–66.

Zhao, L., Peralta-Videa, J.R., Rico, C.M., Sun, Y., Niu, G., Servin, A., Nunez, J.E., Duarte-Gardea, M., Gardea-Torresdey, J.L. 2014. CeO_2 and ZnO nanoparticles change the nutritional qualities of cucumber (*Cucumis sativus*). J. Agric. Food Chem. 62: 2752–59.

Zhao, L., Sun, Y., Hernandez-Viezcas, J.A., Hong, J., Majumdar, S., Niu, G., Duarte-Gardea M., Peralta-Videa, J.R., Gardea-Torresdey, J.L. 2015. Monitoring the environmental effects of CeO_2 and ZnO nanoparticles through the life cycle of corn (*Zea mays*) plants and *in situ* μ-XRF mapping of nutrients in kernels. Environ. Sci. Technol. 49(5): 2921–28.

Zhou, X., Torabi, M., Lu, J., Shen, R., Zhang, K. 2015. Nanostructured energetic composites: synthesis ignition/combustion modeling, and applications. ACS Appl. Mater. Interfaces 6: 3058–74.

Zhu, H.T., Lin, Y.S., Yin, Y.S. 2004a. A novel one-step chemical method for preparation of copper nanofluids. J. Colloid Interface Sci. 277(1): 100–03.

Zhu, H.T., Zhang, C.Y., Yin, Y.S. 2004b. Rapid synthesis of copper nanoparticles by sodium hypophosphite reduction in ethylene glycol under microwave irradiation. J. Cryst. Growth. 270(3-4): 722–728.

Zwieten, M.V., Stovold, G., Zwieten, L.V. 2007. Alternatives to copper for disease control in the Australian Organic Industry. A report for the Rural Industries Research and Development Corporation 1–82.

Chapter 15

Microbially Synthesized Nanoparticles

Exposure Sources and Ecotoxicity

Indarchand Gupta,[1] Alka Yadav,[2] Avinash P. Ingle,[3]
Silvio Silverio da Silva,[3] Chistiane Mendes Feitosa[4] and
Mahendra Rai[2,4,]*

Introduction

Over the last few decades, nanotechnology has made enormous progress and emerged as one of the most promising technologies in a range of fields. The Nanotechnology Consumer Product Inventory was created in 2005 to document the emergence of nanotechnology products in the consumer market (Vance et al. 2015). It is recorded that there was an increase of about 25 times in the production of nano-based products between 2005 and 2010 (Bundschuh et al. 2018), and that production is continuously increasing. According to a conservative forecast, its value will reach $11.3 billion in 2020, and an optimistic forecast predicts a market value of $55.0 billion in 2022 (Inshakova and Inshakov 2017).

Various nanoparticles such as silver, gold, copper, zinc, and silicon, and many oxide nanoparticles, such as zinc oxide and titanium dioxide, are being used for the development of a variety of nano-based products (Vance et al. 2015). All such nanomaterials are considered building blocks of nanotechnology and have a broad range of applications in agriculture, food, renewable energy, textile, cosmetic, biomedicine, and other fields (Raj et al. 2012, Yetisen et al. 2016, Han et al. 2019, He et al. 2019,

[1] Department of Biotechnology, Institute of Science, Aurangabad 431004, Maharashtra, India.
[2] Nanobiotech Laboratory, Department of Biotechnology, SGB Amravati University, Amravati 444602, Maharashtra, India.
[3] Department of Biotechnology, Engineering School of Lorena, University of Sao Paulo, Estrada municipal do Campinho, sn, 12602-810 Lorena, SP, Brazil.
[4] Department of Chemistry, Federal University of Piauí, 64000-040, Teresina, Piauí, Brazil.
* Corresponding author: mkrai123@rediffmail.com

Rai et al. 2019a, Rai et al. 2019b, Rastogi et al. 2019). The novel and unique properties of these nanoparticles, such as smaller size, surface area, surface reactivity, charge and shape, against their bulk counterparts make them more suitable for all the above-mentioned applications (Geonmonond et al. 2018, Lee and Jun 2019). The extensive use of normal and engineered nanomaterials in a wide range of products has raised concerns about their toxicity because the chances of the release of nanomaterials in the environment are also increased (Klaine et al. 2008, Giese et al. 2018).

There is an active debate about the benefits and toxicological concerns of various nanomaterials and nano-based consumer products (Handy and Shaw 2007, Owen and Handy 2007). It concerns mainly the risks to the environment, including terrestrial and aquatic systems (Owen and Handy 2007). However, the scientific community, especially ecotoxicologists, have given a great deal of attention to this concern and are beginning to understand the potential risks to wildlife associated with various nanomaterials (Handy et al. 2008). Biocompatibility and the highly reactive nature of nanomaterials along with other physicochemical properties are considered responsible for their ecotoxicity. For example, Seitz et al. (2014) demonstrated that the crystalline composition of titanium dioxide nanoparticles influences their toxicity to the water flea *Daphnia magna*. Similarly, other parameters such as particle number, surface area, and body burden are considered important to adequately reflect the exposure situation of nanomaterials (Petersen et al. 2015). Moreover, the surface area and crystalline composition of nanoparticles play a crucial role in their toxicity (Seitz et al. 2014).

It was proposed that the evaluation of different routes of exposure is essential for the assessment of risk. However, detection and quantification of nanomaterials released in the environment by any route is extremely challenging owing to the unavailability of specific, sensitive and promising procedures and technologies (Von Der Kammer et al. 2012). Therefore, there is an urgent need to perform extensive studies on toxicities of nanomaterials and to develop effective technologies for their assessment. In addition, it is very important to implement strict guidelines about the use and disposal of nanomaterials. This chapter focuses on various toxicological concerns of nanomaterials. In addition, biocompatibility issues associated with nanomaterials, different routes for the exposure of nanomaterials to the ecosystem, role of various properties (e.g., shape, size, concentration) of nanomaterials in toxicity and other important aspects have been discussed.

Biocompatibility issues of physical, chemical and biological synthesis of nanoparticles

Biocompatibility is defined as the ability of a material to be successfully used as a carrier molecule in drug delivery systems (Gautam and vari Veggel 2013, Adabi et al. 2016). A high level of biocompatibility is observed when the nanomaterial interacts with the body without producing toxic, immunogenic, thrombogenic or carcinogenic responses (Naahidi et al. 2013). The biocompatibility of nanomaterials basically depends on their size, shape, surface area, concentration and functional groups (Markides et al. 2012, Naahidi et al. 2013, Singh et al. 2016) (Fig. 15.1).

Nanoparticles can be synthesized through physical, chemical and biological routes (Singh et al. 2016). Biological synthesis is rapid and ecofriendly, since the nanoparticles do not require a stabilizing agent, as they would with physical and

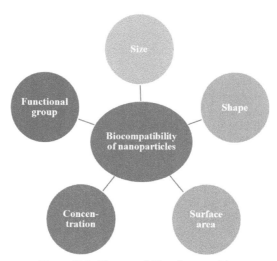

Figure 15.1. Biocompatibility of nanoparticles.

chemical synthesis methods, because plants and microorganisms themselves serve as capping and stabilizing agents (Feng et al. 2013, Singh et al. 2016). Also, biogenic nanoparticles are free of toxic contamination from byproducts that get attached to nanoparticles during the physicochemical synthesis process (Gurunathan et al. 2014) (Fig. 15.2).

Biologically synthesized nanoparticles require fewer steps for synthesis and also possess attached functional groups that make them biologically active (Vadlapudi et al. 2014). On the other hand, in physicochemical synthesis an additional step is needed for attachment of functional groups to the nanoparticles (Vadlapudi et al. 2014, Singh et al. 2016). Unlike chemical and physical methods, the biological route offers rapid synthesis with stable production of nanoparticles of controlled size and shapes (Gautam and vari Veggel 2013, Gurunathan et al. 2014).

Thus, biologically synthesized nanoparticles are mostly used in the production of a variety of nano-based products as they emerge as the safer and better alternative to conventional methods.

Microbial synthesis of nanoparticles

Since biological methods for synthesis of nanoparticles are found to be comparatively rapid, safe and eco-friendly, a number of biological systems including plants and various microorganisms have been successfully used for the synthesis of various kinds of nanoparticles. Researchers differ in their opinions about the use of biological systems. Some studies claim that plants are the most suitable systems, and others claim that microbial systems are more convenient than plants (Ovais et al. 2018). Every biological system has its own advantages and limitations, but in some respects microbial systems are more suitable than plant systems (Singh et al. 2016, Ovais et al. 2018).

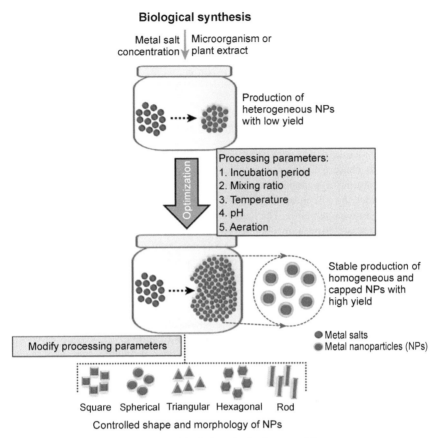

Figure 15.2. Biological synthesis of nanoparticles.

The most important advantage is that any microorganism can be grown in the laboratory under suitable conditions in a very small space (Molnar et al. 2018), whereas in the case of plants specific climatic and geographic conditions are required. Moreover, the growth of plants requires more time than the growth of microorganisms. Another important point is that extensive exploitation of plants may cause harm to the environment.

As mentioned above, it is proposed that nanoparticles produced using plant systems are mostly polydisperse because of the involvement of phenols, flavonoids, terpenoids, and other phytochemicals and also because of their diverse photochemistry (Salunke et al. 2014, Ovais et al. 2016, Ovais et al. 2018). In addition, there is a possibility that the alternation of the phytochemical profile of plant extracts due to seasonal variations may affect the synthesis of nanoparticles as far as the physical and chemical properties are concerned. Microbial synthesis is free of such limitations (Singh et al. 2015). Among microbial systems, fungi are considered most suitable because tolerance to bioaccumulation of metals, high binding ability, and intracellular uptake make them more efficient for biological synthesis of nanoparticles (Qidwai et al. 2018).

Route of exposure to the ecosystem

Nanoparticles released in the environment have diverse fates. They could settle in the soil, remain suspended in the air, or be released into aquatic bodies. In all of these circumstances nanoparticles can interact with the components of the ecosystem and display toxic effects.

Exposure to nanoparticles through the soil

Soil contains a solid matrix with which nanoparticles can interact very easily. Moreover, the aqueous phase containing a considerable amount of natural colloids provides a higher opportunity to nanoparticles for interaction with soil. With respect to toxicity to the ecosystem, the soil phase, i.e., soil and soil water phase, decides the mechanism of soil–nanoparticle interaction. The nanoparticles used in any product can reach the soil through the waste disposal mechanism or through settlement of nanoparticles present in air or from water bodies. All of these factors help to accumulate nanoparticles in the soil. It is well known that soil contains a very diverse microflora and, hence, there is a high probability that nanoparticles reaching the soil can interact with such microflora (Kraas et al. 2017, Kumari et al. 2017). Yang and Watts (2005) reported the reduction in root growth of *Crocus sativus* due to inhibition of microbial growth by alumina nanoparticles (Al_2O_3 NPs). Such incidence raises concerns about the impact of nanoparticle accumulation in soil on food chains. Similarly, it was observed that presence of zinc and zinc oxide nanoparticles in the soil inhibits seed germination and plant root growth (Lin and Xing 2007). If the concentration of nanoparticles in soil is high, that can be hazardous for the soil microflora, which can diminish the soil biodiversity. Additionally, the nanoparticles present in soil can also be trapped by plants growing in the vicinity (Kumari et al. 2017). The nanoparticles thus reach the plant bodies and can interfere with the normal functioning of plant metabolism. The interference in plant metabolism creates an adverse situation for plant growth, thereby affecting plant numbers and diversity. Therefore, it can be summarized that nanoparticles reaching the soil can affect the soil microflora as well as the plants growing in the vicinity.

Exposure to nanoparticles through water

Water is essential to every life form. It is a safe and easy portal for carrying away all types of waste disposal arising from human activities. Water bodies are regarded as a major sink for the disposal of waste and environmental contaminants. Therefore, there is risk of aquatic toxicity to flora and fauna. Nanoparticles present in various nano-based products can be easily released in water bodies. For example, silver nanoparticles used in socks to prevent the development of odor could be easily released in water after washing. Similarly, nanoparticles used in agricultural products were reported to enter water bodies (Prasad et al. 2017). Through these and other routes, nanoparticles can enter rivers, ponds, canals, and other water bodies. The accumulation of nanoparticles in larger water bodies such as rivers can exert a hazardous effect on the flora and fauna of the river, thereby affecting its biodiversity (Wright et al. 2018). The nanoparticles present in water bodies can be absorbed in or on the surface of microorganisms, plants, and animals, affecting their process of homeostasis.

Exposure to nanoparticles through air

There are many reports that suggest that nanoparticles can be found in air circulation (Jing et al. 2015, Stebounova et al. 2018). Nanoparticles can enter the air via the industry that manufactures them. They can also get released into the air through general activities of consumers using nanoparticle-containing products. Air is also an important component of the ecosystem. The nanoparticles present in the air may be inhaled by animals and humans and accumulate in their respiratory system. They tend to reach other body parts through the circulatory system. Hence, nanoparticles in the air will have an impact on living beings.

Nano-ecotoxicity: A major concern

Developments in the field of nanotechnology have redesigned the landscape of modern research. However, fears of their harmful effects on human health and the environment are also rising. After their potential applications, nanomaterials are discharged into the environment and may pose a threat to the ecological system. Hence, there is a need for assessment of factors influencing the toxic action of nanoparticles to the ecosystem. Among the various features of nanoparticles, their size, shape, dissolution, and state of aggregation are considered to influence their potential toxicity (Fig. 15.3).

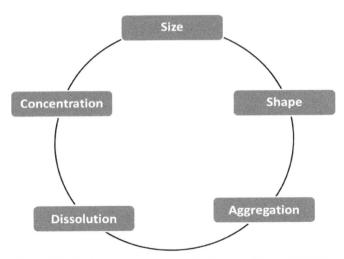

Figure 15.3. Factors influencing the toxicity of nanoparticles to ecosystem.

Influence of size and shape

Size in nanometer-scale implies novel physicochemical characteristics in nanoparticles and, therefore, it will not be wrong to expect nanoparticle size to have an impact on their ecotoxicity. The chemical reactivity and high contact of nanoparticles with body surface is responsible for increased toxicity as compared to their larger counterparts. Many reports suggest that smaller nanoparticles are more toxic than larger nanoparticles. For instance, ZnONPs of diameter 8 nm were reported to display higher growth inhibition in *Staphylococcus aureus* as compared to ZnONPs of diameter 50–70 nm (Jones et al. 2008). Fluorescent nanoparticles showed toxic effect on Japanese medaka (*Oryzias latipes*), in terms of higher accumulation of smaller nanoparticles as compared to larger nanoparticles (Kashiwada 2006). The increase in the size of nanoparticles leads to a decrease in their total surface area, consequently decreasing the specific surface area, superficial reactivity and toxic effects. However, the size-dependent toxicity of nanoparticles is correlated with the number of ions released from them in the aquatic condition. It has been claimed that the smaller nanoparticles release ions more easily than larger ones. Additionally, smaller nanoparticles have easy access to the cellular components. Unlike larger nanoparticles, they can cross the cell membrane and easily pass into the cellular cytoplasm (Gupta et al. 2012).

The effect of shape is yet poorly understood in the ecotoxicity of nanoparticles. There are a few reports that suggest that nanoparticle shape does not have any effect on their toxicity (Visnapuu et al. 2013, Gorka et al. 2015). However, it has been reported that nanosilver triangle plates are more toxic than silver nanospheres (Pal et al. 2007, Sadeghi et al. 2012, Raza et al. 2016). Other reports suggested that silver nanospheres showed more toxicity than silver nanowires and plates (Ashkarran et al. 2013, Gao et al. 2013). In an investigation performed on *Danio rerio* it has been claimed that nanosilver plates were more toxic than silver nanoparticles and silver nanowires (George et al. 2012). Moreover, silver nanoparticles were more toxic than silver nanowires to *D. rerio*, *Lolium multiflorum* (ryegrass), *Oryzias latipes* (juveniles), *D. magna*, and *Pseudokirchneriella subcapitata* (Gorka et al. 2015, Sohn et al. 2015). The reason behind the higher toxicity of nanosilver plates most probably lies in its shape. The edges or sharp corners of nanosilver plates get dissolved more easily to form silver ions and that enhances their toxicity (Moon et al. 2018). All of these studies imply that the shape of nanomaterials under study can alter their toxicity even when those nanomaterials possess the same size and concentration (Moon et al. 2018). Therefore, for evaluating the toxicity data of nanoparticles, their shape should also be considered an important parameter.

Influence of nanoparticle dissolution

The dissolution of nanoparticles, especially in the aquatic environment, plays an important role in deciding their fate. Conventionally, the dissolved fraction of test material is considered to be responsible for exerting biological effects. For nanoparticles, their dissolution to release ions in the aquatic medium is an important parameter for their ecotoxicity studies. The nanoparticles may be poorly, partially or completely soluble depending on their type and environmental conditions (Skjolding et al. 2016). Hence, the dissolution of nanoparticles has an influence on their toxicity.

As discussed earlier, the use of nanoparticles raises major concerns when they reach aquatic bodies. In such an environment, they show toxic impacts on ecosystems that reports suggest are due to release of ions. The ionic form of nanoparticles was reported to be toxic at concentrations of a few micrograms per liter (Kahru and Duboutguier 2010). Smaller nanoparticles can more easily dissolve in the media than larger nanoparticles and, hence, they are more toxic (Raza et al. 2016, Jimeno-Romero et al. 2017, Moon et al. 2018). Moon et al. (2018) reported that smaller particles of 36 nm possessed higher toxicity than particles of 10 and 20 μm. ZnONPs were reported to partially dissolve in water to produce zinc ions; therefore, their toxicity is also considered to be highly dependent on ion released (Mallevre et al. 2014). This could initiate the exposure of ionic zinc to aquatic systems.

Influence of concentration

Any material beyond a certain concentration is harmful, and the same applies to nanoparticles. The dose-dependent toxicity of nanoparticles is mostly studied in aquatic environments. The exposure of aquatic species to nanoparticles has been found to have toxicity depending on their concentration. For instance, with the increase in concentration, silver nanoparticles led to high mortality rate in the embryos of zebrafish. They were also found to retard cell growth due to delayed cell division (Asharani et al. 2008). The higher concentration of nanoparticles leads to generation of reactive oxygen species (ROS). The enhanced ROS generation can further cause cell injury and cell death (Naqvi et al. 2010). As compared to controlled conditions, silver nanoparticles, for instance, have been reported to play an important role in 10-fold increase in ROS accumulation with the increase in their concentration (Hussain et al. 2005).

Nair and Chung (2014) reported that higher concentrations of silver nanoparticles inhibit root growth of rice seedlings. Another study reported the decreased mitotic index of *Alium cepa* with an increase in the dose of silver nanoparticles (Kumari et al. 2009). Silver nanoparticles were found to inhibit the root elongation of *Raphanus sativus* (Zuverza-Mena et al. 2016). Copper oxide nanoparticles were demonstrated to result in dose-dependent cytotoxicity, genotoxicity and oxidative stress (Akhtar et al. 2013). Though some reports are available, more detailed studies are needed to evaluate the concentration-dependent ecotoxicity of nanoparticles.

Influence of nanoparticle aggregation

Aggregation behavior of nanoparticles can also influence their toxicity. In aquatic systems, aggregation of the nanoparticles may result from pH and the existence of organic matter (Handy et al. 2008). There is a difference in toxicity of the same nanoparticles to invertebrates if they are present in well-dispersed and aggregated forms (Lovern and Klaper 2006). Nanoparticle aggregation is one of the most important factors in its toxicity. Aggregation is the product of interaction between the mobility and bioavailability of nanoparticles. The nanoparticles when aggregated, especially in the aquatic system, get deposited on the surface of organisms including microbes, plants, and animals (Wiesner et al. 2006). The aggregation of nanoparticles can lead to bioaccumulation and biomagnification. According to Lovern et al. (2007), the aggregation in various nanoparticles leads to the loss of mobility in zooplankton *D. magna*. Similar reports have been published that suggest the bioaccumulation and biomagnification of silver nanoparticles in a food chain consisting of *Chlorella* sp. (green algae), *Moina macrocopa* (water flea), *Chironomus* spp. (blood worm), and *Barbonymus gonionotus* (silver barb) (Yoo-iam et al. 2014). Titanium dioxide nanoparticles have been shown to accumulate in *D. rerio* through the ingestion of contaminated daphnids (Zhu et al. 2010), whereas the quantum dots were reported to transfer to *Tetrahymena pyriformis* via *Escherichia coli* (Holbrook et al. 2008).

Current challenges and future prospects

There are tremendous challenges associated with the ecotoxicity study of nanomaterials. The study of nano-ecotoxicity is presently in its infancy. The available data still do not give a clear idea about the concentration and form in which nanoparticles should be released into the environment. Hence, there is an urgent requirement for data on safety levels of nanoparticles released into the environment. It should be mandatory to perform studies on the behavior and fate of nanoparticles under different pH and ionic strength. The data obtained will help in the design of standardized methods for the regulatory ecotoxicity testing of nanoparticles to ensure their safe use. Moreover, there is a need for the development of a platform for unbiased, completely documented and transparent discussion on the harmful effects of nanomaterials intended for public use. The participation of concerned researchers, industry personnel and environmentalists across the globe will yield highly fruitful outcomes, which can help in safe use of nanoparticles.

Conclusions

Nanotechnology is a rapidly emerging technology that has enormous potential for human use. Although many products based on this technology are already available in the market and even a large number of potential new applications of nanotechnology are arising, many reports concerning their toxicity are also drawing attention. The reports say that use of nano-based products can release nanoparticles into the air, soil and water bodies, through which they can cause harm to the ecosystem. Hence, there is a need to focus on aspects that are important for the sustainable application of nanotechnology.

It is being found that nanoparticles released in the air, soil, and water can interact with the respective biotic component. In such circumstances, they will disturb the complete ecosystem. The factors that govern their toxicity are mainly influenced by their size, shape, concentration, dissolution, and state of aggregation. Many studies focus mainly on such aspects of nanomaterial use but there is a lack of sufficient data on interaction of emerging nanoparticles with various components of the ecosystem. More attention needs to be paid to those interactions in order to gather suitable data for sustainable and safe use of nanotechnologies.

References

Adabi, M., Naghibzadeh, M., Adabi, M., Zarrinfard, M.A., Esnaashari, S.S., Seifalian, A.M., Faridi-Majidi, R., Tanimowo Aiyelabegan, H., Ghanbari, H. 2016. Biocompatibility and nanostructured materials: applications in nanomedicine. Artif Cells Nanomed. Biotechnol. 45(4): 833–842.

Akhtar, M.J., Kumar, S., Alhadlaq, H.A., Alrokayan, S.A., Abu-Salah, K.M., Ahamed, M. 2013. Dose-dependent genotoxicity of copper oxide nanoparticles stimulated by reactive oxygen species in human lung epithelial cells. Toxicol. Ind. Health 32(5): 809–821. doi:10.1177/0748233713511512.

Asharani, P.V., Wu, Y.L., Gong, Z., Valiyaveettil, S. 2008. Toxicity of silver nanoparticles in zebrafish models. Nanotechnology 19(25): 255102 (8pp) doi:10.1088/0957-4484/19/25/255102.

Ashkarran, A.A., Estakhri, S., Nezhad, M.R.H., Eshghi, S. 2013. Controlling the geometry of silver nanostructures for biological applications. Phys. Procedia. 40: 76–83.

Bundschuh, M., Filser, J., Lüderwald, S., McKee, M.S., Metreveli, G., Schaumann, G.E., Schulz, R., Wagner, S. 2018. Nanoparticles in the environment: where do we come from, where do we go to? Environ. Sci. Eur. 30: 6. https://doi.org/10.1186/s12302-018-0132-6.

Feng J.J., Song, Y.Y., Feng, X.M., Shrestha, N.K., Wisitruangsakul, N. 2013. Biocompatible functional nanomaterials: synthesis, properties and applications. J. Nanomat. Article ID 385939.

Gao, M.J., Sun, L., Wang, Z.Q., Zhao, Y.B. 2013. Controlled synthesis of Ag nanoparticles with different morphologies and their antibacterial properties. Mat. Sci. Eng. C. Mater. 33: 397–404.

Gautam, A., vari Veggel, F.C.J.M. 2013. Synthesis of nanoparticles, their biocompatibility and toxicity behavior for biomedical applications. J. Mat. Chem. B. 1: 5186–5200.

Geonmonond, R.S., Da Silva, A.G.M., Camargo, P.H.C. 2018. Controlled synthesis of noble metal nanomaterials: motivation, principles, and opportunities in nanocatalysis. An. Acad. Bras. Cienc. 90: 719–744.

George, S., Lin, S., Ji, Z., Thomas, C.R., Li, L., Mecklenburg, M., Meng, H., Wang, X., Zhang, H., Xia, T., Hohman, J.N., Lin, S., Zink, J.I., Weiss, P.S., Nel, A.E. 2012. Surface defects on plate-shaped silver nanoparticles contribute to its hazard potential in a fish gill cell line and zebrafish embryos. ACS Nano. 6(5): 3745–59.

Giese, B., Klaessig, F., Park, B., Kaegi, R., Steinfeldt, M., Wigger, H., von Gleich, A., Gottschalk, F. 2018. Risks, release and concentrations of engineered nanomaterial in the environment. Sci. Rep. 8: 1565. doi: 10.1038/s41598-018-19275-4.

Gorka, D.E., Osterberg, J.S., Gwin, C.A., Colman, B.P., Meyer, J.N., Bernhardt, E.S., Gunsch, C.K., DiGulio, R.T., Liu, J. 2015. Reducing environmental toxicity of silver nanoparticles through shape control. Environ. Sci. Technol. 49(16): 10093–10098.

Gupta, I., Duran, N., Rai, M. 2012. Nano-silver toxicity: emerging concerns and consequences in human health. pp. 525–548. *In*: Rai, M., Cioffi, N. (eds.). Nano-Antimicrobials: Progress and Prospects. Springer Verlag, Germany.

Gurunathan, S., Han J., Park, J.H., Kim, J.H. 2014. A green chemistry approach for synthesizing biocompatible gold nanoparticles. Nanoscale Res. Lett. 9(1): 248–259.

Han, X., Xu, K., Taratula, O., Farsad, K. 2019. Applications of nanoparticles in biomedical imaging. Nanoscale 11: 799–819.

Handy, R.D., Owen, R., Valsami-Jones, E. 2008. The ecotoxicology of nanoparticles and nanomaterials: current status, knowledge gaps, challenges, and future needs. Ecotoxicology 17(5): 315–325.

Handy, R.D., Shaw, B.J. 2007. Toxic effects of nanoparticles and nanomaterials: implications for public health, risk assessment and the public perception of nanotechnology. Health Risk. Soc. 9: 125–144.

He, X., Deng, H., Hwang, H.M. 2019. The current application of nanotechnology in food and agriculture. J. Food Drug Anal. 27(1): 1–21.

Holbrook, R.D., Murphy, K.E., Morrow, J.B., Cole, K.D. 2008. Trophic transfer of nanoparticles in a simplified invertebrate food web. Nat. Nanotechnol. 3: 352–355.

Hussain, S.M., Hess, K.L., Gearhart, J.M., Geiss, K.T., Schlager, J.J. 2005. *In vitro* toxicity of nanoparticles in BRL 3A rat liver cells. Toxicol. *In Vitro* 19(7): 975–983.

Inshakova, E., Inshakov, O. 2017. World market for nanomaterials: structure and trends. MATEC Web Conf. 129: 02013.

Jimeno-Romero, A., Bilbao, E., Izagirre, U., Cajaraville, M.P., Marigómez, I., Soto, M. 2017. Digestive cell lysosomes as main targets for Ag accumulation and toxicity in marine mussels, *Mytilus galloprovincialis*, exposed to maltose-stabilised Ag nanoparticles of different sizes. Nanotoxicology 11(2): 168–183.

Jing, X., Park, J.H., Peters, T.M., Thorne, P.S. 2015. Toxicity of copper oxide nanoparticles in lung epithelial cells exposed at the air-liquid interface compared with *in vivo* assessment. Toxicol. *In Vitro* 29(3): 502–511.

Jones, N., Ray, B., Koodali, R.T., Manna, A.C. 2008. Antibacterial activity of ZnO nanoparticles suspensions on a broad spectrum of microorganisms. FEMS Microbiol. Lett. 279: 71–76.

Kahru, A., Dubourtguier, H.C. 2010. From ecotoxicology to nanoecotoxicology. Toxicology 269: 105–119.

Kashiwada, S. 2006. Distribution of nanoparticles in the see-through medaka (*Oryzias latipes*). Environ. Health Perspect. 114: 1697–1702.

Klaine, S.J., Alvarez, P.J.J., Batley, G.E., Fernandes, T.F., Handy, R.D., Lyon, D.Y., Mahendra, S., Mclaughlin, M.J., Lead, J.R. 2008. Nanomaterials in the environment: behavior, fate, bioavailability, and effects. Environ. Toxicol. Chem. 27(9): 1825–1851.

Kraas, M., Schlich, K., Knopf, B., Wege, F., Kägi, R., Terytze, K., Hund-Rinke, K. 2017. Long-term effects of sulfidized silver nanoparticles in sewage sludge on soil microflora, Environ. Toxicol. Chem. 36(12): 3305–3313.

Kumari, M., Mukherjee, A., Chandrasekaran, N. 2009. Genotoxicity of silver nanoparticles in *Allium cepa*. Sci. Total Environ. 407(19): 5243–5246.

Kumari, M., Pandey, S., Mishra, S.K., Nautiyal, C.S., Mishra, A. 2017. Effect of biosynthesized silver nanoparticles on native soil microflora via plant transport during plant–pathogen–nanoparticles interaction, 3 Biotech 7: 345. https://doi.org/10.1007/s13205-017-0988-y.

Lee, S.H., Jun, B.H. 2019. Silver nanoparticles: synthesis and application for nanomedicine. Int. J. Mol. Sci. 20(4): E865. https://doi.org/10.3390/ijms20040865.

Lin, D., Xing, B. 2007. Phytotoxicity of nanoparticles: inhibition of seed germination and root growth. Environ. Pollut. 150: 243–250.

Lovern, S.B., Strickler, J.R., Klaper, R. 2007. Behavioral and physiological changes in *Daphnia magna* when exposed to nanoparticle suspensions (Titanium Dioxide, Nano C60, and C60HxC70Hx). Environ. Sci. Technol. 41(12): 4465–4470.

Lovern, S.B., Klaper, R. 2006. *Daphnia magna* mortality when exposed to titanium dioxide and fullerene (C60) nanoparticles. Env. Toxicol. Chem. 25(4): 1132–1137.

Mallevre, F., Fernandes, T.F., Aspray, T.J. 2014. Silver, zinc oxide and titanium dioxide nanoparticle ecotoxicity to bioluminescent *Pseudomonas putida* in laboratory medium and artificial wastewater. Environ. Pollut. 195: 218–225.

Markides, H., Rotherham, M., El Haj, A.J. 2012. Biocompatibility and toxicity of magnetic nanoparticles in regenerative medicine. J. Nanomat. Article ID 614094.

Molnar, Z., Bódai, V., Szakacs, G., Erdélyi, B., Fogarassy, Z., Sáfrán, G., Varga, T., Kónya, Z., Tóth-Szeles, E., Szűcs, R., Lagzi, I. 2018. Green synthesis of gold nanoparticles by thermophilic filamentous fungi. Sci. Rep. 2018; 8: 3943. doi: 10.1038/s41598-018-22112-3.

Moon, J., Kwak, J.I., An, Y.-J. 2018. The effects of silver nanomaterial shape and size on toxicity to *Caenorhabditis elegans* in soil media. Chemosphere. doi:10.1016/j.chemosphere.2018.09.177.

Naahidi, S., Jafari, M., Edalat, F., Raymond, K., Khademhosseini, A., Chen, P. 2013. Biocompatibility of engineered nanoparticles for drug delivery. J. Control. Release 166(2): 182–194.

Nair, P.M., Chung, I.M. 2014. Physiological and molecular level effects of silver nanoparticles exposure in rice (*Oryza sativa* L.) seedlings. Chemosphere 112: 105–113.

Naqvi, S., Samim, M., Abdin, M.Z., Ahmed, F.J., Maitra, A.N., Prashant, C.K., Dinda, A.K. 2010. Concentration-dependent toxicity of iron oxide nanoparticles mediated by increased oxidative stress. Int. J. Nanomedicine 5: 983–989.

Ovais, M., Khalil, A.T., Ayaz, M., Ahmad, I., Nethi, S.K., Mukherjee, S. 2018. Biosynthesis of metal nanoparticles via microbial enzymes: a mechanistic approach. Int. J. Mol. Sci. 19(12): 4100. doi: 10.3390/ijms19124100.

Ovais, M., Khalil, A.T., Raza, A., Khan, M.A., Ahmad, I., Islam, N.U., Saravanan, M., Ubaid, M.F., Ali, M., Shinwari, Z.K. 2016. Green synthesis of silver nanoparticles via plant extracts: beginning a new era in cancer theranostics. Nanomedicine (Lond.). 11(23): 3157–3177.

Owen, R., Handy, R. 2007. Formulating the problems for environmental risk assessment of nanomaterials. Environl. Sci. Technol. 41: 5582–5588.

Pal, S., Tak, Y.K., Song, J.M. 2007. Does the antibacterial activity of silver nanoparticles depend on the shape of the nanoparticle? A study of the Gram-negative bacterium *Escherichia coli*. Appl. Environ. Microbiol. 73: 1712–1720.

Petersen, E.J., Diamond, S.A., Kennedy, A.J., Goss, G.G., Ho, K., Lead, J., Hanna, S.K., Hartmann, N.B., Hund-Rinke, K., Mader, B., Manier, N., Pandard, P., Salinas, E.R., Sayre, P. 2015. Adapting OECD aquatic toxicity tests for use with manufactured nanomaterials: key issues and consensus recommendations. Environ. Sci. Technol. 49(16): 9532–9547.

Prasad, R.M., Bhattacharyya, A., Nguyen, Q.D. 2017. Nanotechnology in sustainable agriculture: recent developments, challenges, and perspectives. Front Microbiol. 8: 1014. doi: 10.3389/fmicb.2017.01014.

Qidwai, A., Pandey, A., Kumar, R., Shukla, S.K., Dikshit, A. 2018. Advances in biogenic nanoparticles and the mechanisms of antimicrobial effects. Indian J. Pharm. Sci. 80(4): 592–603.

Rai, M., Ingle, A.P., Gupta, I., Pandit, R., Paralikar, P., Gade, A., Chaud, M.V., Santo, C.A.D. 2019a. Smart nanopackaging for enhancement of shelf-life of food. Environ. Chem. Lett. 17: 277–290.

Rai, M., Ingle, A.P., Pandit, R., Paralikar, P., da Silva, S.S. 2019b. Emerging role of nanobiocatalysts in hydrolysis of lignocellulosic biomass leading to sustainable bioethanol production. Cat. Rev. 61(1): 1–26.

Raj, S., Jose, S., Sumod, U.S., Sabitha, M. 2012. Nanotechnology in cosmetics: Opportunities and challenges. J. Pharm. Bioallied. Sci. 4(3): 186–193.

Rastogi, A., Tripathi, D.K., Yadav, S., Chauhan, D.K., Živčák, M., Ghorbanpour, M., El-Sheery, N.I., Brestic, M. 2019. Application of silicon nanoparticles in agriculture. 3 Biotech. 9(3): 90. doi: 10.1007/s13205-019-1626-7.

Raza, M.A., Kanwal, Z., Rauf, A., Sabri, A.N., Riaz, S., Naseem, S. 2016. Size-and shape dependent antibacterial studies of silver nanoparticles synthesized by wet chemical routes. Nanomaterials 6(4): 74.

Sadeghi, B., Garmaroudi, F.S., Hashemi, M., Nezhad, H.R., Nasrollahi, A., Ardalan, S., Ardalan, S. 2012. Comparison of the anti-bacterial activity on the nanosilver shapes: nanoparticles, nanorods and nanoplatess. Adv. Powder Technol. 23: 22–26.

Salunke, G.R., Ghosh, S., Santosh Kumar, R.J., Khade, S., Vashisth, P., Kale, T., Chopade, S., Pruthi, V., Kundu, G., Bellare, J.R., Chopade, B.A. 2014. Rapid efficient synthesis and characterization of silver, gold, and bimetallic nanoparticles from the medicinal plant Plumbago zeylanica and their application in biofilm control. Int. J. Nanomedicine. 9: 2635–2653.

Seitz, F., Rosenfeldt, R.R., Schneider, S., Schulz, R., Bundschuh, M. 2014. Size-, surface- and crystalline structure composition-related effects of titanium dioxide nanoparticles during their aquatic life cycle. Sci. Total Environ. 493: 891–897.

Singh, P., Kim, Y.J., Zhang, D., Yang, D.C. 2016. Biological synthesis of nanoparticles from plants and microorganisms. Trends Biotechnol. 34(7): 588–599.

Singh, R., Shedbalkar, U.U., Wadhwani, S.A., Chopade, B.A. 2015. Bacteriagenic silver nanoparticles: synthesis, mechanism, and applications. Appl. Microbiol. Biotechnol. 99(11): 4579–4593.

Skjolding, L.M., Sørensen, S.N., Hartmann, N.B., Hjorth, R., Hansen, S.F., Baun, A. 2016. Aquatic ecotoxicity testing of nanoparticles-the quest to disclose nanoparticle effects. Angew. Chem. Int. Ed. Engl. 55(49): 15224–15239.

Sohn, E.K., Chung, Y.S., Johari, S.A., Kim, T.G., Kim, J.K., Lee, J.H., Lee, Y.H., Kang, S.W., Yu, I.J. 2015. Acute toxicity comparison of single-walled carbon nanotubes in various freshwater organisms. BioMed. Res. Int. 2015: 323090. http://dx.doi.org/10.1155/2015/323090.

Stebounova, L.V., Gonzalez-Pech, N.I., Peters, T.M., Grassian, V.H. 2018. Physicochemical properties of air discharge-generated manganese oxide nanoparticles: Comparison to welding fumes. Environ. Sci. Nano. 2018(5): 696–707.

Vadlapudi, V., Behara, M., Devamma, M.N. 2014. Green synthesis and biocompatibility of nanoparticles. J. Chem. 7(3): 219–233.

Vance, M.E., Kuiken, T., Vejerano, E.P., McGinnis, S.P., Hochella, Jr. M.F., Rejeski, D., Hull, M.S. 2015. Nanotechnology in the real world: Redeveloping the nanomaterial consumer products inventory. Beilstein. J. Nanotechnol. 6: 1769–1780.

Visnapuu, M., Joost, U., Juganson, K., Künnis-Beres, K., Kahru, A., Kisand, V., Ivask, A. 2013. Dissolution of silver nanowires and nanospheres dictates their toxicity to *Escherichia coli*. Biomed. Res. Int. 2013: 819252. doi: 10.1155/2013/819252.

von der Kammer, F., Ferguson, P.L., Holden, P.A., Masion, A., Rogers, K.R., Klaine, S.J., Koelmans, A.A., Horne, N., Unrine, J.M. 2012. Analysis of engineered nanomaterials in complex matrices (environment and biota): general considerations and conceptual case studies. Environ. Toxicol. Chem. 31(1): 32–49.

Wiesner, M.R., Lowry, G.V., Alvarez, P., Dionysiou, D., Biswas, P. 2006. Assessing the risks of manufactured nanomaterials. Environ. Sci. Technol. 40(14): 4336–4345.

Wright, M.V., Matson, C.W., Baker, L.F., Castellon, B.T., Watkins, P.S., King, R.S. 2018. Titanium dioxide nanoparticle exposure reduces algal biomass and alters algal assemblage composition in wastewater effluent-dominated stream mesocosms. Sci. Total Environ. 626: 357–365.

Yang, L., Watts, D.J. 2005. Particle surface characteristics may play an important role in phytotoxicity of alumina nanoparticles. Toxicol. Lett. 158: 122–132.

Yetisen, A.K., Qu, H., Manbachi, A., Butt, H., Dokmeci, M.R., Hinestroza, J.P., Skorobogatiy, M., Khademhosseini, A., Yun, S.H. 2016. Nanotechnology in textiles. ACS Nano. 10(3): 3042–3068.

Yoo-iam, M., Chaichana, R., Satapanajaru, T. 2014. Toxicity, bioaccumulation and biomagnification of silver nanoparticles in green algae (Chlorella sp.), water flea (Moina macrocopa), blood worm (Chironomus spp.) and silver barb (Barbonymus gonionotus). Chem. Spec. Bioavail. 26(4): 257–265.

Zhu, X., Chang, Y., Chen, Y. 2010 Toxicity and bioaccumulation of TiO_2 nanoparticle aggregates in *Daphnia magna*. Chemosphere 78(3): 209–215.

Zuverza-Mena, N., Armendariz, R., Peralta-Videa, J.R., Gardea-Torresdey, J.L. 2016. Effects of silver nanoparticles on radish sprouts: root growth reduction and modifications in the nutritional value. Front. Plant Sci. 7: 90. http://dx.doi.org/ 10.3389/fpls.2016.00090.

Index

T - #0394 - 071024 - C308 - 234/156/14 - PB - 9780367517106 - Gloss Lamination